Asphalt: Materials Science and Technology

Asphalt: Materials Science and Technology

Edited by
Philip Webster

WILLFORD PRESS

www.willfordpress.com

Published by Willford Press,
118-35 Queens Blvd., Suite 400,
Forest Hills, NY 11375, USA

ISBN: 978-1-68285-783-0

Cataloging-in-Publication Data

Asphalt : materials science and technology / edited by Philip Webster.
    p. cm.
Includes bibliographical references and index.
ISBN 978-1-68285-783-0
1. Asphalt. 2. Bituminous materials. 3. Materials science. 4. Asphalt--Technological innovations. I. Webster, Philip.
TN853 .A76 2020
553.27--dc23

For information on all Willford Press publications
visit our website at www.willfordpress.com

WILLFORD PRESS

Contents

Permissions

List of Contributors

Index

Preface

Asphalt is a semi-solid, black and sticky form of petroleum which is also known as bitumen. It may be found in natural deposits or it could be a refined product called pitch. Pitch is derived from petroleum, coal, plants and tar or it is manufactured. Asphalt is primarily used in road constructions as a binder or glue that is mixed with aggregate particles such as sand, gravel, recycled concrete and crushed stones to create asphalt concrete. The other main uses of asphalt include bituminous waterproofing systems that are designed to protect the commercial and residential buildings from water. This book includes some of the vital pieces of work being conducted across the world, on various topics related to asphalt material. Most of the topics introduced in this book cover new techniques and the applications of such material. It will provide comprehensive knowledge to the readers.

Various studies have approached the subject by analyzing it with a single perspective, but the present book provides diverse methodologies and techniques to address this field. This book contains theories and applications needed for understanding the subject from different perspectives. The aim is to keep the readers informed about the progresses in the field; therefore, the contributions were carefully examined to compile novel researches by specialists from across the globe.

Indeed, the job of the editor is the most crucial and challenging in compiling all chapters into a single book. In the end, I would extend my sincere thanks to the chapter authors for their profound work. I am also thankful for the support provided by my family and colleagues during the compilation of this book.

Editor

Effects of Test Conditions on APA Rutting and Prediction Modeling for Asphalt Mixtures

Hui Wang,[1] **Haoqi Tan,**[2] **Tian Qu,**[2] **and Jiupeng Zhang**[2]

[1]*Key Laboratory of Highway Construction & Maintenance Technology in Loess Region, Shanxi Transportation Research Institute, Taiyuan, Shanxi 030006, China*
[2]*School of Highway, Chang'an University, Xi'an, Shaanxi 710064, China*

Correspondence should be addressed to Jiupeng Zhang; zhjiupeng@163.com

Academic Editor: Ana María Díez-Pascual

APA rutting tests were conducted for six kinds of asphalt mixtures under air-dry and immersing conditions. The influences of test conditions, including load, temperature, air voids, and moisture, on APA rutting depth were analyzed by using grey correlation method, and the APA rutting depth prediction model was established. Results show that the modified asphalt mixtures have bigger rutting depth ratios of air-dry to immersing conditions, indicating that the modified asphalt mixtures have better antirutting properties and water stability than the matrix asphalt mixtures. The grey correlation degrees of temperature, load, air void, and immersing conditions on APA rutting depth decrease successively, which means that temperature is the most significant influencing factor. The proposed indoor APA rutting prediction model has good prediction accuracy, and the correlation coefficient between the predicted and the measured rutting depths is 96.3%.

1. Introduction

Rutting prevention has become one of the most in-demand topics of study with more-extensive research on damage to asphalt pavement. Studies all over the world have established a variety of rutting test methods to analyze and evaluate the antirutting properties of asphalt mixtures. APA (Asphalt Pavement Analyzer) rutting test has gained much international attention in recent years for such advantages as the ability to simulate live load conditions. In 2001, NCAT (National Center for Asphalt Technology) evaluated the applicability of various methods for evaluating the antirutting performance, including APA, HWTD (Hamburg Wheel-tracking Device), FRT (French Pavement Rutting Tester), RLWT, and triaxial repeated load creep tests, and preferentially recommended APA rutting test. Subsequently, numbers of research have conducted the antirutting properties of asphalt mixtures by using APA test.

Xie et al. researched moisture susceptibility through APA and analyzed the results of water-submerged rut testing, from which they presented an index of water submergence stability to assess water resistance of different asphalt mixtures [1]. Han et al. investigated the effects of water on permanent deformation potential by using APA and indicated that APA testing form, preconditioning conditions, and the freezing and thawing cycle times are the main universal reasons that cause a greater extent of deformation in wet rut compared with dry rut [2]. Zhang et al. presented a rule of the effect of different size aggregates on the high-temperature performance of asphalt mixtures on the basis of APA tests and indicated a reasonable percentage of different size aggregates, which could help to design both skeleton and dense-construction asphalt mixtures [3]. Junbiao et al. used digital image-processing to analyze the loading modes of the RLWT (Rotary Loaded Wheel Tester) and APA rutting tests, by comparing the angle between the long axis and the x-axis of coarse aggregates in RLWT testing specimens, APA testing specimens, and untested specimens; they found that the loading modes of RLWT and APA rutting tests differ notably [4]. Cao et al. studied the deformation property of asphalt mixtures at constant and varying temperatures through APA and determined that deformation curves have two phases: delayed elastic deformation and viscous deformation [5]. Xue et al. performed APA tests on SAWI (Stress-Absorbing Waterproof Interlayers) to analyze the fatigue properties and indicated that the fatigue life of SAWI is 3.32 times that of common asphalt concrete [6]. Xiao et al. researched the compaction properties of different asphalt mixtures and analyzed their abilities to resist rutting and water damage by using APA [7, 8]. Rushing and Little assessed rutting sensitivity through APA, triaxial static creep, and triaxial repeated load creep tests and found that the increasing rate of permanent strain and the flow time value determined via triaxial static creep tests has the strongest correlation to APA rutting depth [9]. Ali et al. assessed antirutting properties of WMA (Warm Mix Asphalt) by using APA, pointing out that reducing the production temperature of foamed WMA might result in the increasing potential of permanent deformation and moisture-induced damage [10]. Xie and Shen researched the rutting resistance, moisture susceptibility, and fatigue resistance of rubberized SMA (Stone Matrix Asphalt), analyzing the incorporation of CRM (Crumb Rubber Modifier) by using APA [11]. Malladi et al. researched moisture and rutting susceptibility of WMA by using APA and indicated that WMA performance is on a par with that of HMA and good potential for the widespread use of WMA [12].

Numbers of achievements have been obtained on APA test in the last decades, and most of the previous research focused on the evaluation of asphalt mixture antirutting performance by using APA test. Few attempts have reported the effects of testing conditions on APA rutting depth and are not sufficient in understanding the results of APA rutting tests influenced by test conditions. Therefore, in this paper, the APA experiments were performed under different test conditions for further understanding the influences of load, temperature, and other conditions on rutting depth, and an APA rutting prediction model was established to provide reference for further popularization and application of APA rutting tests.

2. Experimental Design

2.1. Asphalt Mixture Design and Specimen Preparation. Four kinds of AC-20 (Asphalt Concrete whose nominal maximum size of aggregate is 20 mm) asphalt mixtures and two kinds of AC-13 (Asphalt Concrete whose nominal maximum size of aggregate is 13 mm) asphalt mixtures (namely, AC-20 coarse-type modified asphalt, AC-20 coarse-type matrix asphalt, AC-20 fine-type modified asphalt, AC-20 fine-type matrix asphalt, AC-13 modified asphalt, and AC-13 matrix asphalt) were designed according to highway project to research the influence of experimental conditions on APA rutting tests. Shell 70 matrix asphalt and SBS modified asphalt were used in the experiments. Their technical properties are given in Table 1, and Table 2 conforms to the technical requirements published in *Technical Specification for Construction of Highway Asphalt Pavement (JTG F40-2004)*.

Three kinds of aggregate gradations (namely, AC-20 coarse-type, AC-20 fine type, and AC-13) were selected to

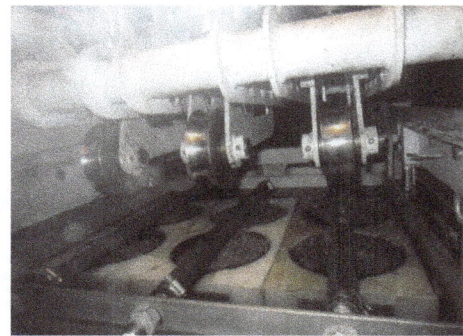

Figure 1: Cylinder specimens in APA testing.

mix with two kinds of asphalt binders, respectively, to prepare six kinds of asphalt mixtures. Optimal asphalt contents of asphalt mixtures were determined by the Marshall method; the results are presented in Table 3.

Cylinder specimens (Φ 150 × 75 mm) were molded by using SGC (Superpave Gyratory Compactor) according to the results of the material composition designs given previously, and the target air voids were 4 and 7%.

2.2. APA Rutting Test Schemes. APA rutting tests using SGC cylinder specimens (as shown in Figure 1) were conducted in both dry and immersion conditions, respectively. Under dry conditions, the specimens with 4% air voids were selected for APA rutting tests, and the test temperatures were 40, 50, and 60°C. The APA rutting test was the force of a concave wheel applied indirectly on the specimen through an inflatable rubber hose. In this paper, standard load was 445 ± 22 N; standard air pressure of the rubber hose was 690 ± 22 kPa; and running frequency of the wheel was 60 Hz, that is, 60 times back and forth per minute. To analyze the effect of overloading on APA rutting, two combinations of load and tire pressure (533 N/827 kPa and 889 N/827 kPa) were increased to test at 60°C.

Under immersion conditions, the specimens with the air voids of 4 and 7% were tested, respectively; test temperatures were 50 and 60°C, respectively; load was 445 N; and tire pressure was 690 kPa. All the specimens were placed in APA test machines at the test temperature for 6–24 h before testing to ensure that the specimens would reach the test temperature and maintain temperature equalizing.

3. Results and Discussion

3.1. Influence of Temperature on APA Rutting under Dry Conditions. Under dry conditions, the APA rutting depth development trends for different types of asphalt mixtures under three different temperatures are shown in Figures 2–4. Results show the high-temperature performances of the asphalt mixtures according to rutting depth-size order, under three different temperatures reaching unanimity, namely, AC-20 fine-type matrix asphalt mixture > AC-13 matrix asphalt mixture > AC-20 coarse-type matrix asphalt mixture > AC-20 fine-type modified asphalt mixture > AC-13 modified

TABLE 1: Technical properties of Shell 70 matrix asphalt.

Test index	Unit	Test result	Technical requirement
Penetration (100 g, 5 s, 25°C)	0.1 mm	76.83	60~80
Penetration index	—	−1.2	−1.5~1.0
Softening point	°C	46.2	≥45
Ductility (5 cm/min, 10°C)	cm	79.5	>25
Ductility (5 cm/min, 15°C)	cm	138.0	>100
Wax content (distillation method)	%	1.9	≤2.2
Flash point	°C	316.3	≥260
Solubility	%	99.58	≥99.5
Density (15°C)	g/cm³	1.036	Measured records
Density (25°C)	g/cm³	1.033	—
Quality loss after RTFOT	%	0.170	±0.8
Residual penetration after RTFOT (100 g, 5 s, 25°C)	0.1 mm	52.83	—
Penetration ratio after RTFOT (25°C)	%	68.76	≥61
Residual ductility after RTFOT (5 cm/min, 10°C)	cm	10.4	>6
Residual ductility after RTFOT (5 cm/min, 15°C)	cm	46.8	>15

TABLE 2: Technical properties of SBS modified asphalt.

Test index	Unit	Test result	Technical requirement
Penetration (100 g, 5 s, 25°C)	0.1 mm	65.67	60~80
Penetration index	—	0.1	≥−0.4
Softening point	°C	83.1	≥55
Ductility (5 cm/min, 5°C)	cm	33.8	≥30
Flash point	°C	272	≥230
Solubility	%	99.7	≥99
Elastic recovery (25°C)	%	99.67	≥65
Segregation of storage stability, difference of softening point after 48 h	°C	1.6	≤2.5
Density (15°C)	g/cm³	1.034	—
Relative density (15°C)	—	1.037	—
Density (25°C)	g/cm³	1.030	—
Relative density (25°C)	—	1.033	—
Quality loss after RTFOT	%	−0.2	±1.0
Residual penetration after RTFOT (100 g, 5 s, 25°C)	0.1 mm	48.7	—
Penetration ratio after RTFOT (25°C)	%	74.16	≥60
Residual ductility after RTFOT (5 cm/min, 5°C)	cm	21.46	≥20

asphalt mixture > AC-20 coarse-type modified asphalt mixture. Thus, it shows that modified bitumen can improve the high-temperature performance of asphalt mixtures. At a test temperature of 40°C, the growth of rutting of different asphalt mixtures is moderate with small degree rutting. At 50°C, rutting is 2-3 times bigger than rutting at 40°C; rutting at 60°C is approximately 1.5 times bigger than rutting at 50°C and 3–5 times bigger than rutting at 40°C. Comparisons of APA rutting test results indicate that temperature appears to significantly affect the development of rutting, which has been confirmed by Zhang et al. [13–15].

Results show that rutting depth changes with the number of loads and the temperature, according to the APA rutting test results under different temperatures. Thus, the temperature-effects regression model shown in (1) was established.

$$\left[\frac{R}{R_0}\right] = \left[\frac{T}{T_0}\right]^{\alpha} \left[\frac{N}{N_0}\right]^{\beta}, \tag{1}$$

where R is the predicted rutting depth (mm); R_0 is the benchmark rutting depth under temperature T_0 and at loading times N_0 on the basis of the rutting test (mm); T, N are the test temperature and the loading times corresponding to the predicted rutting depth; and α, β are the regression coefficients of the materials.

For the asphalt mixtures, many factors, such as asphalt properties, asphalt content, air void, aggregate gradation,

TABLE 3: Design gradations and optimal asphalt contents of asphalt mixtures.

Gradation type	Mass percentage (%) through the following mesh (mm)												OAC (%)	
	26.5	19	16	13.2	9.5	4.75	2.36	1.18	0.6	0.3	0.15	0.075	Matrix asphalt	Modified asphalt
AC-20 C-type	100	92.3	82.9	69.6	58.2	39.9	27.3	18.7	13.7	10.9	6.9	4.7	4.0	4.2
AC-20 F-type	100	95.1	88.9	79.7	68.5	53.3	40.9	31.1	21.6	16.3	8.9	5.4	4.3	4.3
AC-13	100	100	100	97.3	75.4	45.6	27.1	20.3	14.2	10.7	8.2	5.1	5.1	4.9

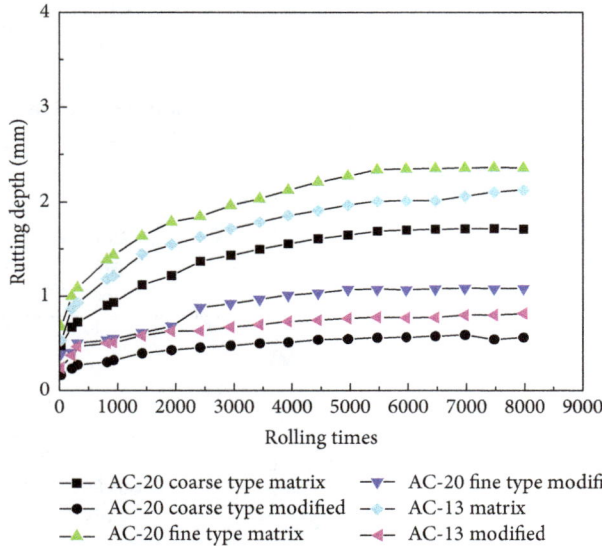

FIGURE 2: APA rutting depths of asphalt mixtures at 40°C under air-dry conditions.

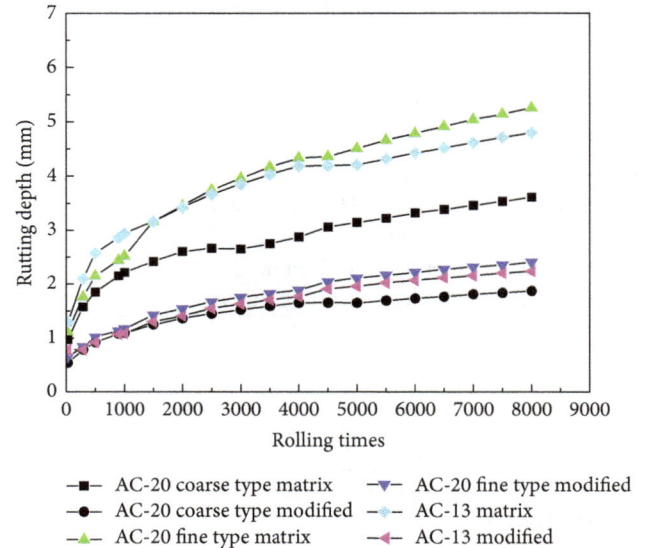

FIGURE 3: APA rutting depths of asphalt mixtures at 50°C under air-dry conditions.

and others could influence rutting depth R and the performance coefficient of the materials α, β. The model would be greatly complicated if these factors were considered in the temperature-effects regression model of rutting depth. Therefore, to simplify the model, R_0 is introduced into formula (1) to standardize the influence of the asphalt mixture type for predicting rutting depth. Benchmark rutting depth R_0 is the result of the APA rutting test under various conditions of temperature T_0 and loading times N_0. Groups of data (324) from 54 times rutting tests were used in the regression analysis. Eight benchmark depths were selected to regress, respectively, and each of them corresponded to different test temperatures and loading times. The regression analyses show the following: a correlation coefficient of $R \geq 0.94$, high correlations exist between both the independent and the dependent variables; standard error ≤ 0.2 mm, the fitting degree of regression equation is good; and significance $F < 0.01$, the established regression equation is very significant. The average values $\alpha = 3.062$ and $\beta = 0.246$ were applied in the regression equation because a negligible difference exists between the regression coefficients α, β obtained from the eight benchmark depths. Thus, (2) can be obtained.

$$\left[\frac{R}{R_0} \right] = \left[\frac{T}{T_0} \right]^{3.062} \left[\frac{N}{N_0} \right]^{0.246}. \tag{2}$$

3.2. Influence of Load on APA Rutting under Dry Conditions. Under dry conditions, the influence from combinations of

different loads and tire pressures on the APA rutting depth of asphalt mixtures is shown in Figure 5. The results show that the orders of asphalt mixtures under three combinations of load and tire pressure (namely, 690 kPa/445 N, 827 kPa/533 N, and 827 kPa/889 N) are the same, from excellent to poor for AC-20 coarse-type modified asphalt, AC-13 modified asphalt, AC-20 fine-type modified asphalt, AC-20 coarse-type matrix asphalt, AC-13 matrix asphalt, and AC-20 fine-type matrix asphalt, respectively.

The testing results are consistent with research results of Gu et al., which indicate rutting depths all tend to increase with increases in load and tire pressure [16, 17]. Little difference exists among rutting test results under combinations of 690 kPa/445 N and 827 kPa/533 N; rutting depths under combinations of 827 kPa/533 N were approximately 1.08 times that of the rutting depths under combinations of 690 kPa/445 N. Comparing combinations of 690 kPa/445 N revealed that the load of 827 kPa/533 N increased 20% and tire pressure increased 20%, however, tire ground pressure increased by only 6%, that is, 42 kPa. A large difference exists between the rutting test results under combinations of 827 kPa/889 N and two other combinations. For matrix asphalt mixtures, rutting deeper than 12 mm appeared under far fewer than 8000 load times. For modified asphalt mixtures, the rutting depths under combinations of 827 kPa/889 N were approximately 2.72 times that of rutting depths under combinations of 690 kPa/445 N and approximately 2.45 times that of rutting

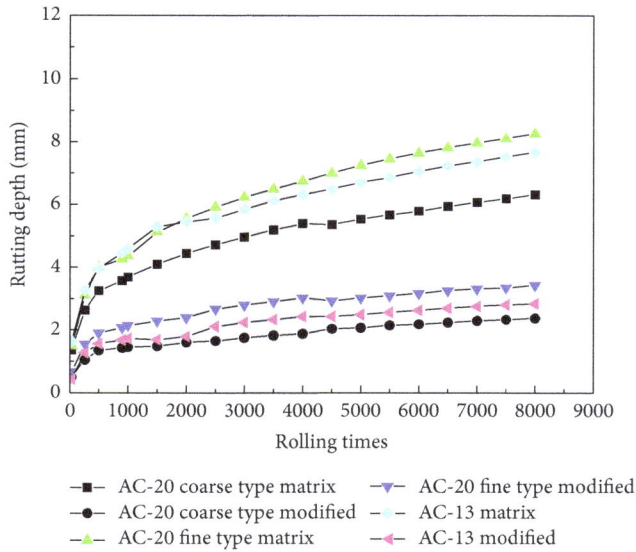

FIGURE 4: APA rutting depths of asphalt mixtures at 60°C under air-dry conditions.

depths under combinations of 827 kPa/533 N. Thus, overloading requires more attention for its significant effect on the rutting depth of asphalt mixtures.

3.3. Development Trend of Immersion APA Rutting. Development trends of APA rutting depth of asphalt mixtures with the air voids of 4% and 7% under different temperature and humidity conditions are shown in Figures 6–9. Cracking, asphalt exfoliating, and stripping appeared at the interface of aggregate and asphalt under the water bath condition in the APA rutting test. A decline in the adhesion of aggregate and asphalt also results in the reduction of the structural stability of the asphalt mixture. The larger rutting appeared under the drier conditions in the APA rutting test.

The rutting depth ratio is defined as the ratio of dry rutting depth and immersion rutting depth, and the water stability of asphalt mixture with greater rutting ratio is better [18]. The rutting depth ratios were calculated according to APA rutting depths under different immersing conditions, and the results are shown in Figures 10–13.

Analyses show that the rutting depth ratio of the modified asphalt mixture is greater which means that the water stability is better considering the viscosity of modified asphalt is greater than that of matrix asphalt, more polar materials exist in the matrix asphalt, and the matrix asphalt has good wettability. The water stabilities of the AC-13 modified asphalt mixture and the AC-13 matrix asphalt mixture are worse than those of the other limestone mixtures because of the use of acidic stone granite. The AC-13 matrix asphalt mixture is obvious, its rutting depth ratio is the smallest, and the specimens were spilled at the test temperature of 60°C.

Comparisons of the rutting depth ratios at different loading times show that the rutting depth ratios increases, whereas the influence of water decreases gradually with the increase in loading times. The air void of asphalt mixtures

decreases and the aggregate skeleton resistance increases with the repeated action of the loading wheel. The increasing amplitude of pore-water pressure is smaller compared to that of the increasing amplitude of the aggregate skeleton resistance, although pore-water pressure also increases. Thus, the influence of water decreases gradually. The results also show that the order of the water stability of the asphalt mixture ordered by rutting depth ratios under 25 or by 4000 loading times is not stable, whereas the order of the water stability of the asphalt mixture ordered by rutting depth ratios under 8000 loading times is more stable. Therefore, 8000 loading times of APA equipment should be ensured when evaluating tests of water stability performance, and test results from loading times under 25 or 4000 are not recommended for evaluating water stability of asphalt mixtures.

3.4. Grey Correlation Analyses of Influencing Factors in APA Rutting Tests. Numbers of researchers analyzed influencing factors of rutting deformation characteristics on asphalt pavement. Peilong et al. analyzed the correlation of influencing factors of rut resistance using grey theory, indicating that rut deformation rate has the maximum grey correlation with rate rut depth among five influencing factors of void ratio, graduation index, rut deformation rate, passing rate at 4.75 mm in middle layer, and filler/asphalt ration [19]. But few attempts have been reported about effects of test conditions on APA rutting using grey theory. Therefore, the correlation degrees of temperature, combinations of load and tire pressure, and water on APA rutting depths were researched by using the grey correlation method on the basis of the tests described previously [20]. The grey correlation method orders correlation degrees and identify the main factor influencing the target value by calculating their target values and influence factors.

Firstly, pinpoint both the reference sequence and the compared sequence when using the grey correlation method for analysis. Assume that the reference sequence is X_0, $X_0 = \{X_0(k) \mid k = 1, 2, \ldots, n\}$; the comparison sequence is X_i ($i = 1, 2, \ldots, m$), $X_i = \{X_i(k) \mid k = 1, 2, \ldots, n\}$. Establish an average value for each of these sequences; namely, each sequence is divided by the average value, and a new sequence is obtained.

$$Y_0 = \{Y_0(k)\} = \{X_0(k) / \mid k = 1, 2, \ldots, n\}. \tag{3}$$

In (3), $\overline{X_0} = \sum_{k=1}^{n} X_0(k)/N$.

$$Y_i = \{Y_i(k)\} = \{X_i(k) / \mid k = 1, 2, \ldots, n\}. \tag{4}$$

In (4), $\overline{X_i} = \sum_{k=1}^{n} X_i(k)/N$, ($i = 1, 2, \ldots, m$).

Y_0 is new reference sequence; Y_i is a new compared sequence. The correlation of the reference curve and the compared curve at time K (indicator and space) is

$$\zeta_i(k)$$

$$= \frac{\min_i \min_k |Y_0(k) - Y_i(k)| + \zeta \max_i \max_k |Y_0(k) - Y_i(k)|}{|Y_0(k) - Y_i(k)| + \zeta \max_i \max_k |Y_0(k) - Y_i(k)|}. \tag{5}$$

In (5), ζ is the distinguish coefficient; its value is between 0 and 1; $\min_i \min_k |Y_0(k) - Y_i(k)|$ is the smallest difference

(a) Matrix asphalt mixture

(b) Modified asphalt mixture

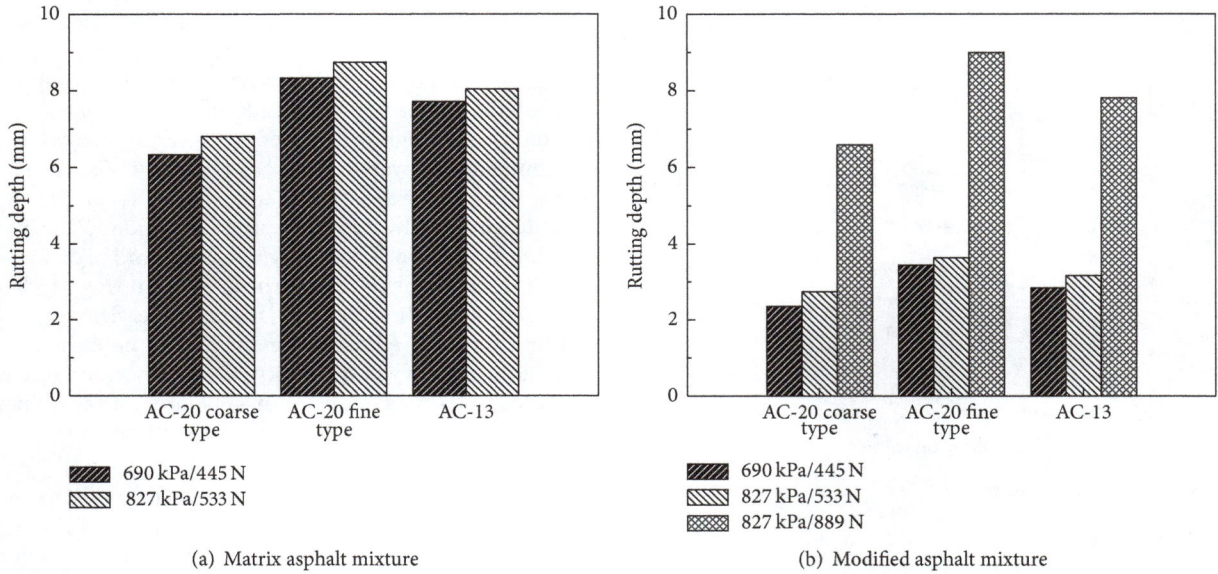

FIGURE 5: APA rutting depths of asphalt mixtures at different combinations of load and tire pressure.

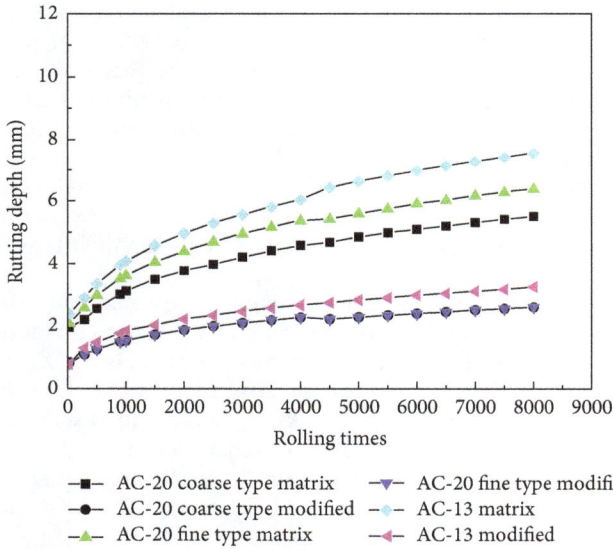

FIGURE 6: APA rutting depths of asphalt mixtures with 4% air voids at 60°C under immersing conditions.

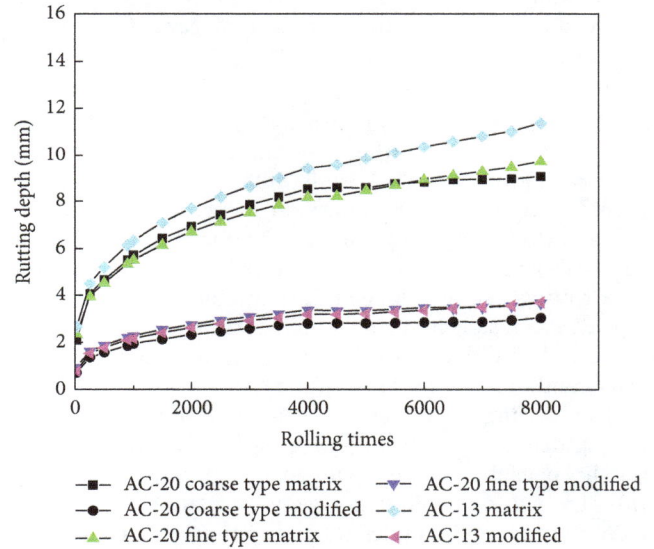

FIGURE 7: APA rutting depths of asphalt mixtures with 7% air voids at 60°C under immersing conditions.

between two poles; $\max_i\max_k|Y_0(k) - Y_i(k)|$ is the biggest difference between two poles; $k = 1, 2, \ldots, n; i = 1, 2, \ldots, m$.

The expression of the correlation degree is $\gamma_i = (1/N)\sum_N^n \zeta_i(k)$; γ_i is the correlation degree of the curve $X_i(Y_i)$ and the reference sequence $X_0(Y_0)$. When the correlation degrees are ordered, the results show that the bigger γ_i leads to the steadier development trend of X_i and X_0 and the greater influence of X_i on X_0. The calculation results of the grey correlation degree of different factors are shown in Figure 14. The Abscissas 1 to 6 corresponded to the AC-20 coarse-type matrix asphalt mixture, AC-20 coarse-type modified asphalt

mixture, AC-20 fine-type matrix asphalt mixture, AC-20 fine-type modified asphalt mixture, AC-13 matrix asphalt mixture, and the AC-13 modified asphalt mixture.

The results show that the order of correlation degree is temperature, load, air void, and immersing condition. Thus, temperature is the closest correlation associated with asphalt mixture rutting depth. The main reason is that the asphalt binder in the asphalt mixture is temperature-sensitive, and the deformation depends significantly on temperature. Under high-temperature conditions, the asphalt is prone to flow, and rutting appears. This finding also verifies that asphalt

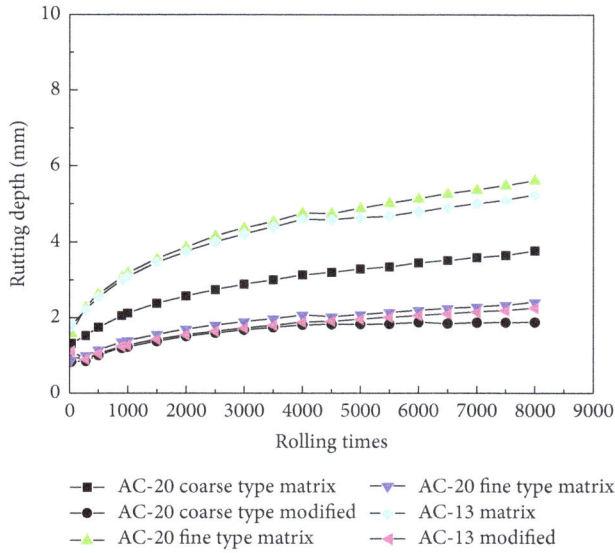

FIGURE 8: APA rutting depths of asphalt mixtures with 4% air voids at 50°C under immersing conditions.

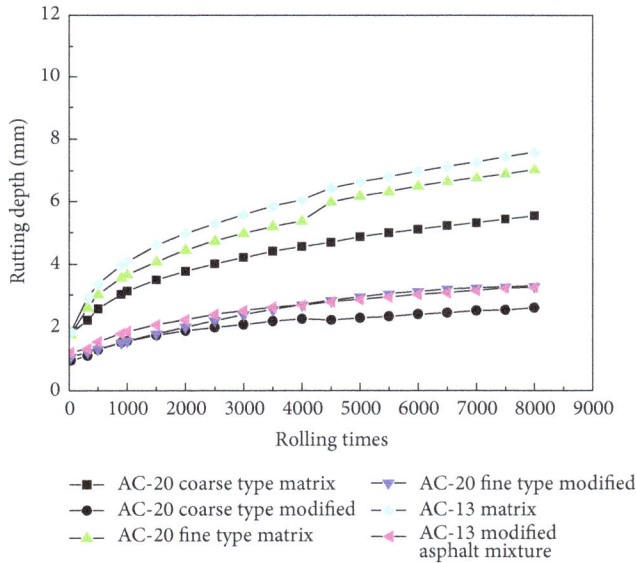

FIGURE 9: APA rutting depths of asphalt mixtures with 7% air voids at 50°C under immersing conditions.

pavement rutting forms mainly under high temperatures. Also, load is closely associated with asphalt mixture rutting depth. Increasing the load might damage the interlock structure between the aggregates in the asphalt mixture, which might affect the temperature of the pavement structure and lead to rutting. The relationship between air void and rutting is that the existence of air voids might cause supplementary compaction under loading. However, the air voids of asphalt mixtures are limited. Hence, the correlation degree of air voids is relatively low. Meanwhile, rutting should be weak. Water damage to the asphalt mixture is mainly attributable to

water entering the interface of the asphalt and the aggregate and causing asphalt and aggregate spalling. However, mixtures with dense gradation are mainly used in this research while the water that could enter the interface of the asphalt to aggregate is limited. Thus, water has a relatively low influence on the stability of mixtures.

4. Indoor Rutting Prediction Modeling and Validation Based on APA

With reference to the model of APA rutting depth-temperature-loading times in formula (1), the indoor APA rutting prediction model accounts for many factors such as temperature, loading time, loading level, and air void and presents these factors in

$$\frac{R}{R_0} = \left(\frac{T}{T_0}\right)^{\alpha}\left(\frac{N}{N_0}\right)^{\beta}\left(\frac{P}{P_0}\right)^{\gamma}\left(\frac{V}{V_0}\right)^{\nu},\qquad(6)$$

where T_0, P_0, N_0, V_0 are the benchmark parameters of the indoor APA rutting prediction model and R_0 is the benchmark parameter showing the characterization of resistance to the permanent deformation of the material, which is obtained from the APA rutting depth under specific conditions. The specific conditions are that the test temperature is the reference temperature T_0; the test loading time is the reference loading time N_0; the test loading level is the reference loading level P_0; and the test air void is the reference air void V_0.

The APA standard test conditions are as follows: $T_0 = 60°C$; $N_0 = 8000$ times; $P_0 = 0.714$ MPa; and $V_0 = 4\%$. The data set (R/R_0, T/T_0, N/N_0, P/P_0, and V/V_0) could be obtained from the rutting depth under other test conditions R divided by the rutting depth under standard test conditions R_0 and other test conditions divided by the standard test conditions. By taking the logarithm of each data in the data set, $\lg(R/R_0)$ is the dependent variable, while $\lg(T/T_0)$, $\lg(N/N_0)$, $\lg(P/P_0)$, and $\lg(V/V_0)$ are the independent variables. The regression index could be obtained after multiple regression analyses of the data set. By applying these results to (6), an APA rutting prediction model which could take into consideration any significant factors could be obtained.

$$\frac{R}{R_0} = \left(\frac{T}{T_0}\right)^{2.960}\left(\frac{N}{N_0}\right)^{0.243}\left(\frac{P}{P_0}\right)^{1.625}\left(\frac{V}{V_0}\right)^{0.413}.\qquad(7)$$

According to the predicted APA rutting depth in (7), the correlation coefficient of the predicted and measured APA rutting depths shown in the Figure 15 reaches 96.3%. Only seven data deviations exist, which are more than 1 mm among the measured and the predicted APA rutting depths in 1188 groups of experimental data. The biggest deviation is 2.392 mm.

5. Conclusions

(1) The orders of rutting depths for 6 kinds of asphalt mixtures under different temperatures and combinations of load and tire pressure are almost the same; that is, AC-20 type matrix asphalt mixture > AC-13

(a) Matrix asphalt mixture

(b) Modified asphalt mixture

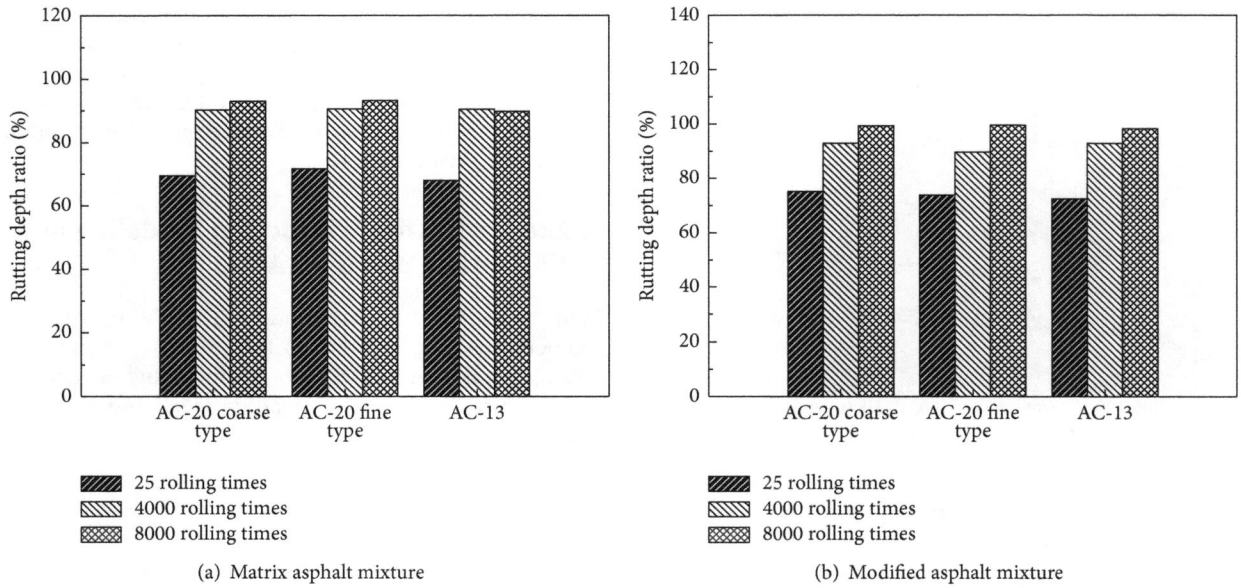

FIGURE 10: Rutting depth ratio of specimens with 4% air void at 60°C under different rolling times.

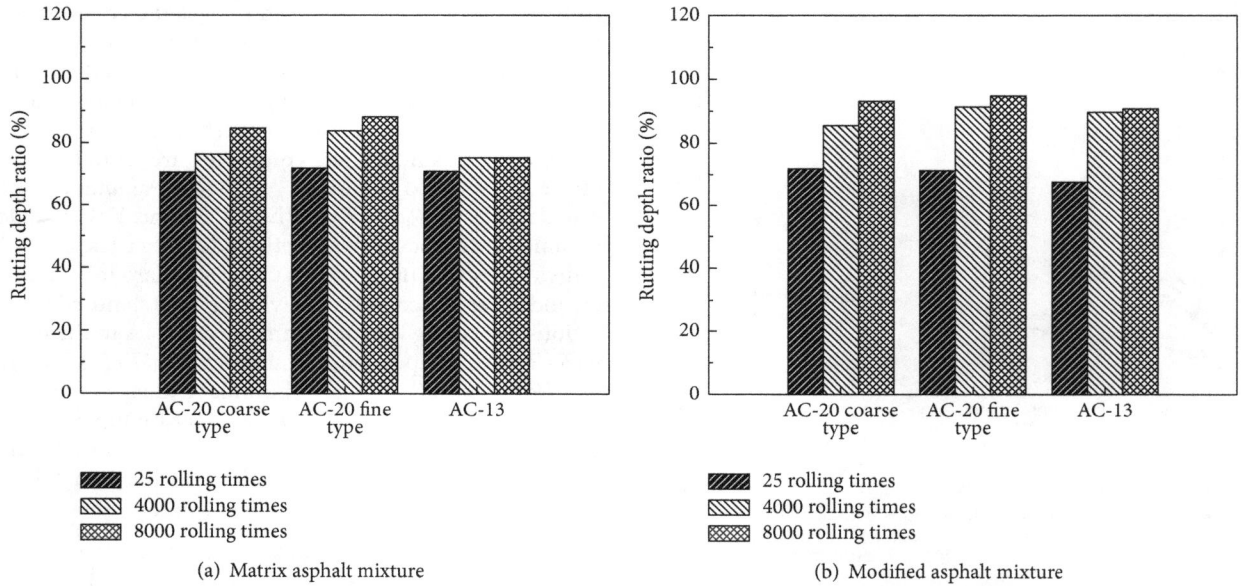

(a) Matrix asphalt mixture

(b) Modified asphalt mixture

FIGURE 11: Rutting depth ratios of specimens with 7% air void at 60°C under different rolling times.

matrix asphalt mixture > AC-20 coarse-type matrix asphalt mixture > AC-20 fine-type modified asphalt mixture > AC-13 modified asphalt mixture > AC-20 coarse-type modified asphalt mixture. Compared with the matrix asphalt mixture, the APA rutting depth of the modified asphalt mixture is smaller. Besides he rutting resistance of the modified asphalt mixture is also better.

(2) The effect of temperature on the development of rutting is nonnegligible. At a test temperature of 40°C, the development of rutting of different asphalt

mixtures is relatively mild, and the rutting depth is small. Rutting depth at a test temperature of 50°C is 2 to 3 times as big as the rutting at 40°C, whereas rutting depth at 60°C is approximately 1.5 times as large as rutting at 50°C. The regression model of APA rutting depth-temperature-loading times was established and verified on the basis of rutting depth under different test temperatures and loading times.

(3) The correlation degrees of temperature, combinations of load and tire pressure, and water on the APA rutting depth were researched through grey

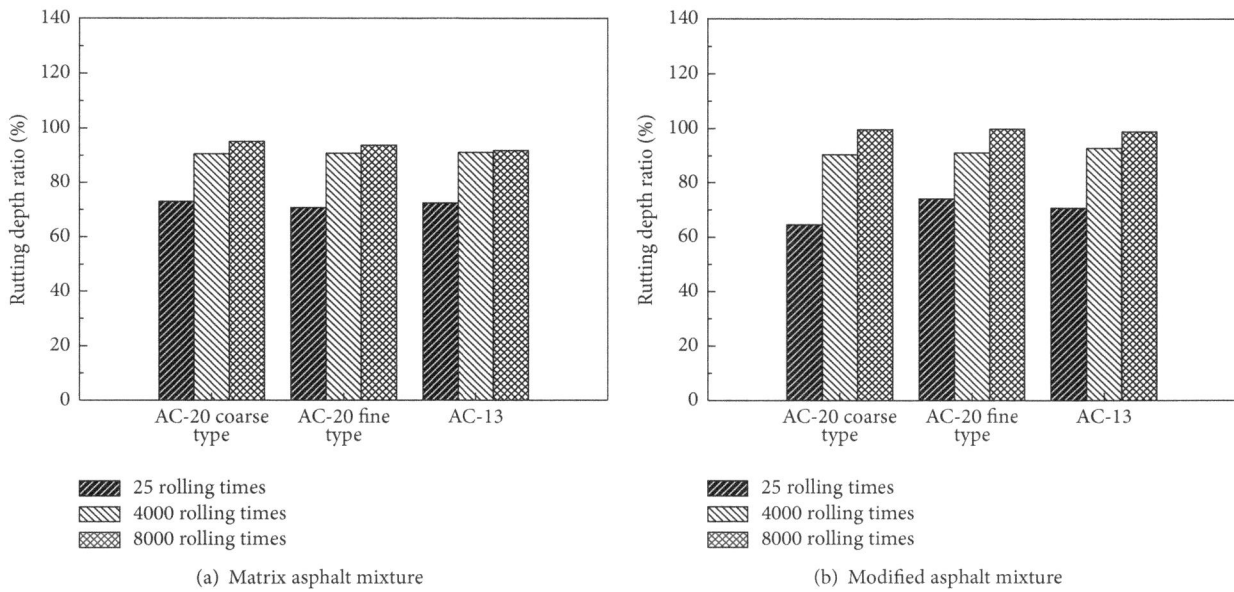

(a) Matrix asphalt mixture

(b) Modified asphalt mixture

FIGURE 12: Rutting depth ratios of specimens with 4% air void at 50°C under different rolling times.

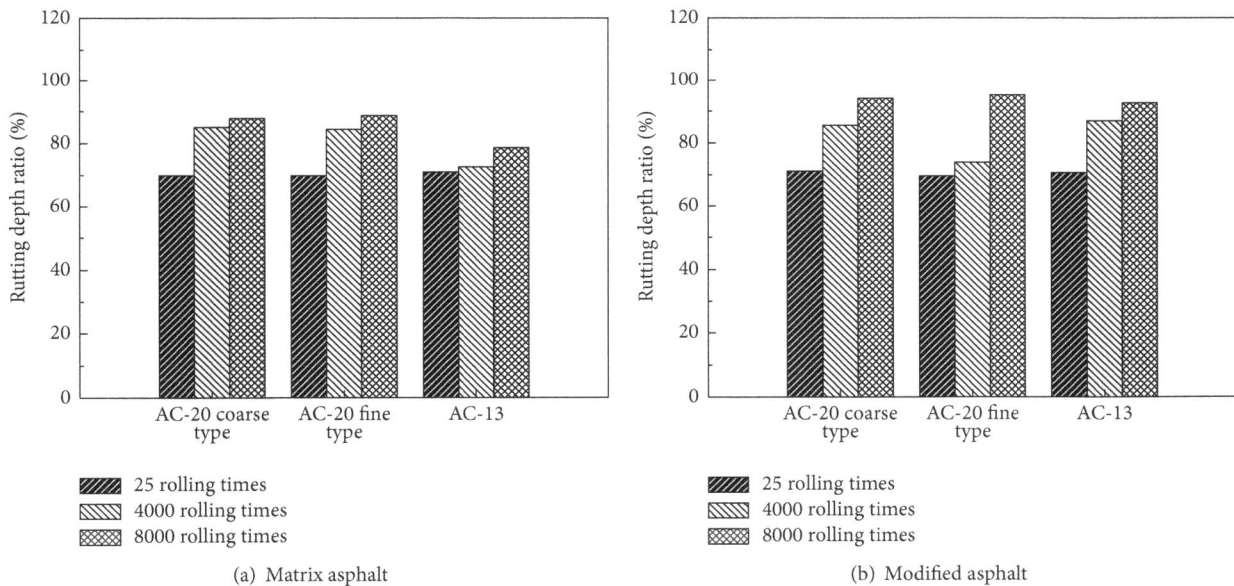

(a) Matrix asphalt

(b) Modified asphalt

FIGURE 13: Rutting depth ratios of specimens with 7% air void at 50°C under different rolling times.

correlation method. The results show that temperature is the most significant influencing factor.

(4) The indoor APA rutting prediction model considering several factors such as temperature, loading times, loading level, and air void was established. The prediction of this model is precise and convincing. Rutting depth under other conditions could be predicted on the basis of APA rutting depth under these benchmark conditions.

Conflicts of Interest

The authors declare that there are no conflicts of interest regarding the publication of this paper.

Acknowledgments

This work was supported by the National Natural Science Foundation of China (no. 51408043), the Department of Science & Technology of Shaanxi Province (no. 2016KJXX-69), the Open Foundation of Key Laboratory of Highway

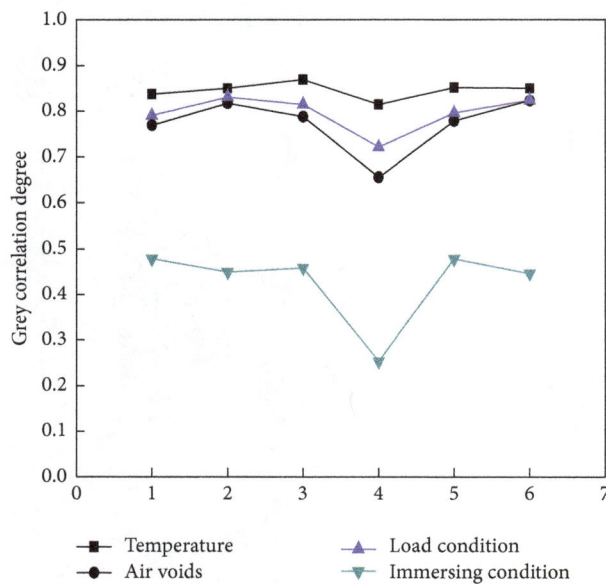

FIGURE 14: Grey correlation degrees of temperature, air void, load, and immersing condition.

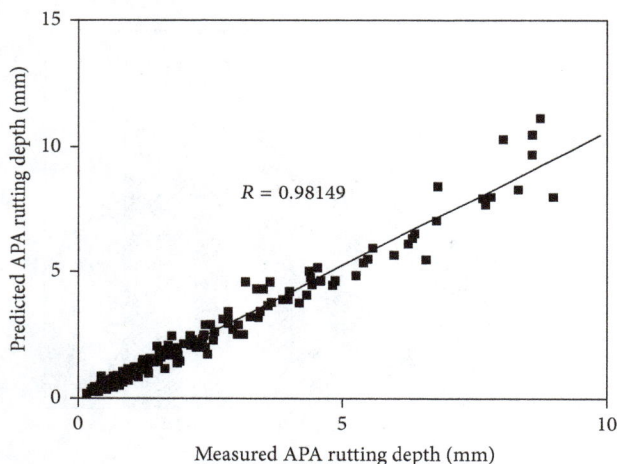

FIGURE 15: Comparison of measured and predicted APA rutting depths.

Construction & Maintenance Technology in Loess Region, Ministry of Transport, China (KLTLR-Y11-7), and the Special Fund for Basic Scientific Research of Central College of Chang'an University (nos. 310821153502 and 310821173501). The authors gratefully acknowledge their financial support.

References

[1] J. Xie, Y.-Z. Li, and L.-G. Shao, "Laboratory evaluation on the moisture susceptibility of hot asphalt mixtures by asphalt pavement analyzer," *Journal of Hunan University of Science and Technology (Natural Science Edition)*, vol. 20, no. 2, pp. 53–57, 2005.

[2] H.-F. Han, W.-M. Lu, and G.-P. He, "Effects of water on permanent deformation potential of hot-mix asphalt," *China Journal of Highway and Transport*, vol. 16, no. 4, pp. 4–8, 2003.

[3] C. Zhang, Z. Xu, R. Yang, and Y. Qiu, "High temperature performance and gradation design in skeleton and dense construction asphalt mixture," *Journal of Tongji University*, vol. 36, no. 5, pp. 603–608, 2008.

[4] Z. Junbiao, X. Ke, and Z. Xiaoning, "Comparative analysis of loading mode between RLWT and APA rutting test," *Journal of Highway and Transportation Research and Development*, vol. 7, pp. 331–336, 2006.

[5] L.-P. Cao, L.-J. Sun, and Z.-J. Dong, "Deformation property of asphalt mixture at varying temperature," *Journal of Building Materials*, vol. 12, no. 5, pp. 625–630, 2009.

[6] Z.-J. Xue, J.-N. Wang, and Y.-Q. Tan, "Performance of reflection crack resistance of stress-absorbing waterproof interlayer," *Journal of South China University of Technology (Natural Science Edition)*, vol. 36, no. 10, pp. 81–85, 2008.

[7] F. Xiao, S. Amirkhanian, H. Wang, and H. Wang, "Performance characteristics comparisons of various asphalt mixture technologies," *Journal of Testing and Evaluation*, vol. 43, no. 4, pp. 841–852, 2014.

[8] F. Xiao, D. A. Herndon, S. Amirkhanian, and L. He, "Aggregate gradations on moisture and rutting resistances of open graded friction course mixtures," *Construction and Building Materials*, vol. 85, pp. 127–135, 2015.

[9] J. F. Rushing and D. N. Little, "Static creep and repeated load as rutting performance tests for airport HMA mix design," *Journal of Materials in Civil Engineering*, vol. 26, no. 9, Article ID 04014055, 2014.

[10] A. Ali, A. Abbas, M. Nazzal, A. Alhasan, A. Roy, and D. Powers, "Effect of temperature reduction, foaming water content, and aggregate moisture content on performance of foamed warm mix asphalt," *Construction and Building Materials*, vol. 48, pp. 1058–1066, 2013.

[11] Z. Xie and J. Shen, "Performance properties of rubberized stone matrix asphalt mixtures produced through different processes," *Construction and Building Materials*, vol. 104, pp. 230–234, 2016.

[12] H. Malladi, D. Ayyala, A. A. Tayebali, and N. P. Khosla, "Laboratory evaluation of warm-mix asphalt mixtures for moisture and rutting susceptibility," *Journal of Materials in Civil Engineering*, vol. 27, no. 5, Article ID 04014162, 2015.

[13] J. Zhang, J. Pei, and B. Wang, "Analysis of asphalt pavement rutting in continuously changing temperature field and early warning for high temperature," *Journal of Tongji University*, vol. 39, no. 2, pp. 242–258, 2011.

[14] X. Huang, Y. Yang, H. Li et al., "Asphalt pavement short-term rutting analysis and prediction considering temperature and traffic loading conditions," *Journal of South China University of Technology (English Edition)*, vol. 3, no. 3, pp. 1–24, 2009.

[15] N.-X. Zheng, S.-S. Niu, and X.-Q. Xu, "Temperature, axle load and axle load frequency model of rutting prediction of heavy-duty asphalt pavement," *China Journal of Highway and Transport*, vol. 22, no. 3, pp. 7–13, 2009.

[16] X.-Y. Gu, Q.-Q. Yuan, and F.-J. Ni, "Analysis of factors on asphalt pavement rut development based on measured load and temperature gradient," *China Journal of Highway and Transport*, vol. 25, no. 6, pp. 30–36, 2012.

[17] L. Zhou, N. Fujian, and Z. Yanjing, "Impact of Environment Temperature and Vehicle Loading on Rutting Development in Asphalt Concrete Pavement," *Journal of Highway and Transportation Research and Development*, vol. 28, no. 3, pp. 42–47, 2011.

[18] M. W. Witczak, K. Kaloush, T. Pellinen, and et al., "Simple performance test for Superpave mix design," National Cooperative

Highway Research Program (NCHRP) Report 465, Transportation Research Board, National Research Council, Washington, D.C., USA, 2002.

[19] L. Peilong, S. Yanjun, and Z. Zhengqi, "Analysis of Rutting Deformation Characteristics on Asphalt Pavement and It's Influencing Factors," *Subgrade Engineering*, vol. 3, pp. 37–40, 2011.

[20] D. Julong, "Grey Control System," *Journal of Huazhong University of Science and Technology*, vol. 10, no. 3, pp. 9–18, 1982.

2

Performance Characteristics of Silane Silica Modified Asphalt

Xuedong Guo, Mingzhi Sun, Wenting Dai, and Shuang Chen

School of Transportation, Jilin University, Changchun 130022, China

Correspondence should be addressed to Wenting Dai; daiwt@jlu.edu.cn

Academic Editor: Ying Li

At present there are many kinds of fillers and modifier used for modified asphalt, but the effect of modifier differs in thousands of ways; most of them can increase the high temperature performance of asphalt, but the modified effect of low-temperature crack resistance, water stability, and antifatigue performance is different. Aiming at the subsistent problems, this paper innovatively puts forward the idea of taking the silane silica (nanosilica modified with silane coupling agent) as filler to develop one kind of modified asphalt concrete which has excellent comprehensive performance based on the idea of enhance the whole performance of asphalt concrete and interface consolidation strength between aggregate and asphalt at the same time. The best mixing amount of silane silica and the production process of modified asphalt were conducted by contrasting the test date as penetration, viscosity, and softening point; the aging of asphalt and modified asphalt was analyzed by TG test, the superiority of silane silica modified asphalt is more clearly understandable by chemical analysis results. Meanwhile it is proved that silane silica has positive effect to improve the mixture of high and low temperature performance, water stability, and aging resistance through a series of road performance tests.

1. Introduction

Asphalt has complicated chemical composition which exhibits both viscous and elastic properties that heavily depend on both time and temperature. Researchers and engineers have been trying to use various kinds of modifiers to modify and improve the asphalt materials' performance; up to now the main kinds of modifiers had polymers (such as SBS, rubber, and resin [1–3]), antistripping agent (such as lime ash, alkaline saponin, and coupling agent [4]), fiber (such as steel fiber, wood fiber, basalt fiber, polyester fiber, glass fiber, and carbon fiber [5–7]), and filler (such as carbon black, sulfur, diatomite, and silica [8–14]). The effects of various modifiers are different; in general a certain kind of modifier has only one or a few positive effects on high and low temperature performance, aging performance, or water stability. Therefore, it is of great significance to develop a modified asphalt of comprehensive performance, and the research work has a long way to go.

In recent years, nanotechnology has been widely used in many fields as a new technology with broad prospects and rich creativity. Many researchers have used nanomaterials for cement, but the nanomaterial modified asphalt is relatively late compared to modified concrete. Because of the very small size and huge surface area, nanomaterials can achieve the same effect of the ordinary material only by adding relatively little amount. The research on nanoclay by You et al. [8] and Jahromi and Khodaii [9] showed that it can significantly improve asphalt softening point, reduce needle penetration and ductility, improve the antiaging performance by adding little dosage of nanoclay, and improve the rheological properties of asphalt by enhancing the toughness and elasticity of asphalt at the same time. The research by Yao et al. [13] showed that the 2%–4% nanosilica can reduce the rutting depth to nearly half. However, it has not been able to get a good solution for the problem that inorganic nanomaterials have poor dispersal ability to mix together with the organic asphalt and inorganic materials are easy to agglomerate. This paper is proposed to solve the above problems by using one kind of nanosilica which is modified with silane coupling agent (hereinafter called silane silica).

The situation that silane silica used in rubber, paint, plastic, and other industry applications is getting better and better [15–20]. As new materials, more and more researchers pay attention to it, but it is almost in a blank stage that the silane silica is used as modifier to improve the adhesion between asphalt and aggregate and the road performance of asphalt mixture. As a superiority modifier due to its potentially beneficial properties (e.g., strong adsorption, huge surface area, good stability, and excellent dispersal ability), silane silica has not really been recognized in the using process of the asphalt pavement. The objective of this study is to evaluate the modified effect and process with silane silica in order to improve the overall road performance of asphalt mixture and enhance the adhesion between asphalt and aggregate.

2. Raw Material Property and Mechanism of Action

2.1. Mechanism Analysis of Modified Material. The special properties of nanoscale modified materials due to the mesoscopic size are the hot spots in the research of composite materials. Due to the large surface of inorganic nanomaterials and the poor compatibility between inorganic and organic phase, ordinary nanosilica particles in asphalt will cluster easily, and it is hard to achieve uniform and stable dispersion through high speed stirring. Grafting modification of nanosilica by using silane coupling agent can not only realise inorganic material surface modification but also reduce the surface energy of inorganic materials and also make distribution of inorganic materials in organic matrix materials more uniform.

In addition, silane contains two different chemical functional groups by chemical bonding theory analysis, and one end reacts with hydroxyl groups of inorganic materials (such as silica particles, aggregate) to form a covalent bond; the other end can form stationary phase with organic functional groups of asphalt molecular through chemical reaction. So it can be considered that the silane has a bridging effect on the organic-inorganic system. From the standpoint of the mechanism of silane, particles of nanosilica modified by silane can better dissolve and disperse into asphalt, and the strength and toughness of asphalt binder can improve greatly. Besides, the material disposed by silane can form an organic coupling agent film on the surface and aggregate change from hydrophily into lipophilicity which provides the basis for the consolidation enhancement of the interface between asphalt and aggregate. Therefore, the mechanical strength and the impact toughness of the composite modified asphalt are much better than the matrix asphalt, and it will have good high temperature performance, antiaging, antifatigue, and other properties.

2.2. Raw Material

2.2.1. Silane Silica. The hydrophobic silane silica which is selected to modify asphalt is silica which reacts with silane coupling agent. The contact angle is reaching 70° to 150° of hydrophilic silica after surface modification and the surface has grafted polysiloxane which successfully changes the hydrophilic silica into excellent hydrophobic silane silica. The main technical indicators of silane silica are shown in Table 1.

TABLE 1: Physical properties of silane silica.

Project	Index
Model	Silane silica
Appearance	White powder
Whiteness/%	≥93
Silica content (dry goods)/%	≥98
PH value	≥7
BET (specific surface area)/m^2/g	≥220
DBP mL/100 g (oil absorption capacity)	≥270
Particle size/nm	≤100
Cauterant (dry goods) (1000°C, 2 h)/%	≤7.0
Heating reduction (105°C, 2 h)/%	≤6.0

2.2.2. Asphalt. Panjin number 90 was chosen to be the base asphalt, and the main technical indexes of asphalt were able to meet the requirements of the standard.

2.2.3. Aggregate. The aggregate is from Jiutai stone material factory which is alkaline aggregate.

3. Test to Find the Optimum Mixing Amount of Silane Silica

3.1. Needle Penetration and Penetration Index. 90# asphalt was used to modify with silane silica; the preparation method of modified asphalt was as follows: First control asphalt was heated to molten state; then silane silica was added into control asphalt at concentrations of 0.5%, 1%, 1.5%, 2%, 3%, 4%, and 5% by weight of the control asphalt and mixed in the high-speed shear machine. During the mixing, the asphalt samples were kept at constant temperature of 160°C and blended using a shear rate of 4000 r/min for an hour. Finally we obtain modified asphalt with different dosage of silane silica added. The optimum mixing amount of silane silica added in asphalt is determined according to the basic index of the asphalt such as penetration, viscosity, and softening point. The penetration and PI value of asphalt under different temperature and content according to the specification of asphalt penetration test are measured in Table 2.

The test data of Table 2 showed that the denseness of modified asphalt has trend to become bigger with the increasing of silane silica content.

Penetration index with different silane silica content is as shown in Figure 1.

The curve of penetration index showed in Figure 1 has two peaks, respectively, in 0.5% and 3%. When 3% silane silica is added, PI index is close to zero; at this point, the sensitivity of the modified asphalt is the least sensitive, which can improve the high temperature stability and low temperature cracking resistance of asphalt mixture, and it can be considered that the 3% mixing amount is the best. While the mixing amount is more than 3%, PI values show a downward trend. The flow ability of asphalt will become worse rapidly at this time when the trial tested is more than 5%, and it is not conducive to the mixture.

TABLE 2: Penetration and penetration index PI.

Volume	15° penetration	25° penetration	30° penetration	PI
0%	31.7	81.3	133.9	−0.26
0.5%	25.1	82.6	124.4	−1.06
1.00%	29.2	65.1	113.2	0.239
1.50%	27.8	73.3	128	−0.61
2%	25.5	73.9	123.3	−0.867
3.0%	30	74	116.7	0.116
4%	20.1	57.2	90	−0.566
5%	19	64.3	87.1	−0.816

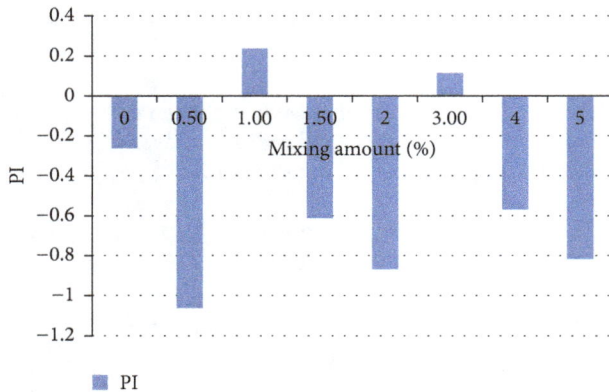

FIGURE 1: Penetration index value of modified asphalt in different mixing amount.

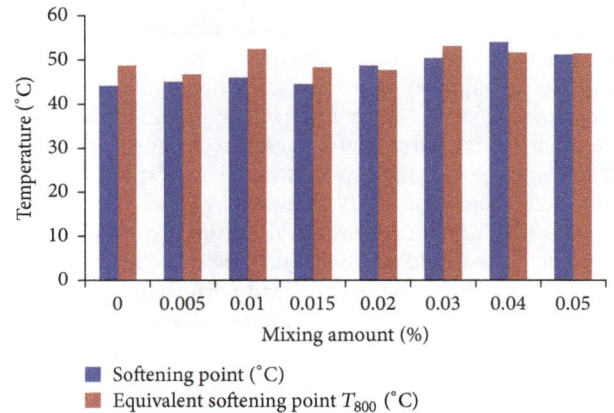

FIGURE 2: Softening point and equivalent softening point of modified asphalt in different mixing amount.

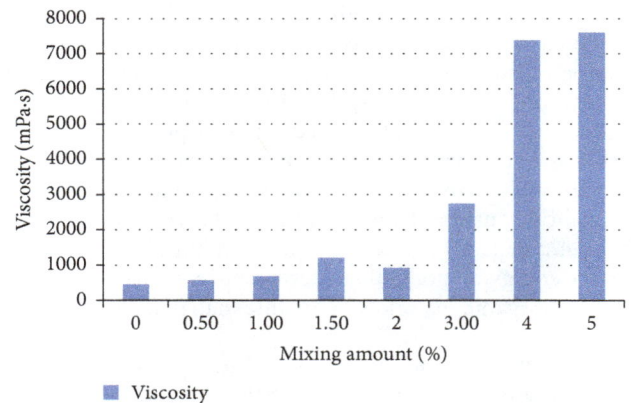

FIGURE 3: Rotational viscosity of modified asphalt in different mixing amount.

3.2. Softening Point and Equivalent Softening Point.

The softening point and equivalent softening point of modified asphalt are calculated and the results are shown in Figure 2.

The requirement of the softening point in our specification for Grade A is more than or equal to 44°C, and the test results all meet the primary requirements. Softening point test value increased with the mixing amount increase, but when more than 4% softening point began to decline. Equivalent softening points appear in two peaks in 1% and 3%; it also began to decline when the dosage was more than 3%.

3.3. Rotational Viscosity.

The degree of asphalt fluidity is generally estimated by the viscosity. The purpose of rotational viscosity is to check the pumping and handling ability of asphalt storage, mixing, and compaction. It can be used for nonmodified and modified asphalt binders. In addition, six kinds of samples with different mixing amount were used in the test. The viscosity of 135°C is shown in Figure 3.

Rotational viscosity of asphalt 135°C shown in Figure 3 had upward trend in 0% to 4%, mainly due to the full reaction between of asphalt and silane silica, while the viscosity decreased when silane silica mixing amount is more than 4%; this is because mixture contains excess modified material, and it cannot fully react; in other words silane silica maintains the original state in asphalt. Viscosity of asphalt should not be too large, and the standard in Superpave specification should be less than 3000 mPa·s, and it meets the specification

requirements when the silane silica content is less than 3%. In addition, the mixing process of silane silica modified asphalt suggests that those chemical reactions and physical dispersion are likely to happen, and a new network structure might be formed. Furthermore, silane silica holds the potential to strengthen the control asphalt and to improve the recovery ability when stress is applied.

3.4. Equivalent Brittle Point.

The calculation result of the equivalent brittle point is shown in Figure 4.

When the mixing amount is less than 3%, equivalent brittle point is in wave state, and the 1% and 3% appear in peak value; when it is more than 3%, equivalent brittle point showed rising trend. The main reason is that the excess part of modified material cannot react with asphalt and will play a negative effect. Equivalent brittle point in specification requirements for the A-level is ≤ -14.4°C, and mixing amount will not be satisfied with the requirements if it is more than 4%.

3.5. The Optimum Mixing Amount of Silane Silica.

By the comparison and analysis of the test data, if silane silica content is more than or equal to 4%, the penetration index,

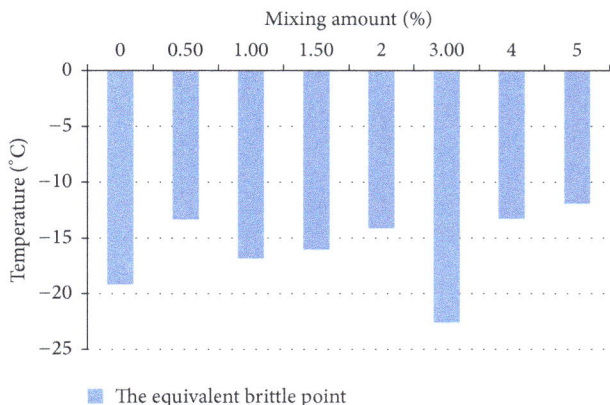

FIGURE 4: Equivalent brittle point of modified asphalt in different mixing amount.

viscosity, and equivalent brittle point of asphalt cannot meet the specification. Therefore, it is considered that content more than or equal to 4% cannot meet the requirements. A number of indices showed that performance of modified asphalt at 0% to 3% range had rising phase for the curve. PI value, equivalent softening point, and equivalent brittle point reached the peak when the mixing amount is 1% and 3%; meanwhile viscosity at 3% is the best. In summary, the optimum mixing amount of silane silica modified asphalt is 3% so as to meet the requirements of the specifications and achieve the best effect.

4. TG Analysis

Thermal gravimetric analysis method was used to determine the weight of asphalt (burnout) with the curve of the temperature and determine the asphalt temperature stability. DTA and thermal gravimetric curves of asphalt samples were tested by TG209F3 thermal gravimetric analyzer, the gas environment of test instrument was nitrogen, the heating rate was 40°C/10.0 (K/min), and the temperature range was from 0 to 300°C. The experiment mainly used nitrogen providing oxygen-free environment for the specimen to avoid reaction between oxygen and specimen. According to the TG analysis data, the relationship curve between temperature and DTA was drawn mainly to measure the physical and chemical changes in the asphalt under the sustained changing temperature environment. In addition, the curves declining represented the endothermic reaction happened, and the curves rising represented the exothermic reaction that happened.

The Tg and DTA analysis results are shown in Figures 5 and 6.

The difference between first endothermic and exothermic peaks is not great between control asphalt and silane silica modified asphalt, but due to the fact that temperature test is only 300°C, the temperature of next endothermic and exothermic peak cannot be determined. An exothermic peak was observed at 100°C from all the samples. Then the peak area of the control asphalt is much larger than the

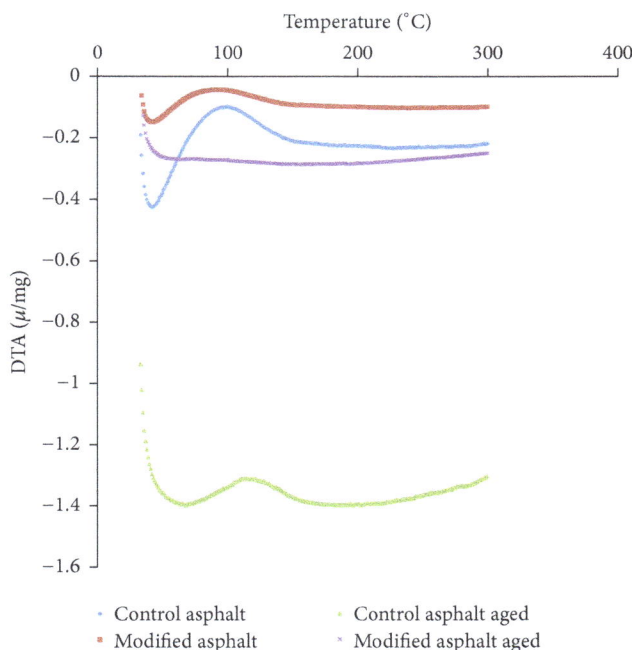

FIGURE 5: DTA analysis result of four kinds of samples.

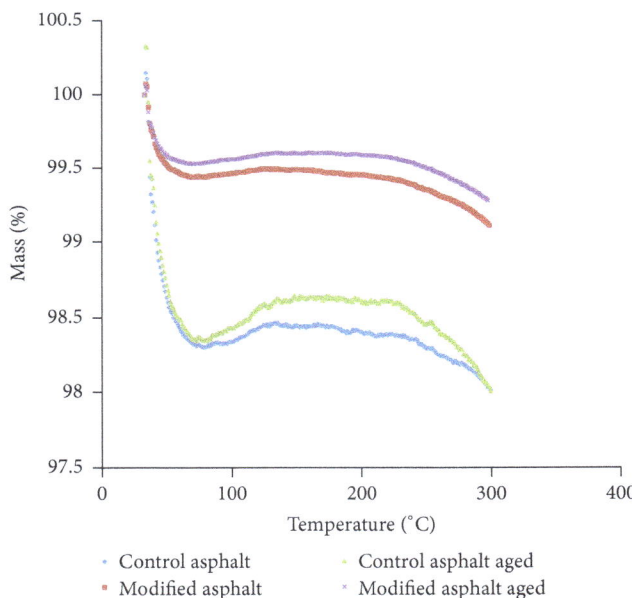

FIGURE 6: TG analysis result of four kinds of samples.

modified asphalt, and it can be judged that the heat effect of control asphalt is more significant and physical and chemical reactions during the temperature rising process are severer. So we can conclude that the temperature stability of modified asphalt is better.

The first endothermic peak of aging control asphalt was 75°C, while that in aging modified asphalt was 125°C, and the first exothermic peak in the former was 190°C, while the latter has not reached the exothermic peak at 300°C, so it can be known that the properties of aging modified asphalt are much better than aging control asphalt, because, after aging asphalt showed slower physical and chemical reaction activity, we can also evidently see that peak area of aging asphalt compared to the unaged asphalt is much smaller from Figure 5.

Asphalt structure is composed by four fractions of saturate, aromatic, resin, and asphaltine. Among them, light weight fractions in which the number of carbon atom is less than 4 will sublimate with the increasing of temperature, and the weight of asphalt will reduce. In comparison with the burning loss of control asphalt and modified asphalt, the amount of control asphalt was 1.6% at 105°C, and the loss of modified asphalt was 0.5%. There is little difference with endothermic peak and exothermic peak of two kinds of asphalt, while the peak area is much smaller than that of the control asphalt. In consideration of different peak areas, it can also be judged that the physical and chemical reactions of matrix asphalt are more active, so the amount of burning loss is even greater. Besides, the loss of the control asphalt at 300°C is 2.2%, while the burning loss of the silane silica modified asphalt at 300°C is only 0.9%; it can be seen that the high temperature stability of modified asphalt is better than that of the control asphalt. Meanwhile by comparison the burning loss at 300°C it is obviously seen that aging resistance of the modified asphalt is also better than the matrix asphalt.

By comparison with the burning loss of aging control asphalt and aging modified asphalt, the amount of aging control asphalt was 1.4% at 125°C, while the amount of aging modified asphalt was 0.4%. At 300°C, the burning loss of aging control asphalt was 2.0%, while the amount of the aging modified asphalt at 300 was only 0.72%. So we can conclude that the high temperature stability and aging resistance of modified asphalt were significantly improved after aging.

In summary, in both unaged modified asphalt and aging modified asphalt, the chemical compositions and chemical properties by differential thermal analysis are better than the control asphalt. TG test showed that the performance of asphalt such as viscosity, tensile strength, high temperature stability, and antiaging performance had been improved significantly by mixing with silane silica. That is due to not only the coupling effect that silane has on asphalt, but also the activation and modification effect silica on asphalt which can promote the chemical reaction between asphalt and silane, and silane also will improve the compatibility between silica and asphalt, strengthen the interaction of the interface between them, and improve the tear strength of asphalt.

5. The Road Performance Test

A series of road performances were conducted to determine the superiority of asphalt mixture modified by silane silica, and it can provide theoretical support for the application of the modified asphalt and mixture.

The effect of modified asphalt on the high temperature stability of the mixture was verified by the rutting test. The effect of modified asphalt on the water stability of mixture was verified by water boiling test, Marshall immersion test, freeze-thaw split test, and Marshall test, and the effect of modified

TABLE 3: Results of rutting test.

Mixture type	Dynamic stability number (time/mm)
Control asphalt mixture	2700.16
Modified asphalt mixture	3391.96

TABLE 4: Results of immersion Marshall test.

Mixture type	Control asphalt mixture	Modified asphalt mixture
Stability (kN)	11.27	14.03
Flow value (mm)	2.51	2.79
Water immersion 48 h stability (kN)	7.36	10.39
Water immersion 48 h flow value (mm)	1.18	1.59
Residual stability (%)	65.3	74.05

TABLE 5: Result of vacuum water saturated Marshall test.

Mixture type	Control asphalt mixture	Modified asphalt mixture
Stability (kN)	11.27	14.03
Flow value (mm)	2.51	2.79
Vacuum saturation 48 h stability (kN)	6.82	8.99
Vacuum saturated 48 h flow value (mm)	1.82	1.52
Residual stability (%)	60.51	64.08

asphalt on antiaging property of modified asphalt was verified by using the short-term aging test.

5.1. Research on High Temperature Stability. The rutting test results of asphalt mixture with 3% mixing amount and control asphalt mixture are shown in Table 3.

From Table 3, we can see that the modified asphalt mixture can improve the dynamic stability by 25.6%. The effect of the addition of silica silane on the high temperature stability of the mixture is obvious.

5.2. Research on Water Stability. The immersion Marshall test results are shown in Table 4.

The stability of the modified asphalt mixture is higher than the control asphalt mixture. The additional silane silica in the asphalt mixture improves the stability about 24.4%, while after water immersion the value is up to 41.2%. The residual stability of the modified asphalt mixture is also higher than the control asphalt mixture.

Vacuum saturated Marshall test results are in Table 5.

After modification of the stability of the mixture after vacuum saturation was improved by about 31.8%, the residual stability of modified asphalt mixture also increased from 60.51% to 64.08%.

The freeze-thaw split test results were shown in Table 6.

After modification of the splitting strength of the mixture was improved by about 22%, after freezing and thawing cycles

TABLE 6: Result of freeze-thaw split test.

Mixture type	Control asphalt mixture	Modified asphalt mixture
Nonfreezing and thawing splitting strength (KN)	11.6	14.16
Freezing and thawing splitting strength (KN)	9.45	11.91
Freeze-thaw splitting tensile strength ratio (%)	81.47	84.11

TABLE 7: Result of boiling test.

Test type	Boiling, 3 min	Boiling, 20 min
Control asphalt	Grade 2	Grade 1
Modifier asphalt	Grade 5	Grade 5

TABLE 8: Result of aging test.

Mixture type	Control asphalt mixture	Modified asphalt mixture
The residual stability of the water immersion before and after aging (%)	65.3/62.6	74.05/72.7
The residual stability of the vacuum saturation before and after aging (%)	60.51/55.7	64.08/62.3
Tensile strength ratio of freeze-thaw splitting strength before and after aging (%)	81.47/77.3	84.11/83.3

the value was up to 26%. The splitting tensile strength ratio of modified asphalt was 84.11%; in contrast this value of control asphalt mixture was 81.47.

The boiling test was used to verify the modification effect of adhesion between asphalt and stone, and the test results are as shown in Table 7.

From Table 7 it can be seen that the adhesion of asphalt and stone is very poor without adding silane silica. By the boiling test, the membrane of control asphalt dropped significantly after 1 min boiling, and the aggregate has been exposed more than half after 3 min, while the effect of silane silica added in asphalt is obvious, and after boiling for 20 min adhesion reached grade 5. Through the experiment, we can see that the silane silica added in asphalt greatly improved the adhesion and enhanced the consolidation strength of the oilstone interface.

The above four test results showed that silane silica modified asphalt mixture had excellent effect to improve the water stability of asphalt mixture.

5.3. Research on Antiaging Performance. In accordance with the asphalt mixture test procedures T0734 short-term aging method, asphalt mixture was heated in the 135°C ± 1°C oven for 4 h ± 5 min, mixture must be stirred once an hour during the aging process.

Then the specimens were formed to measure the index of the residual stability of the water immersion, the residual stability of the vacuum saturation, and tensile strength ratio of freeze-thaw splitting strength before and after aging. The test results are as shown in Table 8.

From the test results in Table 8, the water stability of the two samples was decreased after aging, while the decreasing amplitude of modified asphalt mixture was less than that of control asphalt mixture, and the indexes of modified asphalt after aging are better than asphalt unaged which indicates that modified asphalt mixture significantly improved antiaging properties.

6. Conclusions

Silane in silane silica with adsorption and invasion function can reduce the surface tension and enhance the hydrophobicity of the asphalt and also form key joints and improve the adhesion between asphalt and aggregate. Silica with activation and modification function can improve the tear strength of the asphalt. The objective of this study is to evaluate the modified effect and process with silane silica in order to improve the overall road performance of asphalt mixture and enhance the adhesion between asphalt and aggregate. The specific conclusions are as follows:

(1) Silane silica was innovatively used as a new type of modified filler.

(2) The production process of modified asphalt is as follows: First control asphalt was heated to molten state; then silane silica was added into control asphalt at certain concentrations by weight of the control asphalt and mixed in the high-speed shear machine. During the mixing, the asphalt samples were kept at constant temperature of 160°C and blended using a shear rate of 4000 r/min for an hour.

(3) Based on the limited laboratory work done in this study, it can be concluded that the addition of 3% silane silica is the optimum mixing amount to improve the performance characteristics of modified asphalt in various conditions.

(4) TG test showed that tear strength, high temperature resistance, and aging resistance of the modified asphalt are obviously improved. The superiority of modified asphalt was proved from the point of chemical analysis.

(5) The water stability of modified asphalt mixture has been significantly improved; meanwhile the antiaging performance and high temperature stability of the modified asphalt mixture are also improved obviously.

Conflict of Interests

The authors declare that there is no conflict of interests regarding the publication of this paper.

Acknowledgment

The research work described herein was funded by the National Nature Science Foundation of China (NSFC) (Grant no. 51178204). This financial support is gratefully acknowledged.

References

[1] B. Sengoz and G. Isikyakar, "Evaluation of the properties and microstructure of SBS and EVA polymer modified bitumen," *Construction and Building Materials*, vol. 22, no. 9, pp. 1897–1905, 2008.

[2] M. S. Cortizo, D. O. Larsen, H. Bianchetto, and J. L. Alessandrini, "Effect of the thermal degradation of SBS copolymers during the ageing of modified asphalts," *Polymer Degradation and Stability*, vol. 86, no. 2, pp. 275–282, 2004.

[3] K.-D. Jeong, S.-J. Lee, S. N. Amirkhanian, and K. W. Kim, "Interaction effects of crumb rubber modified asphalt binders," *Construction and Building Materials*, vol. 24, no. 5, pp. 824–831, 2010.

[4] L. Xin, *Research of Modified Asphalt Binder based on Asphalt Aggregate Interface Enhancement*, Jilin University, Changchun, China, 2012.

[5] M. J. Khattak, A. Khattab, H. R. Rizvi, and P. Zhang, "The impact of carbon nano-fiber modification on asphalt binder rheology," *Construction and Building Materials*, vol. 30, no. 5, pp. 257–264, 2012.

[6] W. Ning, *Research on the Performance of the Basalt Fiber and Modified Asphalt*, China University of Geosciences, 2013.

[7] T. Jiyu, *Composition and Performance Test of Fiber Asphalt Mixture*, Zhengzhou University, Zhengzhou, China, 2013.

[8] Z. You, J. Mills-Beale, J. M. Foley et al., "Nanoclay-modified asphalt materials: preparation and characterization," *Construction and Building Materials*, vol. 25, no. 2, pp. 1072–1078, 2011.

[9] S. G. Jahromi and A. Khodaii, "Effects of nanoclay on rheological properties of bitumen binder," *Construction and Building Materials*, vol. 23, no. 8, pp. 2894–2904, 2009.

[10] Z. Zhang and G. Hu, "Research on properties of low temperature cracking Silica modified asphalt mixture," *Journal of Beijing University of Technology*, vol. 32, no. 11, pp. 1007–1010, 2006.

[11] S. Chen and Y. Yu, "Effect of silica on properties of SBS modified asphalt," *Petroleum Asphalt*, vol. 21, no. 1, pp. 18–22, 2007.

[12] M. Lin, *Research the Properties of PCF-Diatomite Composite Modified Asphalt*, Jilin University, Changchun, China, 2014.

[13] H. Yao, Z. You, L. Li et al., "Rheological properties and chemical bonding of asphalt modified with nanosilica," *Journal of Materials in Civil Engineering*, vol. 25, no. 11, pp. 1619–1630, 2012.

[14] N. I. M. Yusoff, A. A. S. Breem, H. N. M. Alattug, A. Hamim, and J. Ahmad, "The effects of moisture susceptibility and ageing conditions on nano-silica/polymer-modified asphalt mixtures," *Construction and Building Materials*, vol. 72, pp. 139–147, 2014.

[15] K. Chrissafis, K. M. Paraskevopoulos, G. Z. Papageorgiou, and D. N. Bikiaris, "Thermal and dynamic mechanical behavior of bionanocomposites: fumed silica nanoparticles dispersed in poly(vinyl pyrrolidone), chitosan, and poly(vinyl alcohol)," *Journal of Applied Polymer Science*, vol. 110, no. 3, pp. 1739–1749, 2008.

[16] H. Kun, C. Lan, and L. Xuan, "Modified silica and silicone powder on properties of styrene butadiene rubber," *Rubber Industry*, vol. 62, no. 1, pp. 21–26, 2015.

[17] Y. Zhu, "Preparation of rubber with silane silica nano filler for rubber," *Rubber Reference Method*, vol. 37, no. 6, pp. 7–11, 2007.

[18] Q. Liu, H. Zhao, and H. Hu, "Surface modification of silica reinforced SSBR," *Performance Synthetic Rubber Industry*, vol. 37, no. 2, pp. 144–148, 2014.

[19] L. L. Song, X. Jie, and G. Yu, "Rubber filler mutual effect of styrene butadiene rubber (SBR)/silica composite performance," *Science China Chemistry*, vol. 11, pp. 1–5, 2014.

[20] E. Hesami, D. Jelagin, N. Kringos, and B. Birgisson, "An empirical framework for determining asphalt mastic viscosity as a function of mineral filler concentration," *Construction and Building Materials*, vol. 35, pp. 23–29, 2012.

3

Fatigue Evaluation of Recycled Asphalt Mixture Based on Energy-Controlled Mode

Tao Ma,[1] Kai Cui,[1,2] Yongli Zhao,[1] and Xiaoming Huang[1]

[1]*School of Transportation, Southeast University, Nanjing, Jiangsu 210096, China*
[2]*State Engineering Laboratory of Highway Maintenance Technology, Changsha University of Science and Technology, Changsha 410114, China*

Correspondence should be addressed to Tao Ma; matao@seu.edu.cn

Academic Editor: Meor Othman Hamzah

The fatigue properties of asphalt mixtures are important inputs for mechanistic-empirical pavement design. To understand the fatigue properties of asphalt mixtures better and to predict the fatigue life of asphalt mixtures more precisely, the energy-controlled test mode was introduced. Based on the implementation theory, the laboratory practice for the energy-controlled mode was realized using a four-point-bending fatigue test with multiple-step loading. In this mode, the fatigue performance of typical AC-20 asphalt specimens with various reclaimed asphalt pavement (RAP) contents was tested and evaluated. Results show that the variation regulation of the dissipated energy and accumulative energy is compatible with the loading control principle, which proves the feasibility of the method. In addition, the fatigue life of the asphalt mixture in the energy-controlled mode was between that for the stress-controlled and strain-controlled modes. The specimen with a higher RAP content has a longer fatigue life and better fatigue performance.

1. Introduction

With the service life of asphalt pavement increasing, maintenance and rehabilitation are inevitable. In addition, more and more reclaimed asphalt pavement (RAP) materials have been created. Meanwhile, with the construction cost and energy consumption increasing, the recycling of RAP in new asphalt pavement has attracted more and more attention, especially in China [1, 2]. Many studies have proved that the successful conduction of hot recycling technology of RAP can generate recycled asphalt pavement with the same performance as that of new asphalt pavement [3, 4]. However, studies also indicate that the addition of RAP can affect the fatigue durability of recycled asphalt pavement [5, 6]. Thus, many researchers are still focusing on how to evaluate and improve performance, especially the fatigue durability of the recycled asphalt mixture.

The fatigue properties of asphalt pavement determine its design life. Asphalt pavement structural defects, especially fatigue cracks, are caused by the failure of the asphalt mixture's antifatigue properties. Therefore, fatigue properties

are the key issue that researchers have focused on [7]. Traditionally, the stress- or strain-controlled loading mode is used for laboratory tests. One of the two modes is selected, according to the pavement thickness. Generally, 12.7 cm is regarded as the boundary thickness. If the surface layer of the asphalt pavement is thicker than 12.7 cm, the stress-controlled mode would be selected; otherwise, the strain-controlled mode would be selected [8]. However, with the fabrication of new pavement materials, alteration of pavement design theory, and the appearance of new failure modes, the rationale behind this classification standard has been widely questioned [9].

Fatigue tests for different gradations and different mixture components were conducted, based on the stress-controlled and the strain-controlled mode, respectively [10–12]. Results indicated that, in the strain-controlled mode, fatigue life increases as the initial modulus decreases. However, in the stress-controlled mode, the fatigue life increases as the initial modulus increases. Thus, it can be seen that the control mode plays a critical role in the evaluation of asphalt mixture fatigue properties. In view of this, many researchers only use one control mode to evaluate asphalt mixture properties and construct fatigue damage models considering modulus, dissipated energy, and accumulative energy [13–16]. Some studies concentrated on the differences and relations between the two control modes, trying to unify the evaluation indexes [17–20] or to convert one to another [21]. However, these studies are still confined to either the stress-controlled mode or the strain-controlled mode.

Previous research [22–26] has shown that the fatigue process of the bottom surface layer of the asphalt pavement is more related to the energy than to the stress in the stress-controlled and the strain in the strain-controlled modes. The dissipated energy of each loading cycle remains approximately constant [27, 28]. Thus, the energy-controlled loading mode was proposed in this study, and the method to achieve it utilized a four-point bending (4PB) fatigue test device. Thereafter, fatigue properties of the asphalt mixture with 0%, 25%, and 50% RAP contents were evaluated. The test results could contribute to the establishment of a new estimation index for fatigue properties of asphalt mixtures.

2. Theory

During one fatigue loading cycle in a 4PB test, the dissipated energy of a beam could be calculated by the following equation:

$$E = \int_0^{2\pi/\omega} F \sin(\omega t) L \sin(\omega t - \varphi) \, dt, \tag{1}$$

where F is the force applied to the beam [N], L is the measured beam deflection [m], φ is the phase angle between force and displacement [rad], and ω is the frequency of applied force and displacement [rad/s].

F could be determined by the product of the flexural modulus and the deflection of the beam, so (1) could also be expressed as

$$E = \int_0^{2\pi/\omega} SL \sin(\omega t) L \sin(\omega t - \varphi) \, dt$$
$$= SL^2 \int_0^{2\pi/\omega} \sin(\omega t) \sin(\omega t - \varphi) \, dt = SL^2 \times \pi \sin\varphi, \tag{2}$$

where S is the modulus of the beam.

In (2), it should be noted that SL^2 equals the energy applied to the beam in the energy-controlled mode. The phase angle φ is a property of the material itself, which corresponds to the proportion of dissipated energy. Changes during the test are so small that they could be neglected when calculating the dissipated energy. During the fatigue test, as long as SL^2 remains constant, the energy applied to the beam could be regarded as a constant. In this way, the energy-controlled mode can be achieved theoretically.

Under the strain-controlled mode, the modulus S and bottom tensile strain ε of the beam can be continuously renewed by software using real-time data from the displacement sensor, which makes it possible to control the applied energy. Because the deflection L is directly proportional to

the bottom tensile strain of the beam ε, ε can be used as the controlled index. The control theory is as follows:

$$S_0 \times \varepsilon_0^2 = S_n \times \varepsilon_n^2, \tag{3}$$

where S_0 is initial flexural modulus of the beam, ε_0 is the initial bottom tensile strain of the beam, S_n is the flexural modulus of the beam at Cycle N, and ε_n is the bottom tensile strain of the beam at Cycle N.

Because of the inherent limitation of the 4PB fatigue test device, the energy-controlled mode cannot be achieved directly. In the strain-controlled mode, when the strain was controlled, the modulus S_n at Cycle N could be output in real time. Accordingly, the force applied to the beam F_n was adjusted to $F_0 \times S_n/S_0$, to guarantee that the bottom tensile strain of the beam ε_n at Cycle N would be equal to the initial tensile strain ε_0. Referring to (3) and the theory of the strain-controlled mode, we propose a laboratory operational method, based on a 4PB device, called multiple-step loading: according to the modulus S_n at Cycle N, shift the bottom tensile strain of the beam ε_n to $\sqrt{S_0/S_n} \times \varepsilon_0$ and then shift the force applied to the beam F_n to $\sqrt{S_0/S_n} \times F_0$. Thus, the energy applied to the beam can remain constant as $F_0 L_0$. Using this method, the energy-controlled mode could be achieved. In the experiment, each time the force and the strain are adjusted, the fatigue test machine should be restarted. Thus, the higher the precision requirement is, the longer the break time that will be taken. However, the fatigue damage might mitigate during the intermittence due to the asphalt self-healing effect [29–31], which would reduce the accuracy. In this study, a 15% reduction of the beam modulus was used for the loading process, and the fracture of the beam was set as the standard of failure.

3. Operation Method for the Energy-Controlled Mode

A base asphalt with a penetration grade of 70 was used as a binder. Aggregates used for the mixture design were 100% limestone, and the mineral powder was limestone powder. RAP materials were milled from the surface layer of an expressway that had been used for 10 years and showed serious pavement defects such as cracking, raveling, and rutting. These materials were subsequently then processed in a plant. Next, the RAP materials were obtained from the mix plant and used for laboratory tests. The gradation of RAP materials is shown in Table 1. The basic properties of the extracted RAP binder are shown in Table 2. A dense graded asphalt mixture with a nominal maximum aggregate size of 19 mm (AC20) was used as the design type of the new asphalt mixture and recycled asphalt mixture. The designed gradation of AC20 is shown in Figure 1. By using the designed gradation as the target gradation, three AC20 with RAP contents of 0%, 25%, and 50% were prepared, and their gradations are also shown in Figure 1. All the prepared mixtures were subjected to fatigue tests with different controlled modes, and their fatigue properties were analyzed.

The operational procedures for the energy-controlled mode were as follows.

TABLE 1: Gradation of RAP.

Sieving size/mm	16	13.2	9.5	4.75	2.36	1.18	0.6	0.3	0.15	0.075
Percentage passing/%	100	96.5	77.2	48.3	35.9	20.5	15.1	10.7	6.1	4.3

TABLE 2: Basic properties of binders with different RAP contents.

Type of binder	Penetration/0.1 mm	Softening point/°C	25°C ductility/cm
Extracted RAP binder	31.3	60.6	7.7
Virgin binder	68.0	48.3	130.7
25% RAP binder	60.5	52.1	90.6
50% RAP binder	48.8	55.0	45.5

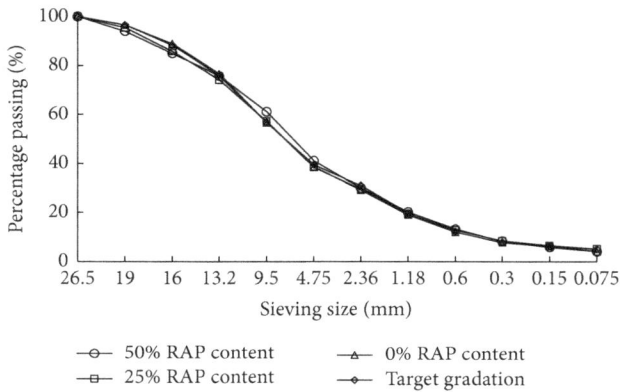

FIGURE 1: Gradations of asphalt mixtures with different RAP contents.

Step 1 (specimen preservation). Put the specimen into the environmental chamber for preservation for no less than 4 h at the test temperature ± 0.5°C.

Step 2 (specimen placement). Install the specimen on the 4PB test device and fix it with clamps. Place the LVDT and make sure it is in contact with specimen surface. Then, adjust its position to the middle of the beam. The initial readout of the LVDT should be as close to zero as possible.

Step 3 (parameter decision). The beam should be tested under a sine wave load, $200\mu\varepsilon$ initial strain, and a loading frequency of 10 Hz.

Step 4 (initial strain decision). Specimens with different RAP contents have different moduli. Therefore, to ensure that the energy applied to each specimen is equal, the initial strain value should be different. The beam is first loaded under $200\mu\varepsilon$ for 500 cycles, and the stiffness modulus in the 500th cycle is the initial modulus S_0. The initial strain ε_0 for each beam can then be calculated.

Step 5. Create a new test file on the computer, set the target strain as ε_0, and then begin loading. When the modulus of the beam decreases to $0.85S_0$, stop the test.

Step 6. Repeat Step 5 with the strain target as $1.0847\varepsilon_0$. Because of the asphalt self-healing effect, the modulus may

increase during the break. Generally, the modulus will drop again to the level before the break after about 200 cycles. Continue to load, until the modulus drops to $0.7225S_0$.

Step 7. Repeat Step 5 with the target strains shown in Table 3. Keep loading until the beam fractures.

Step 8. Add up the cycle number of each loading level. Then, the fatigue life and a diagram of the flexural stiffness modulus varying with cycle number can be obtained.

At the beginning of loading Step 1, since the initial modulus S_0 was obtained with an initial strain ε_0, the energy applied to the beam at the beginning of the first loading process could be expressed as

$$E_0 = a \times S_0 \times \varepsilon_0{}^2, \tag{4}$$

where E_0 is the energy applied to the beam and a is a constant.

The force applied to the beam could be expressed as

$$F_0 = b \times S_0 \times \varepsilon_0, \tag{5}$$

where b is a constant.

A 15% reduction in beam modulus was used for the loading process. The modulus dropped to $0.85S_0$ at the end of loading Step 1, and the energy applied to the beam dropped to $0.85E_0$. The force applied on the beam dropped to $0.85F_0$. Then, the device was stopped and prepared for loading Step 2. To achieve the energy-controlled mode, the energy applied to the beam at the beginning of loading Step 2 should be the same as E_0. According to (3), if ε_2 is equal to $1.0847\varepsilon_0$ ($1.0847\varepsilon_0 = \sqrt{S_0/0.85S_0} \times \varepsilon_0$), the energy applied to the beam at the beginning of loading Step 2 could be E_0, the same as the first loading process. The force applied to the beam becomes $0.9219F_0$ ($0.9219F_0 = b \times 0.85S_0 \times 1.0847\varepsilon_0$). Repeating the calculation, the values in Table 3 can be obtained step by step.

4. Comparison between Different Control Modes

4.1. Fatigue Test Results under the Energy-Controlled Mode. The fatigue test in the energy-controlled mode was conducted using a multiple-step loading method, at a temperature of 15°C. The size and manufacturing method of the beam, test

TABLE 3: Parameter-controlled process under the energy-controlled mode.

Loading step	Stiffness	Tensile strain	Load	Energy
1	$S_0 \rightarrow 0.85S_0$	ε_0	$F_0 \rightarrow 0.85F_0$	$E_0 \rightarrow 0.85E_0$
2	$0.85S_0 \rightarrow 0.7225S_0$	$1.0847\varepsilon_0$	$0.9219F_0 \rightarrow 0.7836F_0$	$E_0 \rightarrow 0.85E_0$
3	$0.7225S_0 \rightarrow 0.0.6141S_0$	$1.1765\varepsilon_0$	$0.8500F_0 \rightarrow 0.7225F_0$	$E_0 \rightarrow 0.85E_0$
4	$0.6141S_0 \rightarrow 0.5220S_0$	$1.2761\varepsilon_0$	$0.7836F_0 \rightarrow 0.6661F_0$	$E_0 \rightarrow 0.85E_0$
5	$0.5220S_0 \rightarrow 0.4437S_0$	$1.3841\varepsilon_0$	$0.7225F_0 \rightarrow 0.6141F_0$	$E_0 \rightarrow 0.85E_0$
6	$0.4437S_0 \rightarrow 0.3771S_0$	$1.5012\varepsilon_0$	$0.6661F_0 \rightarrow 0.5662F_0$	$E_0 \rightarrow 0.85E_0$
7	$0.3771S_0 \rightarrow 0.3206S_0$	$1.6283\varepsilon_0$	$0.6141F_0 \rightarrow 0.5220F_0$	$E_0 \rightarrow 0.85E_0$

conditions, and devices all followed the standard T0739-2011 test method from the Chinese specification JTG E20-2011 [32]. After each loading level, data such as fatigue life, dissipated energy, and modulus can be obtained. Summarizing these data makes it possible to draw the variation curves for the entire fatigue life. The variation regulation of dissipated energy, accumulative dissipated energy, and modulus is illustrated in Figures 2, 3, and 4, respectively.

As shown in Figures 2 and 3, when loading in the energy-controlled mode, there is a highly linear relationship between the accumulative dissipated energy and loading times, although the force applied and the dissipated energy during each loading level decreased. The energy applied to each beam is the same, so the curves of accumulative dissipated energy of different beams roughly coincide with each other. This confirms the accuracy of the test method. From Figure 4, it can be seen that the modulus decreases with an increase in loading times, and, as in the strain-controlled mode and the stress-controlled mode, the fatigue process of the energy-controlled mode can be divided into three periods: initial, stable, and fracture period. In addition, fatigue properties were evaluated by loading times in the energy-controlled mode. Figures 2, 3, and 4 show that AC-20 with 50% RAP content performs best, followed by 25% RAP content, and 0% performs the worst. In addition, the poorer the fatigue properties were, the faster the material modulus reduced. The loading times of the initial period and fracture period vary little between different control modes, and the fatigue life depends greatly on the stable period. In this respect, there was no significant difference among the three control modes.

4.2. Comparison of Fatigue Test Results under Different Control Modes.
Fatigue tests under the strain-controlled and stress-controlled modes were conducted at a temperature of 15°C, for comparison with the energy-controlled mode. In the strain-controlled mode, the target strain was set at 450$\mu\varepsilon$, and the test stopped when the modulus decreased by 50%. The accumulative dissipated energy variation regulation during the fatigue test is shown in Figure 5. The modulus variation regulation is shown in Figure 6.

Results showed that AC-20 with 25% RAP content had the longest fatigue life, 0% RAP content was second, and 50% had the shortest life. Figure 6 indicates that the mixture had a higher modulus when the RAP content was higher. Thus, the dissipated energy is larger if the mixture has a higher RAP content, when tested at the same strain level. Consequently,

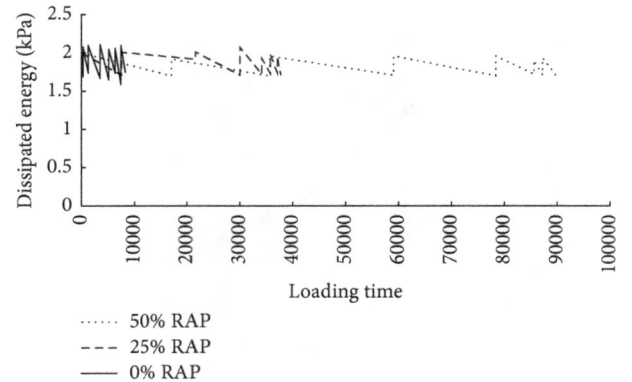

FIGURE 2: Dissipated energy variation under the energy-controlled mode at 15°C.

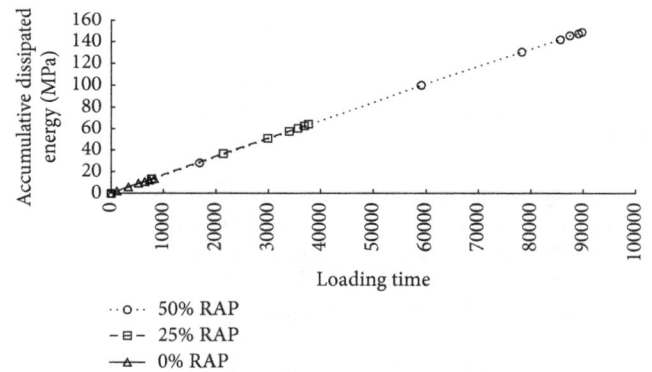

FIGURE 3: Accumulative dissipated energy variation under the energy-controlled mode at 15°C.

in Figure 5, the slope of the 50% RAP content is the largest. Compared with the energy-controlled mode, the growth of accumulative dissipated energy slowed down in later stages in the strain-controlled mode, resulting from a greater decrease in force. Hence, the slope of the curves decreases as loading times increase in Figure 5. Figure 6 illustrates the attenuation process of the modulus. The modulus of the 50% RAP content reduced faster, possibly because the energy applied to the beam was higher. Moreover, the fatigue life of 50% RAP content was the shortest. However, it was not found that its fatigue properties were the poorest, because the energy applied to the beam was higher, which may result in the

TABLE 4: Comparison of fatigue life under different control modes.

	Energy-controlled	Stress-controlled	Strain-controlled
50% RAP content	89870	39540	13590
25% RAP content	37750	28450	62450
0% RAP content	8270	1040	39100

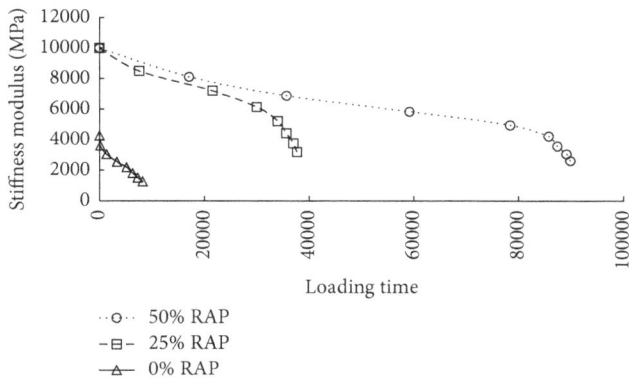

FIGURE 4: Flexural stiffness variation under the energy-controlled mode at 15°C.

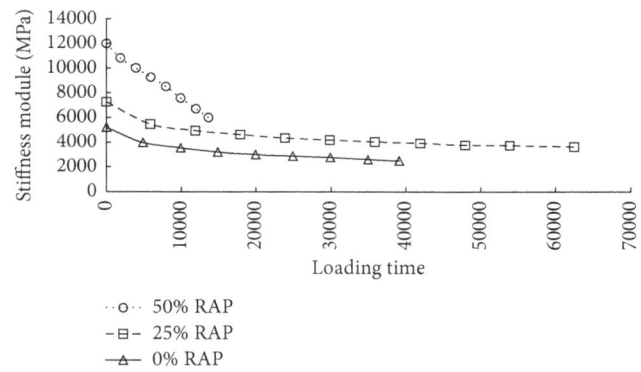

FIGURE 6: Flexural stiffness variation under the strain-controlled mode at 15°C.

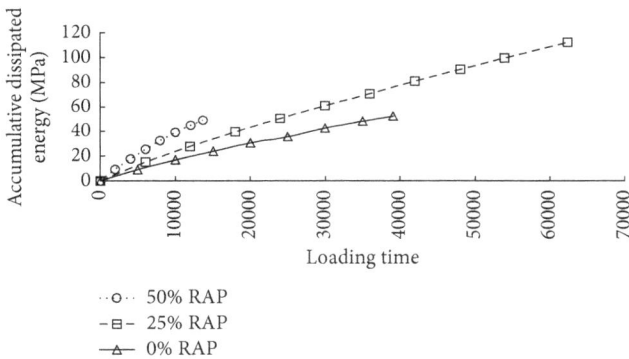

FIGURE 5: Accumulative dissipated energy variation under the strain-controlled mode at 15°C.

shortening of the initial period and hastening of the damage accumulation. Even when put in a pavement, the material was able to bear a greater vehicle load with less strain, because of the higher modulus of the 50% RAP content. However, this could hardly be seen from a strain-controlled fatigue test.

The force applied to the beam was 0.952 kN in the stress-controlled mode. Given the limitations of the test machine, neither the modulus nor the deflection could be obtained. Only the fatigue life of the three control modes could be measured. The results are listed in Table 4.

It is obvious that the higher the modulus was, the longer the fatigue life would be in the stress-controlled mode. On the contrary, a higher modulus would shorten fatigue life in the strain-controlled mode. Under the energy-controlled mode, the modulus did not affect the fatigue life. In this way, the model can better simulate the actual fatigue process of asphalt pavement, making the evaluation of asphalt mixture fatigue properties more objective.

5. Conclusions

(1) The working process of the test machine was analyzed using dissipated energy variation regulation during fatigue testing in the strain-controlled mode. Based on the process, a theory for the energy-controlled loading mode was proposed. A four-point-bending fatigue test device and a multiple-step loading method were used to conduct the operation steps for the energy-controlled mode.

(2) The fatigue life in the energy-controlled mode is between that of the strain-controlled and stress-controlled modes. The thickness of the asphalt pavement ceased to be the determining factor for selecting control modes. Instead, the energy-controlled mode can be used for any thickness. The dissipated energy of each loading mode should remain constant.

(3) Fatigue properties of different RAP contents were tested under the energy-controlled mode. It was indicated that the AC-20 mixture with 50% RAP content had the longest fatigue life and the best fatigue properties. In conclusion, the higher the RAP content was, the better the fatigue properties would be. The laboratory test results matched the theoretical predictions well, while the consistency between theory and field test results should be further investigated.

Conflicts of Interest

The authors declare that there are no conflicts of interest regarding the publication of this paper.

Acknowledgments

This work is financially supported by Natural Science Foundation of China (no. 51378006 and no. 51378123), Natural Science Foundation of Jiangsu (BK20161421), Fundamental Research Funds for the Central Universities (no. 2242015R30027), and Open Fund of State Engineering Laboratory of Highway Maintenance Technology (Changsha University of Science & Technology, kfj160104).

References

[1] T. Ma, X. Huang, Y. Zhao, Y. Zhang, and H. Wang, "Influences of preheating temperature of RAP on properties of hot-mix recycled asphalt mixture," *Journal of Testing and Evaluation*, vol. 44, no. 2, pp. 762–769, 2016.

[2] T. Ma, Y. Zhao, X. Huang, and Y. Zhang, "Using RAP material in high modulus asphalt mixture," *Journal of Testing and Evaluation*, vol. 44, no. 2, pp. 781–787, 2016.

[3] I. L. Alqadi, M. Elseifi, and S. H. Carpenter, "Reclaimed asphalt pavement—a literature review," *Cracking of Asphalt Concrete Pavements*, 2007.

[4] T. W. Kennedy and W. O. Tam, "Optimizing Use of Reclaimed Asphalt Pavement With the Superpave System," *Asphalt Paving Technol*, vol. 67, pp. 311–328, 1998.

[5] M. Tusar, E. Nielsen, F. Batista, M. L. Antunes, K. Mollenhauer, and S. Vansteenkiste, "Optimization of reclaimed asphalt in asphalt plant mixing, re-road end of life strategies of asphalt pavements," *European Commission DG Research*, 2012.

[6] R. Ghabchi, M. Barman, D. Singh, M. Zaman, and M. A. Mubaraki, "Comparison of laboratory performance of asphalt mixes containing different proportions of RAS and RAP," *Construction and Building Materials*, vol. 124, pp. 343–351, 2016.

[7] C. Shannon, A. Mokhtari, H. Lee, S. Tang, C. Williams, and S. Schram, "Effects of high reclaimed asphalt-pavement content on the binder grade, fatigue performance, fractionation process, and mix design," *Journal of Materials in Civil Engineering*, vol. 29, no. 2, Article ID 04016218, 2017.

[8] J. Xie and Z. Guo, "Researching on fatigue model of asphalt mixtures," *Journal of Highway and Transportation Research and Development*, vol. 24, no. 5, pp. 21–25, 2007.

[9] S. Lu, "Research on the selection of the fatigue test control mode of asphalt mixture," *Shandong Jiaotong Keji*, vol. 1, pp. 60–63, 2016.

[10] H. Wang, Z. X. Dang, L. Li, and Z. P. You, "Analysis on fatigue crack growth laws for crumb rubber modified (CRM) asphalt mixture," *Construction and Building Materials*, vol. 47, pp. 1342–1349, 2013.

[11] M. Sabouri, T. Bennert, J. S. Daniel, and Y. R. Kim, "A comprehensive evaluation of the fatigue behaviour of plant-produced rap mixtures," *Road Materials and Pavement Design*, vol. 16, pp. 29–54, 2015.

[12] S. Tang, R. C. Williams, and A. A. Cascione, "Reconsideration of the fatigue tests for asphalt mixtures and binders containing high percentage RAP," *International Journal of Pavement Engineering*, vol. 18, no. 5, pp. 443–449, 2017.

[13] L. Cong, J. Peng, Z. Guo, and Q. Wang, "Evaluation of fatigue cracking in asphalt mixtures based on surface energy," *Journal of Materials in Civil Engineering*, vol. 29, no. 3, Article ID D4015003, 2017.

[14] E. Fallon, C. McNally, and A. Gibney, "Evaluation of fatigue resistance in asphalt surface layers containing reclaimed asphalt," *Construction and Building Materials*, vol. 128, pp. 77–87, 2016.

[15] X. Luo, R. Luo, and R. L. Lytton, "Energy-based crack initiation criterion for viscoelastoplastic materials with distributed cracks," *Journal of Engineering Mechanics*, vol. 141, no. 2, Article ID 04014114, 2015.

[16] M. Sabouri and Y. R. Kim, "Development of a failure criterion for asphalt mixtures under different modes of fatigue loading," *Transportation Research Record*, vol. 2447, pp. 117–125, 2014.

[17] E. Masad, V. T. F. Castelo Branco, D. N. Little, and R. Lytton, "A unified method for the analysis of controlled-strain and controlled-stress fatigue testing," *International Journal of Pavement Engineering*, vol. 9, no. 4, pp. 233–246, 2008.

[18] A. Bhasin, V. F. Castelo Branco, E. Masad, and D. N. Little, "Quantitative comparison of energy methods to characterize fatigue in asphalt materials," *Journal of Materials in Civil Engineering*, vol. 21, no. 2, pp. 83–92, 2009.

[19] M. Pasetto and N. Baldo, "Unified approach to fatigue study of high performance recycled asphalt concretes," *Materials and Structures*, vol. 50, article no. 113, 2017.

[20] T. Hou, B. S. Underwood, and Y. R. Kim, "Fatigue performance prediction of north carolina mixtures using the simplified viscoelastic continuum damage model," *Asphalt Paving Technology: Association of Asphalt Paving Technologists-Proceedings of the Technical Sessions*, vol. 79, pp. 35–80, 2010.

[21] H. Zhang, L. Shan, Y. Tan, Y. Feng, and H. He, "Comparison of asphalt fatigue characteristics under different control modes," *Asphalt Pavements*, vol. 1, pp. 1281–1289, 2014.

[22] D. Lei, P. Zhang, J. He, P. Bai, and F. Zhu, "Fatigue life prediction method of concrete based on energy dissipation," *Construction and Building Materials*, vol. 145, pp. 419–425, 2017.

[23] H. Omrani, A. Tanakizadeh, A. R. Ghanizadeh, and M. Fakhri, "Investigating different approaches for evaluation of fatigue performance of warm mix asphalt mixtures," *Materials and Structures/Materiaux et Constructions*, vol. 50, no. 2, article no. 149, 2017.

[24] A. Modarres and A. Aloogar, "Comparison between the fatigue response of hot and warm mix asphalts based on the dissipated energy approach," *International Journal of Pavement Engineering*, vol. 18, no. 1, pp. 60–72, 2017.

[25] J. Ji, Z. Suo, S. Yang, Y. Xu, and S. Xu, "Application of the dissipated energy theory to estimate the fatigue characteristics of hot and warm recycled sma mixtures," *International Symposium on Frontiers of Road and Airport Engineering*, pp. 1–9, 2015.

[26] A. M. Bahadori, A. Mansourkhaki, and M. Ameri, "A phenomenological fatigue performance model of asphalt mixtures based on fracture energy density," *Journal of Testing and Evaluation*, vol. 43, no. 1, pp. 133–139, 2015.

[27] A. Wu, *Fatigue Performance Evaluation of Asphalt Mixture*, Southeast University, Jiangsu, China, 2015.

[28] T. Ma, Y. Zhang, D. Zhang, J. Yan, and Q. Ye, "Influences by air voids on fatigue life of asphalt mixture based on discrete element method," *Construction and Building Materials*, vol. 126, pp. 785–799, 2016.

[29] X. Luo and R. L. Lytton, "Characterization of healing of asphalt mixtures using creep and step-loading recovery test," *Journal of Testing and Evaluation*, vol. 44, no. 6, pp. 2199–2210, 2016.

[30] Y. Hou, L. Wang, T. Pauli, and W. Sun, "Investigation of the asphalt self-healing mechanism using a phase-field model,"

Journal of Materials in Civil Engineering, vol. 27, no. 3, Article ID 04014118, 2015.

[31] W. Huang, P. Lin, and M. Huang, "Comparison of the fatigue performance of asphalt mixtures considering self-healing," *International Symposium on Frontiers of Road and Airport Engineering*, pp. 109–121, 2015.

[32] Ministry of Transport of the People's Republic of China, *Standard Test Methods of Bitumen and Bituminous Mixtures for Highway Engineering (JTG E20-2011)*, China Communications Press, Beijing, China, 2011.

4

Fatigue Performance of Recycled Hot Mix Asphalt: A Laboratory Study

Marco Pasetto[1] and Nicola Baldo[2]

[1]Department of Civil, Environmental and Architectural Engineering, University of Padua, Via Marzolo 9, 35131 Padua, Italy
[2]Polytechnic Department of Engineering and Architecture, University of Udine, Via del Cotonificio 114, 33100 Udine, Italy

Correspondence should be addressed to Nicola Baldo; nicola.baldo@uniud.it

Academic Editor: Katsuyuki Kida

The paper introduces and analyses the results of an experimental trial on the fatigue resistance of recycled hot mix asphalt for road pavements. Based on the gyratory compaction and the indirect tensile strength test, the mix design procedure has optimized nine different mixes, considering both conventional limestone and Reclaimed Asphalt Pavement (RAP), the latter used at different quantities, up to 40% by weight of the aggregate. A standard bitumen and two polymer modified binders were used for the production of the mixes. The fatigue study was carried out with four-point bending tests, each one performed at 20°C and 10 Hz. The empirical stiffness reduction method, along with the energy ratio approach, based on the dissipated energy concept, was adopted to elaborate the experimental data. Unaged and aged specimens were checked, to analyse the ageing effects on the fatigue performance. In comparison with the control mixes, produced only with limestone, improved fatigue performance was noticed for the mixtures prepared with RAP, especially when made with polymer modified binders, under both aged and unaged conditions. Both the approaches adopted for the experimental data analysis have outlined the same ranking of the mixes.

1. Introduction

As a consequence of the continuous increase of the traffic loads recorded in the last decades, significant efforts have been made by the scientific community and the highway engineers to develop high performance hot mix asphalt. For example, asphalt rubber mixtures [1–4] and polymer modified asphalt concrete [5–11] are just few of the several technologies proposed and widely investigated over the years.

However, the scientific research has been focused also on different themes, for instance, the recycling of waste materials for the construction of road pavements, in order to save natural aggregates, which nowadays are less available [12–22]. Hence, the study of high performance hot mix asphalt, characterized by aggregate structures made with recycled materials, gives the possibility of meeting the demand of high mechanical resistance as well as that of sustainability of the road construction.

Among the different waste materials investigated for recycling in the road pavements, the Reclaimed Asphalt Pavement (RAP) aggregates are widely considered a fundamental resource to implement the sustainability philosophy in the road construction [23–27]. Several researchers have demonstrated the beneficial effects given by the use of RAP aggregates in the lithic skeleton of the hot mix asphalt, for example, a higher resistance to permanent deformation in comparison with traditional bituminous mixtures [23–25, 28, 29]. However, the effect of the RAP material on the fatigue performance of the mix has to be still completely understood; in fact, some researchers have found a reduction of the fatigue life [25, 29], basically associated with a brittle behaviour of the asphalt concrete, due to the high stiffness of the aged binder of the RAP grains [28]. Contrary findings were observed in other investigations [14, 23], primarily in case of use of a rejuvenator agent, to improve the response of the aged RAP bitumen [24, 26]. Recent studies have even verified a comparable fatigue life between RAP asphalt concrete and high modulus hot mix asphalt [29, 30]. Currently such topic deserves to be still investigated, in order to achieve a deep understanding of the fatigue performance of RAP asphalt concrete.

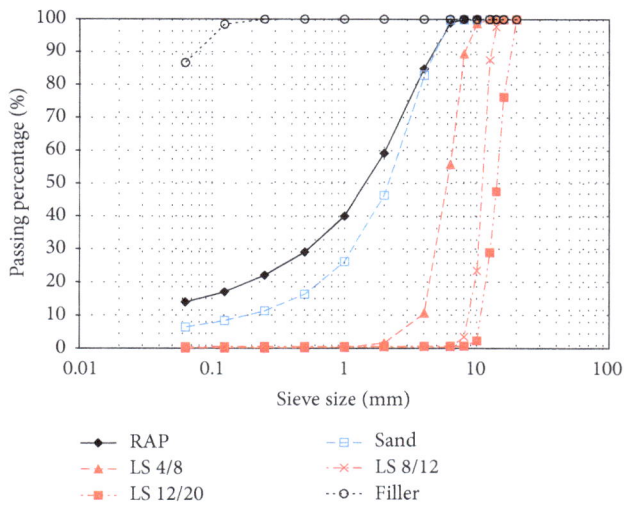

FIGURE 1: Grading curves of the aggregates.

In the present laboratory research, the fatigue performance of polymer modified hot mix asphalt, for base courses of flexible pavements, made with Reclaimed Asphalt Pavement (RAP) aggregates, has been studied by means of the four-point bending test (4PBT), using the strain control mode. The fatigue data have been analysed and elaborated by means of the empirical stiffness reduction approach and the energy ratio method based on the dissipated energy concept.

2. Materials and Methods

2.1. Materials

2.1.1. Aggregates and Bitumen. The RAP aggregates utilized in the study consist of a waste material obtained from demolition work of highway bituminous pavements, at the conclusion of their service life; such flexible structures were all situated in the Northern Italy and had similar composition. However, also natural aggregates (namely, crushed limestone, sand, and filler) have been used for the production of the hot mix asphalt considered in the laboratory study.

Both types of aggregates were provided by private companies from the Northeastern Italian area. The natural aggregates have been supplied in four different grading fractions: 0/4, 4/8, 8/12, and 12/20 mm. Figure 1 shows the grading curves of the conventional aggregates (EN 933-1) and that of the RAP material. The grading curve of the RAP has been obtained testing such material after the extraction of the bitumen (white curve, centrifugation method; EN 12697-1).

Three different types of bitumen were employed in the experimental investigation: two bituminous binders modified with SBS (styrene–butadiene–styrene) polymers, along with a conventional, not modified, bitumen. Based on the data provided by the manufacturers, the pair of SBS modified bitumen types are characterized by a different polymer concentration (in the present research the codes "Hard" and "Soft" have been associated with the modified bitumen with the highest and the lowest polymer dosage, respectively).

The major concern related to the use of the RAP material for the production of hot mix asphalt is due to the higher brittleness of the aged RAP binder, which could lead to a worsening of the fatigue performance of the mixes. In order to avoid such issue, soft bitumen can be adopted as a virgin binder to be added to the RAP mixes [23–26]. Nevertheless, some researchers have verified a significant improvement of the fatigue life of RAP mixes, even if low penetration grade bitumen is used, provided that a proper mix design procedure is adopted [29, 30].

In this experimental study, the feasibility to go further the concept of high performance recycled hot mix asphalt, using RAP aggregates along with polymer modified bitumen, has been investigated; the aim was to overcome the fatigue performance of the conventional polymer modified asphalt concrete.

2.1.2. Hot Mix Asphalt. Nine different types of hot mix asphalt were considered in the laboratory investigation; each mix has been designed to accomplish the acceptance requisites of base courses asphalt concrete for road pavement.

The conventional bitumen, the soft modified binder, and the hard modified one have been used for the production of three different types of hot mix asphalt, marked with the code CM (conventional mixtures), SM (soft modified), and HM (hard modified), respectively.

Each one of the three types of mixes, in turn includes three different types of asphalt concrete, prepared with an increasing RAP percentage and therefore characterized by a different aggregate structure. The mixes BC/R0, BC/R2, and BC/R4 were produced with conventional bitumen and a RAP quantity of 0% (R0), 20% (R2), and 40% (R4), respectively. Similarly, the mixes SM/R0, SM/R2, and SM/R4, as well as HM/R0, HM/R2, and HM/R4, were prepared with a RAP content of 0%, 20%, and 40%, but utilizing the soft modified binder and the hard one, for the SM mixes and the HM mixtures, respectively. Such experimental programme has been considered to analyse the effect of different types of modified bitumen on the fatigue performance of asphalt concrete made with RAP aggregates. Other researchers have verified the feasibility to produce hot mix asphalt characterized by higher RAP percentages, up to 50% [23, 25, 28, 29], 60% [30], and 100% [24, 26]; however, for such mixes, it has been recommended to use rejuvenator agents.

In the present investigation it has been decided to avoid the use of such agents, as a consequence of a particular request of the local road agencies (Northeastern area of Italy); therefore, the maximum amount of RAP materials has been fixed at 40% by weight of the aggregate. Indeed, the present study has been focused on the improvement of the fatigue performance of hot mix asphalt containing significant amount of RAP materials, rather than on increasing the RAP quantity up to very high contents.

2.2. Methods

2.2.1. Mix Design Phase. A conventional trial and error procedure [19] was used in order to design the grading curves of the bituminous mixes, optimizing the combination

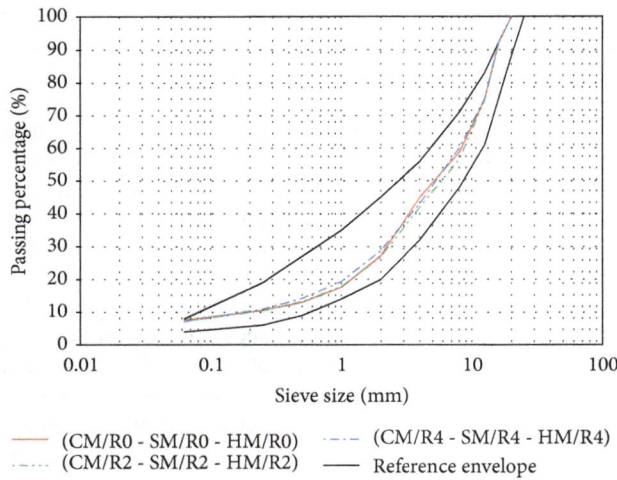

FIGURE 2: Design grading curves of the hot mix asphalt.

TABLE 1: Mix design acceptance requisites.

Parameter	Threshold
V_a at 10 revs	10–14%
V_a at 100 revs	3–5%
V_a at 180 revs	>2%
ITS_{dry}	>0.6 MPa
TSR	>75%

total bitumen content results are equal to the sum of the DBC and the bitumen of RAP.

The RAP material has been preheated for 2 hours at 90°C before being mixed with the limestone and the virgin bitumen, in a heated lab mixer, for one minute.

According to the CIRS specifications [32], the mixing temperatures were 150°C, 160°C, and 170°C for the conventional bitumen, the soft modified binder, and the hard modified one, respectively.

2.2.2. Fatigue Tests. The four-point bending equipment was used for the fatigue investigation, according to the main specifications described in Annex D of the European EN 12697-24 Standard; therefore, a continuous sinusoidal waveform, characterized by a frequency of 10 Hz, has been adopted in each test. The bending tests have been performed under the strain control mode, using three different strain levels, within the range 200–600 μm/m, in order to properly characterize the fatigue response of each mixture. The fatigue trials were conducted at 20°C, namely, the most significant temperature to investigate the fatigue performance of asphalt concrete in Italy. The stress and strain data, along with the phase angle and dissipated energy, have been determined by the equipment for each loading cycle. A laboratory compacting roller has been used to prepare slabs ($300 \times 400 \times 50$ mm), for each type of asphalt concrete, following the main specifications of the EN 12697-33 Standard. The slabs were subsequently accurately sawed, in order to obtain the beam specimens ($400 \times 50 \times 60$ mm) required for the fatigue investigation.

Some of the beam specimens were conditioned in an oven at 85°C for 5 days [34]; such protocol produces an accelerated long-term ageing, which allows taking into consideration the effect of seasoning on the fatigue resistance of the hot mix asphalt.

The fatigue life of hot mix asphalt is currently determined using primarily the conventional stiffness reduction approach [23–26, 29, 30]. However, such approach relies on the empirical assumption that the fatigue failure of the asphalt specimen occurs at the achieving of the 50% of the initial stiffness modulus. In the present study the fatigue life has been estimated by means of such empirical approach, in order to allow a comparison with results of previous investigations, but also using the energy ratio approach [35–37], which relies on more rational assumptions.

2.2.3. Stiffness Reduction Approach. In the fatigue test under strain control mode, strain is fixed at a constant value and therefore the stress is continuously decreasing during the test.

of the particle size with respect to the reference grading envelope for base courses used in the Northeastern Italian area [31]. Figure 2 shows the design grading curves of the mixes referred to the considered reference envelope. The control hot mix asphalt (CM/R0, SM/R0, and HM/R0) has been produced only with limestone aggregates, to allow a proper performance comparison with the mixes made with RAP aggregate.

For each type of hot mix asphalt the design bitumen content (DBC) was determined using an integrated mix design method, based on the volumetric properties and the indirect tensile strength (ITS), with respect to the CIRS-Italian Ministry of Infrastructure Specifications [32]. Cylindrical specimens, characterized by a 150 mm diameter and a 60 mm height, were prepared for the mix design phase, by means of a gyratory compactor. The key compaction parameters, namely, pressure, speed, and angle of rotation, have been fixed at 600 kPa, 30 revs/minute and 1.25°, respectively. The ITS tests were carried out at 25°C, according to the main indications of EN 12697-23 Standard. Both dry and wet specimens were tested, to determine the stripping resistance of the mixes. The wet specimens were soaked for 15 days in water, using a thermostatic bath. The ratio between the ITS of wet (ITS_{wet}) and dry (ITS_{dry}) specimens represents the Tensile Strength Ratio (TSR); such parameter has been determined for each mix investigated.

According to the CIRS design method [32], the volumetric and mechanical acceptance requisites reported in Table 1 have to be satisfied in order to identify the design bitumen content. The volumetric properties, quantified by means of the residual air voids percentage, have been verified at three different compaction levels (10, 100, and 180 revs), whereas the mechanical resistance was tested at the design number of gyrations (100 revs).

The conventional hypothesis of total blending between the bitumen of RAP and the new virgin binder has been considered in the mix design phase [33]. The DBC regards the new bitumen used for the hot mix asphalt production; therefore, for the mixes prepared with RAP aggregates, the

Such reduction, achieved at a high number of loading cycles, down to extremely low values, rarely causes a macrocrack and the complete physical failure of the specimen [38]. Therefore, the conventional failure criterion, largely accepted in the literature, is defined by a 50% reduction of the initial stiffness [38–41].

In this research, the conventional fatigue curves were determined on the basis of the initial value of strain ε_0 and the number of loading repetitions N_f, at which a 50% decrement of the initial stiffness is recorded. According to previous relevant studies [42–44], the strain value evaluated at the 100th cycle has been assumed as the initial strain (EN 12697-24, Annex D); in fact at this stage of the bending test, the hot mix asphalt specimen is still not damaged and therefore it can be considered in its initial state.

The fatigue data have been interpolated by a power law model, according to the following expression:

$$\varepsilon_0 = aN_f^{\ b}, \tag{1}$$

where a and b are interpolation coefficients associated with the type of hot mix asphalt.

2.2.4. Energy Ratio Approach.
The conventional failure criteria, even if established by significant changing of the mechanical parameter involved (namely, the stiffness modulus), in comparison to the initial physical state of the sample, are basically empirical and do not reliably describe the condition of internal damage in the hot mix asphalt. In order to achieve a more rational approach, in the study of the fatigue life, Hopman et al. [35] and Pronk [37] proposed a failure criterion related to the energy dissipated during fatigue bending tests, defining the failure condition as the number of cycles N_1 at which a macrocrack is developed for coalescence of microcracks. Hence, N_1 is associated with the onset of that macrocrack, which under further loading cycles propagates in the specimen [36]. The use of such failure criterion, for the analysis of the results of fatigue bending tests, permits comparing the fatigue response between mixes in an equivalent damage state, related to the onset of macrocracks. Such criterion is recognized in the literature as more physically significant and reliable [37]. From the analytical point of view, this approach is based on the definition of the energy ratio R_n, computed as the ratio between the energy dissipated from the beginning of the test, up to the nth cycle and that dissipated at the nth cycle:

$$R_n = \frac{\pi \sum_{i=0}^{n} \sigma_i \varepsilon_i \sin \phi_i}{\pi \sigma_n \varepsilon_n \sin \phi_n}, \tag{2}$$

where σ represents the stress, ε the strain, ϕ the phase angle, i the generic ith cycle, and n the nth cycle. In order to identify N_1, the evolution of the energy ratio with respect to the number of cycles has to be accurately studied. With regard to the strain control mode, N_1 can be identified in correspondence with the experimental point in the R_n curve, at which a nonlinear trend is detected. Instead, the identification of N_1 for the stress control mode is associated with the achievement of the peak point of the R_n curve. Artamendi

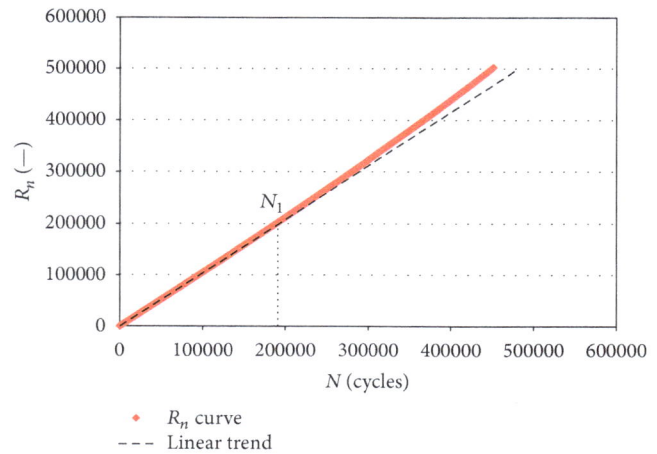

FIGURE 3: Determination of N_1 for mix HM/R2 at 400 μm/m.

and Khalid [42] have already discussed the subjectivity of the N_1 identification procedure, for the strain control tests. Figure 3 represents a typical example, for the mix HM/R2, of the N_1 identification in case of controlled strain. The accurate identification of such number of cycles, N_1, is potentially affected by the subjectivity of the researcher.

In a similar way to that followed with the empirical approach, it is possible to elaborate new fatigue curves, using a power model as that of (1), introducing the N_1 concept in substitution of the empirical N_f.

3. Results and Discussion

3.1. Materials Characterization.
The particle morphology of the aggregates has been characterized in terms of Flakiness Index (EN 933-3). The results were within the range 7–10% for the limestone, depending on the particle size; instead, a value equal to 8% was obtained for the RAP. The mechanical resistance of the aggregates was verified by means of the Los Angeles test (EN 1097-2), which has outlined similar results for the different materials: 23% for the natural aggregates (4/8 mm fraction) and 25% for RAP. Finally, the particle density of the aggregates was evaluated (EN 1097-6); values within the range 2.784–2.790 Mg/m^3 have been determined for the natural aggregates, whereas the RAP has presented a lower value, equal to 2.746 Mg/m^3.

The cold extraction test, carried out on the RAP, has outlined a bitumen percentage of 5.3% by weight of the aggregate. The bituminous binder of the RAP has been recovered using the ABSON method (EN 12697-3) and then characterized in terms of penetration at 25°C (EN 1426), softening point (Ring & Ball Method; EN 1427), Fraass breaking point (EN 12593), and viscosity at 160°C (EN 13702-2); the results are presented in Table 2, along with the values obtained for the conventional bitumen and the polymer modified binders.

According to the experimental data reported in Table 2, the RAP bitumen can be considered very aged and hard. Both the polymer modified binders presented lower penetration, as well as higher values for the softening point and the viscosity,

TABLE 2: Bitumen characterization.

Properties	Standard	RAP binder	Conventional bitumen	Hard modified	Soft modified
Penetration (0.1 x mm), 100 g, 5 s at 25°C	EN 1426	18	52	43	50
Softening point (°C), R&B method	EN 1427	64	48	98	92
Fraass breaking point (°C)	EN 12593	—	−18	−18	−19
Viscosity at 160°C	UNI EN 13702-2	0.256	0.13	1.33	0.62

TABLE 3: Aggregate composition of the hot mix asphalt.

Aggregate type	Fraction (mm)	Hot mix asphalt								
		CM/R0	CM/R2	CM/R4	HM/R0	HM/R2	HM/R4	SM/R0	SM/R2	SM/R4
Sand	0/4	47	25	8	47	25	8	47	25	8
Limestone	4/8	7	9	12	7	9	12	7	9	12
Limestone	8/12	7	9	5	7	9	5	7	9	5
Limestone	12/20	34	34	34	34	34	34	34	34	34
Filler	—	5	3	1	5	3	1	5	3	1
RAP	0/4	0	20	40	0	20	40	0	20	40

TABLE 4: Physical and mechanical properties of the hot mix asphalt.

Asphalt concrete	DBC (%)	Va, 10 revs (%)	Va, 100 revs (%)	Va, 180 revs (%)	Bulk density (Mg/m³)	ITS dry (MPa)	ITS wet (MPa)	TSR (%)
CM/R0	5.00	13.9	4.9	3.4	2.457	1.04	0.89	86
CM/R2	3.94	13.6	4.5	3.3	2.473	1.54	1.36	88
CM/R4	2.88	11.2	3.5	2.3	2.484	1.95	1.78	91
SM/R0	5.00	14.0	4.9	3.6	2.455	1.23	1.08	88
SM/R2	3.94	13.9	4.7	3.5	2.470	1.67	1.50	90
SM/R4	2.88	11.4	3.6	2.4	2.480	2.19	1.97	90
HM/R0	5.00	14.0	5.0	3.8	2.451	1.62	1.46	90
HM/R2	3.94	13.8	4.8	3.7	2.465	2.08	1.92	92
HM/R4	2.88	11.3	3.8	2.6	2.477	2.32	2.16	93

compared to the conventional bitumen. The hard modified bitumen was characterized by the highest viscosity.

3.2. Mix Design Results. Table 3 reports the aggregate structure of the hot mix asphalt considered in the study.

For each type of bitumen used in the investigation, three different aggregate compositions have been designed. The greater the RAP quantity, the lower the filler percentage that is needed in order to optimize the mix. Each mix is characterized by the same amount of the coarser limestone grains, namely, the fraction 12/20. Instead, for the other two limestone fractions (4/8 and 8/12 mm), the contents are different, depending on the RAP percentage.

The results of the mix design phase are summarised in Table 4. For each type of hot mix asphalt studied, the most relevant properties are presented: DBC (by weight of the aggregate); air voids (Va) at 10, 100, and 180 revs; bulk density at 100 revs; indirect tensile strength (ITS) for dry and wet conditions, at 100 revs.

The bitumen percentage shown in Table 4 represents the virgin binder used in order to optimize the hot mix asphalt.

For all the mixes, the air voids requisites prescribed by CIRS mix design procedure [32] have been totally satisfied,

depending on the specific DBC. The dry ITS and TSR resulted in much higher values than the CIRS requisites, for all the hot mix asphalt. Even the wet ITS values were always greater than the dry ITS requirement, so demonstrating a very good moisture resistance of the designed mixes.

The total bitumen contents resulted within the range usually used for base courses, namely, 4.5–5.5% by weight of the aggregate [32], for all the hot mix asphalt. As it was expected, the greater the RAP amount used in the mix, the smaller the new bitumen content necessary to fulfil the CIRS requirements; such laboratory data are consistent with the hypothesis of a complete blending between RAP bitumen and new binder.

No workability difficulties were noted during the mixing phase, neither for the compaction of the hot mix asphalt made with RAP, independently from the type of binder.

3.3. Fatigue Analysis Based on the Stiffness Reduction Approach. Based on the stiffness reduction approach, the elaboration of the fatigue data has allowed obtaining the fatigue curves shown in Figures 4 and 5, for the unaged and aged hot mix asphalt, respectively. The regression coefficients

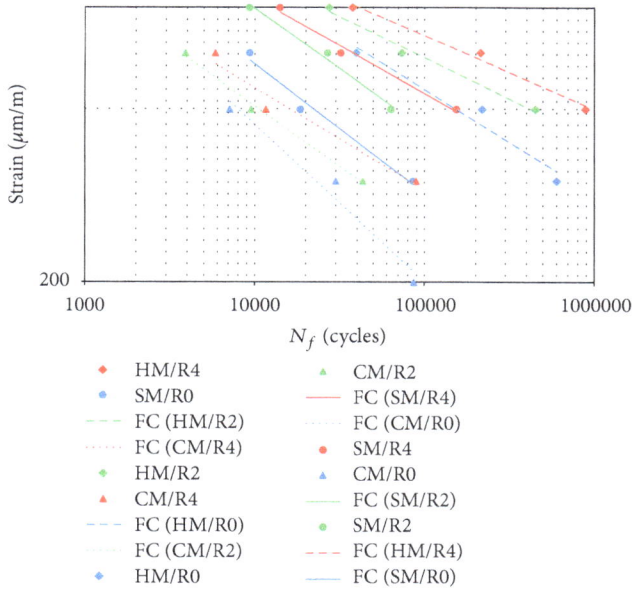

FIGURE 4: Fatigue life N_f versus initial strain: unaged mixes.

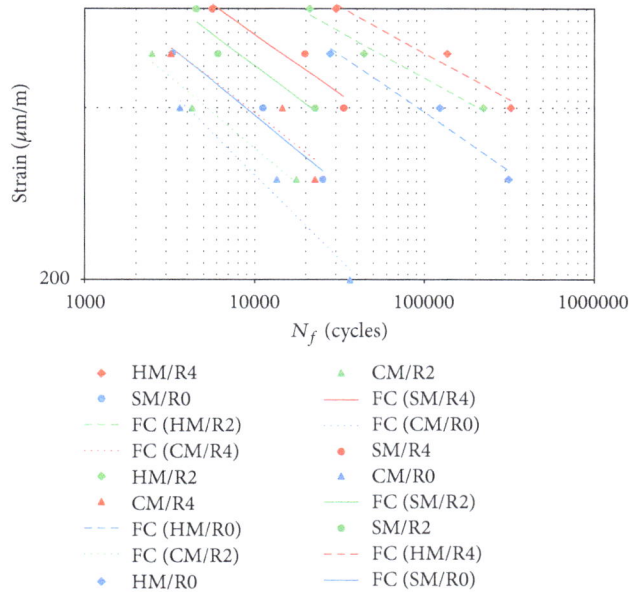

FIGURE 5: Fatigue life N_f versus initial strain: aged mixes.

TABLE 5: Regression coefficients of the fatigue curves: N_f approach for unaged mixtures.

Mixture	a ($\mu m/m$)	b (-)	ε (10^6) ($\mu m/m$)	R^2 (-)
CM/R0	4658.4	−0.273	107	0.9661
CM/R2	2793.9	−0.210	154	0.9943
CM/R4	2221.2	−0.177	193	0.9587
SM/R0	3741.8	−0.223	172	0.9791
SM/R2	4113.8	−0.209	229	0.9864
SM/R4	2866.5	−0.166	289	0.9858
HM/R0	3508.2	−0.182	284	0.953
HM/R2	2515.0	−0.142	364	0.9880
HM/R4	2322.1	−0.127	402	0.9867

TABLE 6: Regression coefficients of the fatigue curves: N_f approach for aged mixtures.

Mixture	a ($\mu m/m$)	b (-)	ε (10^6) ($\mu m/m$)	R^2 (-)
CM/R0	4638.4	−0.295	79	0.9685
CM/R2	3388.0	−0.250	107	0.9679
CM/R4	3340.2	−0.232	135	0.8637
SM/R0	3670.4	−0.244	126	0.9632
SM/R2	3835.8	−0.227	167	0.9176
SM/R4	3821.9	−0.212	204	0.9171
HM/R0	4155.0	−0.205	245	0.9582
HM/R2	3069.6	−0.166	310	0.9768
HM/R4	3351.3	−0.165	343	0.9550

and, moreover, they are shifted towards higher loading cycles values (Tables 5 and 6). Then, it is possible to observe the higher tensile strain ε (10^6) obtained for the RAP mixes, particularly for the greater RAP content (40%); such response has been observed for all the binders used in the investigation (Tables 5 and 6).

Hence, the use of the polymer modified binders, rather than RAP, considering each of them individually, enhances the fatigue life of the mixes. Nevertheless, the achievement of the highest fatigue resistance is feasible only using simultaneously the modified binders along with the greater RAP percentage. The highest fatigue life, 402 $\mu m/m$, was obtained for HM/R4 (Table 5). Such value is much higher (42%) than that determined for the control mix produced with natural materials (HM/R0). However, the increase in the fatigue resistance, associated with the utilization of RAP aggregates, has been nonlinear with the RAP quantity; in fact the improvement was higher using the 40% of RAP rather than the 20%.

This general trend can be observed also for the aged hot mix asphalt (Table 6). However, as it was expected [11], the ageing process reduces the fatigue life of all the mixes.

3.4. Fatigue Analysis Based on the Energy Ratio Approach. Figures 6 and 7 show the fatigue curves processed on the basis of the energy ratio approach, for the unaged and aged hot mix asphalt, respectively. Tables 7 and 8 present the coefficients

and the coefficient of determination R^2 are reported in Tables 5 and 6. The tensile strain associated with a fatigue resistance of 1,000,000 loading repetitions, namely, ε (10^6), has been calculated according to the indications of the Standard EN 12697–24, Annex D.

As it was expected, the mixes produced with the polymer modified binders, especially in case of the hard one, have shown a fatigue life much higher than that of the hot mix asphalt made with the conventional bitumen, independently of the RAP percentage. In fact, the fatigue curves of the polymer modified mixes are characterized by a lower slope

TABLE 7: Regression coefficients of the fatigue curves: N_1 approach for unaged mixtures.

Mixture	a (μm/m)	b (-)	ε (10^6) (μm/m)	R^2 (-)
CM/R0	3315.6	−0.254	99	0.9809
CM/R2	3081.3	−0.227	134	0.9923
CM/R4	2552.4	−0.195	173	0.9978
SM/R0	4160.3	−0.242	147	0.9763
SM/R2	3166.5	−0.197	208	0.9750
SM/R4	2811.0	−0.172	261	0.9849
HM/R0	4359.3	−0.207	250	0.9763
HM/R2	2902.9	−0.162	310	09841
HM/R4	2362.3	−0.135	366	0.9949

TABLE 8: Regression coefficients of the fatigue curves: N_1 approach for aged mixtures.

Mixture	a (μm/m)	b (-)	ε (10^6) (μm/m)	R^2 (-)
CM/R0	2984.6	−0.267	75	0.9528
CM/R2	2936.7	−0.248	95	0.9782
CM/R4	3016.8	−0.233	121	0.9216
SM/R0	3599.2	−0.250	114	0.9968
SM/R2	2918.9	−0.214	152	0.9501
SM/R4	2761.8	−0.195	187	0.9306
HM/R0	4631.1	−0.221	219	0.9951
HM/R2	3280.9	−0.182	265	0.9986
HM/R4	3165.5	−0.168	311	0.9964

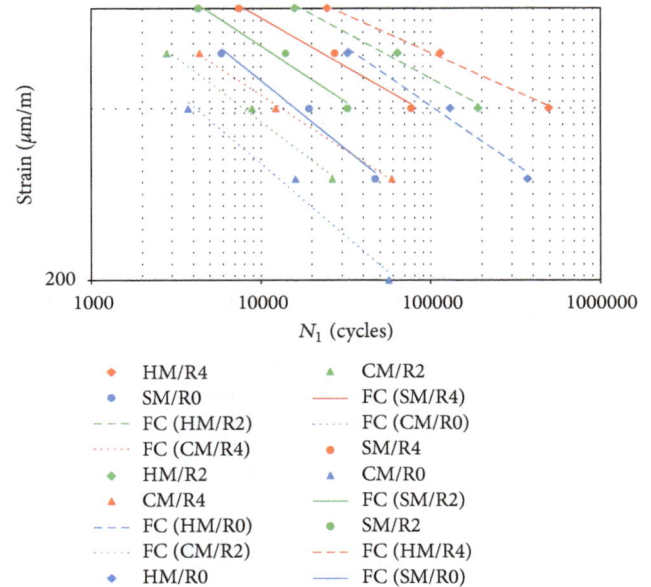

FIGURE 6: Fatigue life N_1 versus initial strain: unaged mixes.

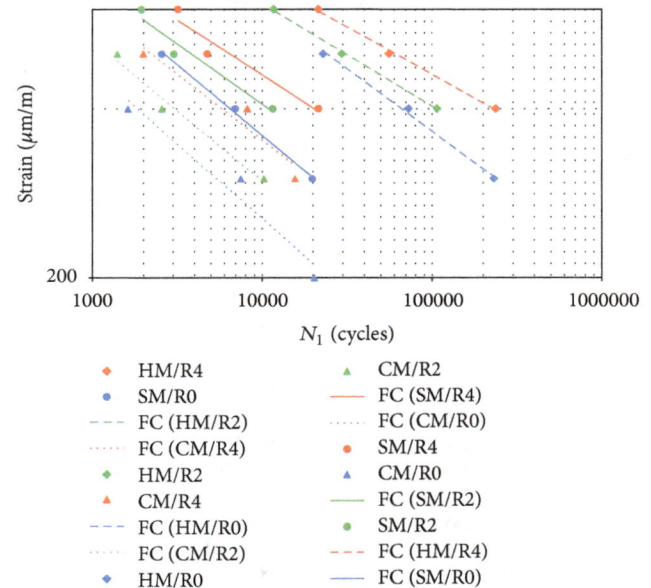

FIGURE 7: Fatigue life N_1 versus initial strain: aged mixes.

of interpolation and determination, along with the value of ε (10^6).

Even if the comparison of values of ε (10^6) associated with the various asphalt concrete, by means of the empirical method and the energy ratio approach [37] leads to an analogous ranking of the mixes, it can be noted that the analysis of the fatigue data with the energy ratio method outlines a more conservative evaluation of the fatigue resistance of the hot mix asphalt. In fact, the values of tensile strain for 1,000,000 loading repetitions, computed using the energy ratio approach, are lower than those determined by means of the empirical method.

Compared to the control bituminous mixtures (CM/R0, SM/R0 and HM/R0), the hot mix asphalt produced with RAP aggregates was characterized by higher ε (10^6) values; improvements from 35% to 74% in case of standard bitumen, from 42% to 78% using the soft polymer modified binder, and from 24% to 47% with the hard one have been obtained, depending on the RAP percentage. The highest fatigue life, independently of the type of bitumen, was obtained for the hot mix asphalt prepared with 40% RAP (CM/R4, SM/R4 and HM/R4).

The improved fatigue resistance presented by the RAP mixes (Tables 7 and 8) could be due to the good affinity between the bitumen added to the mixes (both the modified and the standard one) and that of RAP; such affinity probably allows very strong adhesion between the thin film of virgin binder and the RAP grains. Improved fatigue behaviour for RAP mixtures has been already documented by other research groups [14, 24]. They have considered the RAP particles and the virgin bitumen as a layered system in which the aged binder of the RAP assumes the role of an intermediate stiff layer between the stiffer lithic grains of the RAP and the softer new bitumen that coats the RAP particles. On the basis of such assumption, the aged binder of the RAP allows reducing the stress concentration within the hot mix asphalt, which in turn improves the fatigue life.

Nevertheless, the greater ε (10^6) values were achieved by the mixes produced with RAP and polymer modified binders;

the highest fatigue life has been obtained by HM/R4. Such additional increment of the fatigue life determined for the RAP mixtures prepared with PMB could be justified by the flexibility given by the polymers of the modified binders.

According to the CIRS technical specifications [32], the greatest RAP percentage admissible for road base hot mix asphalt is up to 30%; furthermore, the use of a polymer modified bitumen is not taken in consideration for RAP mixtures. Hence, the feasibility to produce hot mix asphalt with significant RAP content (up to 40%), characterized by a fatigue life greater than that of high performance polymer modified bituminous mixtures, made with limestone materials, can be considered one step further towards sustainability and high resistance to loading cycles, in the Italian road pavement technology.

According to the results of the current study, the RAP effect on the fatigue resistance of the asphalt concrete analysed can be evaluated very positively; however, a straight generalization of such experimental results has to be very carefully considered. Indeed, the integration of such waste material in the structure of a bituminous mixture is very challenging, in relation to some important elements, for example, the RAP particle size distribution, the RAP bitumen percentage and its ageing state, the particular type of hot mix asphalt being investigated, and the rheological properties of the new binder to be added to the mixture.

The ranking of the mixes, in terms of fatigue life, has been completely confirmed also for the hot mix asphalt which underwent the accelerated ageing process (Table 8), as already observed with the empirical approach; it is still the mix HM/R4 to achieve the highest fatigue life, in terms of ε (10^6) values (311 μm/m). Nevertheless, it can be noticed that the higher the RAP content, the greater the fatigue life reduction due to the ageing phenomenon, which affects the ductility of the mixes; such reduction falls within the range from 12% to 30%, depending on the RAP content and the type of bitumen. However, for each type of binder, the reductions in the fatigue life due to ageing were very similar between the mixes with 40% RAP and those made with 20% RAP. For instance, for the hard modified mixtures, such reductions were equal to 15% and 14%, at 40% RAP and 20% RAP, respectively.

Instead, the type of binder affects noticeably the fatigue life reduction; in fact, it can be observed that for the hard modified mixes, the ageing produces lower detrimental effects (up to 15%), with respect to the soft modified asphalt (up to 28%) and the unmodified mixtures (up to 30%). Probably the polymer modification of the bitumen allows guaranteeing better ductility of the mixes.

4. Conclusions

The experimental study described in the present paper discusses the fatigue performance of hot mix asphalt for road base courses produced with RAP and polymer modified binders.

The fatigue data, obtained by means of four-point bending tests carried out at 10 Hz and 20°C, under strain control mode, have been analysed according to both the conventional stiffness reduction approach and the energy ratio method; the latter focused on the dissipated energy concept.

Compared to the control mixes, prepared with a full limestone aggregate structure, the hot mix asphalt containing RAP has presented higher resistance to loading cycles, especially at 40% RAP, also under postageing conditions.

The improvements in the fatigue life observed for the RAP mixes have been even more relevant using the polymer modified binders, particularly in the case of the hard modified one.

Although the experimental fatigue data have outlined an extremely positive effect of RAP on the fatigue performance of the hot mix asphalt tested, a broad generalization of the RAP influence to different asphalt concrete has to be very carefully taken in consideration.

The fatigue analysis conducted using the empirical approach and the energy ratio method has led to the same qualitative ranking of the mixes considered; however quantitative differences have been observed that cannot be neglected.

The energy ratio approach has identified a fatigue life that has been always lower than that determined by the empirical method, for all the cases studied; therefore, the conventional stiffness reduction approach should be carefully used to avoid an overestimation of the fatigue resistance of the mixes.

The ageing phenomenon affects the fatigue performance of all the hot mix asphalt involved, particularly those prepared with RAP and conventional bitumen; however a modification from 20% to 40% RAP in the aggregate structure involves just a small reduction of the fatigue life (about 1%, in terms of ε (10^6) values).

Competing Interests

The authors declare that they have no conflict of interest.

References

[1] W. Cao, "Study on properties of recycled tire rubber modified asphalt mixtures using dry process," *Construction and Building Materials*, vol. 21, no. 5, pp. 1011–1015, 2007.

[2] C.-T. Chiu and L.-C. Lu, "A laboratory study on stone matrix asphalt using ground tire rubber," *Construction and Building Materials*, vol. 21, no. 5, pp. 1027–1033, 2007.

[3] Y. Liu, S. Han, Z. Zhang, and O. Xu, "Design and evaluation of gap-graded asphalt rubber mixtures," *Materials and Design*, vol. 35, pp. 873–877, 2012.

[4] C. C. Wong and W.-G. Wong, "Effect of crumb rubber modifiers on high temperature susceptibility of wearing course mixtures," *Construction and Building Materials*, vol. 21, no. 8, pp. 1741–1745, 2007.

[5] N. Ali, S. Zahran, J. Trogdon, and A. Bergan, "A mechanistic evaluation of modified asphalt paving mixtures," *Canadian Journal of Civil Engineering*, vol. 21, no. 6, pp. 954–965, 1994.

[6] Y. Becker, M. P. Méndez, and Y. Rodríguez, "Polymer modified asphalt," *Vision Tecnologica*, vol. 9, no. 1, pp. 39–50, 2001.

[7] U. Isacsson and X. Lu, "Testing and appraisal of polymer modified road bitumens-state of the art," *Materials and Structures*, vol. 28, no. 3, pp. 139–159, 1995.

[8] B. V. Kök and H. Çolak, "Laboratory comparison of the crumb-rubber and SBS modified bitumen and hot mix asphalt," *Construction and Building Materials*, vol. 25, no. 8, pp. 3204–3212, 2011.

[9] A. Othman, L. Figueroa, and H. Aglan, "Fatigue behavior of styrene-butadiene-styrene modified asphaltic mixtures exposed to low-temperature cyclic aging," *Transportation Research Record*, no. 1492, pp. 129–134, 1995.

[10] H. Özen, A. Aksoy, S. Tayfur, and F. Çelik, "Laboratory performance comparison of the elastomer-modified asphalt mixtures," *Building and Environment*, vol. 43, no. 7, pp. 1270–1277, 2008.

[11] P. Marco and B. Nicola, "Fatigue performance of asphalt concretes made with steel slags and modified bituminous binders," *International Journal of Pavement Research and Technology*, vol. 6, no. 4, pp. 294–303, 2013.

[12] P. Ahmedzade and B. Sengoz, "Evaluation of steel slag coarse aggregate in hot mix asphalt concrete," *Journal of Hazardous Materials*, vol. 165, no. 1–3, pp. 300–305, 2009.

[13] B. A. Asmatulaev, R. B. Asmatulaev, A. S. Abdrasulova, B. L. Levintov, M. F. Vitushchenko, and O. A. Stolyarskiiv, "Using blast-furnace slag in road construction," *Steel in Translation*, vol. 37, no. 8, pp. 722–725, 2007.

[14] B. Huang, G. Li, D. Vukosavljevic, X. Shu, and B. K. Egan, "Laboratory investigation of mixing hot-mix asphalt with reclaimed asphalt pavement," *Transportation Research Record*, no. 1929, pp. 37–45, 2005.

[15] A. Kavussi and M. J. Qazizadeh, "Fatigue characterization of asphalt mixes containing electric arc furnace (EAF) steel slag subjected to long term aging," *Construction and Building Materials*, vol. 72, pp. 158–166, 2014.

[16] E. A. Oluwasola, M. R. Hainin, and M. M. A. Aziz, "Evaluation of asphalt mixtures incorporating electric arc furnace steel slag and copper mine tailings for road construction," *Transportation Geotechnics*, vol. 2, pp. 47–55, 2015.

[17] M. Pasetto and N. Baldo, "Fatigue performance of asphalt concretes with RAP aggregates and steel slags," in *Proceedings of the 7th RILEM International Conference on Cracking in Pavements*, vol. 4, pp. 719–727, RILEM Bookseries, Delft, Netherlands, 2012.

[18] M. Pasetto and N. Baldo, "Resistance to permanent deformation of road and airport high performance asphalt concrete base courses," *Advanced Materials Research*, vol. 723, pp. 494–502, 2013.

[19] M. Pasetto and N. Baldo, "Influence of the aggregate skeleton design method on the permanent deformation resistance of stone mastic asphalt," *Materials Research Innovations*, vol. 18, no. 3, pp. S96–S101, 2014.

[20] Y. Xue, S. Wu, H. Hou, and J. Zha, "Experimental investigation of basic oxygen furnace slag used as aggregate in asphalt mixture," *Journal of Hazardous Materials*, vol. 138, no. 2, pp. 261–268, 2006.

[21] Y. Xue, H. Hou, S. Zhu, and J. Zha, "Utilization of municipal solid waste incineration ash in stone mastic asphalt mixture: pavement performance and environmental impact," *Construction and Building Materials*, vol. 23, no. 2, pp. 989–996, 2009.

[22] S. Wu, Y. Xue, Q. Ye, and Y. Chen, "Utilization of steel slag as aggregates for stone mastic asphalt (SMA) mixtures," *Building and Environment*, vol. 42, no. 7, pp. 2580–2585, 2007.

[23] C. Celauro, C. Bernardo, and B. Gabriele, "Production of innovative, recycled and high-performance asphalt for road pavements," *Resources, Conservation and Recycling*, vol. 54, no. 6, pp. 337–347, 2010.

[24] J. R. M. Oliveira, H. M. R. D. Silva, C. M. G. Jesus, L. P. F. Abreu, and S. R. M. Fernandes, "Pushing the asphalt recycling technology to the limit," *International Journal of Pavement Research and Technology*, vol. 6, no. 2, pp. 109–116, 2013.

[25] P. A. A. Pereira, J. R. M. Oliveira, and L. G. Picado-Santos, "Mechanical characterisation of hot mix recycled materials," *International Journal of Pavement Engineering*, vol. 5, no. 4, pp. 211–220, 2004.

[26] H. M. R. D. Silva, J. R. M. Oliveira, and C. M. G. Jesus, "Are totally recycled hot mix asphalts a sustainable alternative for road paving?" *Resources, Conservation and Recycling*, vol. 60, pp. 38–48, 2012.

[27] A. Stimilli, A. Virgili, F. Giuliani, and F. Canestrari, "In plant production of hot recycled mixtures with high reclaimed asphalt pavement content: a performance evaluation in," in *Proceedings of the 8th RILEM International Symposium on Testing and Characterization of Sustainable and Innovative Bituminous Materials*, vol. 11, pp. 927–939, RILEM Bookseries, Ancona, Italy, 2016.

[28] B. Colbert and Z. You, "The determination of mechanical performance of laboratory produced hot mix asphalt mixtures using controlled RAP and virgin aggregate size fractions," *Construction and Building Materials*, vol. 26, no. 1, pp. 655–662, 2012.

[29] R. Miró, G. Valdés, A. Martínez, P. Segura, and C. Rodríguez, "Evaluation of high modulus mixture behaviour with high reclaimed asphalt pavement (RAP) percentages for sustainable road construction," *Construction and Building Materials*, vol. 25, no. 10, pp. 3854–3862, 2011.

[30] G. Valdés, F. Pérez-Jiménez, R. Miró, A. Martínez, and R. Botella, "Experimental study of recycled asphalt mixtures with high percentages of reclaimed asphalt pavement (RAP)," *Construction and Building Materials*, vol. 25, no. 3, pp. 1289–1297, 2011.

[31] *Capitolato Speciale d'Appalto Tipo per la Manutenzione e la Costruzione delle Infrastrutture Stradali*, Veneto Strade, Venezia, Italy, 2012 (Italian).

[32] CIRS—Ministero delle Infrastrutture e dei Trasporti, *Capitolato Speciale d'Appalto Tipo per Lavori Stradali*, CIRS—Ministero delle Infrastrutture e dei Trasporti, Italy, 2001 (Italian).

[33] I. L. Al-Qadi, M. Elseifi, and S. H. Carpenter, "Reclaimed asphalt pavement—a literature review," Tech. Rep. FHWA-ICT-07-001, Federal Highway Administration (FHWA), Washington, DC, USA, 2007.

[34] I. Artamendi, B. Allen, and P. Phillips, "Influence of temperature and aging on laboratory fatigue performance of asphalt mixtures," in *Proceedings of the Advanced Testing and Characterization of Bituminous Materials Conference*, pp. 185–194, Rhodes, Greece, 2009.

[35] P. C. Hopman, P. A. Kunst, and A. C. Pronk, "A renew interpretation method for fatigue measurements, verification of Miner's rule," in *Proceedings of the 4th Eurobitume Symposium*, Madrid, Spain, 1989.

[36] G. M. Rowe, "Performance of asphalt mixtures in the trapezoidal fatigue test," *Journal of Association of Asphalt Paving Technologists*, vol. 62, pp. 344–384, 1993.

[37] A. C. Pronk, "Comparison of 2 and 4 point fatigue tests and healing in 4 point dynamic test based on the dissipated energy concept," in *Proceedings of the 8th International Conference on Asphalt Pavement*, vol. 2 and 4, Seattle, Wash, USA, 1997.

[38] H. A. Khalid, "Comparison between bending and diametral fatigue tests for bituminous materials," *Materials and Structures/Materiaux et Constructions*, vol. 33, no. 231, pp. 457–465, 2000.

[39] W. Van Dijk and W. Visser, "The energy approach to fatigue for pavement design," *Journal of Association of Asphalt Paving Technologists*, vol. 46, pp. 1–40, 1977.

[40] A. A. Tayebali, G. M. Rowe, and J. B. Sousa, "Fatigue response of asphalt-aggregate mixtures," *Journal of Association of Asphalt Paving Technologists*, vol. 61, pp. 333–360, 1992.

[41] K. A. Ghuzlan and S. H. Carpenter, "Traditional fatigue analysis of asphalt concrete mixtures," in *Proceedings of the Transportation Research Board Annual Meeting*, Washington, DC, USA, January 2003.

[42] I. Artamendi and H. Khalid, "Characterization of fatigue damage for paving asphaltic materials," *Fatigue and Fracture of Engineering Materials and Structures*, vol. 28, no. 12, pp. 1113–1118, 2005.

[43] H. Di Benedetto, M. A. Ashayer Soltani, and P. Chaverot, "Fatigue damage for bituminous mixtures: a pertinent approach," *Journal of Association of Asphalt Paving Technologists*, vol. 65, pp. 142–158, 1996.

[44] M. Pasetto and N. Baldo, "Unified approach to fatigue study of high performance recycled asphalt concretes," *Materials and Structures*, vol. 50, article no. 113, 2017.

5

Assessment Model and Virtual Simulation for Fatigue Damage Evolution of Asphalt Mortar and Mixture

Danhua Wang,[1] **Xunhao Ding** ⓘ**,**[2] **Linhao Gu,**[2] **and Tao Ma** ⓘ[2]

[1]*School of Computer Engineering, Nanjing Institute of Technology, 1 Hongjin Road, Nanjing, Jiangsu 211167, China*
[2]*School of Transportation, Southeast University, 2 Sipailou, Nanjing, Jiangsu 210096, China*

Correspondence should be addressed to Tao Ma; matao@seu.edu.cn

Guest Editor: Quantao Liu

Focused on the fatigue performance of the asphalt mortar, this study proposed an assessment model for fatigue damage evolution based on the continuum mechanics. From the perspective of the material scale rather than the macrostructure, the proposed damage model was set by concentrating on the stress-strain state of a tiny point which could characterize the material performance accurately. By the mechanical formula derivation and based on the four-point bending fatigue tests, the damage evolution law was determined and then the proposed model was verified. Based on the finite element method (FEM), a commercial software named ABAQUS was utilized to develop the random mixtures consisting of coarse aggregates, mortar, and voids. Eventually, combined with the damage model and virtual simulation of bending tests, the factors influencing the fatigue resistance of the whole asphalt mixtures were analyzed further.

1. Introduction

Asphalt mixture is a three-phase structure consisting of aggregate, void, and mortar [1–3]. The asphalt mortar, acting as a bonding layer between adjacent particles, has been highlighted by many researchers to play an important role in the performance of the whole mixtures [4, 5]. Different form the mixtures, the asphalt mortar consists of the binder, mineral powder, and the fine aggregates only.

Tests conducted by Tan et al. [6] have pointed out that the fine aggregate properties play an important role in the viscoelastic performance of the asphalt mortar. To obtain better viscous properties and low-temperature performance, abundant angular fine aggregates are necessary. Hasan et al. [7] designed and carried out the creep tests of the asphalt mortar. In their tests, four kinds of binder were used including one base binder and three modified ones. It is found that the test environmental temperature along with the binder types all affected the occurrence of the mortar creep in porous asphalt. Permeability loss was more significant when the base binder was used. Wang et al. [8] utilized the rolling thin film oven to reveal the influence of mineral powder on the aging characters of asphalt mortar.

The results showed that when the mass ratio of mineral powder to mortar was less than 1.5, the mineral powder contents had a positive effect on asphalt aging; when the mass ratio of mineral powder to mortar was more than 1.5, increasing the contents of mineral powder could only accelerate asphalt aging. Thus, it is necessary to determine the best ratio of mineral powder during pavement construction. Similar findings have been put forward by many others showing the mortar effects in the whole mixtures [9–11].

Apart from the research studies based on the laboratory tests, many numerical methods such as discrete element method (DEM) and finite element method (FEM) were introduced to illustrate the mechanical behavior of the asphalt mortar. These computational techniques accelerate the related studies and reveal the materials' mechanism from a new perspective [12–15]. Based on these numerical methods, many researchers have done virtual tests to evaluate the performance of the asphalt mortar and concluded some meaningful findings [16–18].

In the road engineering, the main concern is about how to evaluate the fatigue properties of the materials and establish the corresponding mathematical model to predict the long-term performance of the pavement structure. These fatigue damage evolution models were necessary guarantees for the precise simulations. Only by importing the model into numerical simulation firstly, the outputs from the numerical software can be convinced and valuable. To reveal the long-term behavior of pavements under cyclic vehicle load, many researchers proposed their own damage models [19–23]. However, most of the proposed damage evolution models were developed from the macrostructure scale and could only predict the behavior of the designed composition and structures. The methods characterizing the damage evolution from the material scale still need to be further studied, especially for the asphalt mortal in the road field.

2. Objective and Scope

The objective of this study is to establish the fatigue damage evolution model for the asphalt mortar. It is for the mortar material rather than the test structures. To achieve this, the four-point bending fatigue tests and the FEM simulation were utilized together. Prior to processing, several assumptions were made as follows:

(1) The macrofracture of the aggregate, mortar, and mixtures was not included here, and only the damage evolution was taken into considerations. The damage and fracture are two different stages for the materials [24]. This helps to determine the study scope of this paper, and the damage evolution of the materials was concentrated on only here.

(2) The damage of the asphalt mortar dominated within the whole mixtures while the coarse aggregates were regarded as the variation of the boundary conditions. It is believed that the damage mainly occurs within the asphalt mortar rather than the aggregates. Because the properties of aggregates are not changed mostly under the cyclic loading, few of the particles will break. Oppositely, the interfaces between aggregates and mortar and the inner area of the mortar tend to generate microcracks more easily which leads to a performance decline. And this is the main cause of the fatigue damage. So, the damage of the asphalt mortar rather than the mixtures was studied here.

(3) Due to the low temperature of the designed tests, the viscoelastic characteristics of asphalt mixtures could be neglected temporary, and the mixtures were regarded as elastic materials [25]. Thus, the fatigue damage evolution models of the asphalt mortar could be derived based on the linear elastic theory.

The rest of this paper is organized as follows. Section 3 mainly introduces the materials, apparatus, simulation methods, and the derivation of the fatigue damage evolution formulas. In Section 4, the damage evolution law and the major factors on the damage process are analyzed based on the laboratory tests. Moreover, the validation of the proposed model is verified and is applied within the FEM simulation to predict structure performance further. Finally, some research findings are summarized in Section 5.

3. Methodology and Experimental

3.1. Materials. The asphalt binder, mineral powder, and aggregates were prepared for this study. Prior to laboratory tests, the base properties of materials were measured firstly to meet the requirements of the Chinese Technical Specification for Construction of Highway Asphalt Pavement [26]. The summaries of the material properties are shown in Tables 1–3. Two types of the specimens are needed in the following process: (a) asphalt mixtures consisting of asphalt binder, mineral powder, and aggregates and (b) asphalt mortar consisting of asphalt binder, mineral powder, and the fine part of aggregates. The testing gradation of the asphalt mixture is shown in Figure 1(a) meeting the requirements of Chinese standard [26]. Since the aggregates smaller than 2.36 mm are regarded as part of asphalt mortar, the specific content of the fine aggregates within mortar can be determined based on the mixture gradation as shown in Figure 1(b). This recalculation of the mortar gradation is necessary to keep the other compositions same as the prepared mixtures. Only the coarse aggregates should be extracted.

3.2. Laboratory Test. According to the Chinese specification named Standard Test Methods of Bitumen and Bituminous Mixtures for Highway Engineering (T0739-2011) [27], the four-point bending beam fatigue tests based on the universal testing machine (UTM) were selected to evaluate the fatigue performance of the asphalt mortar and asphalt mixture. The fatigue tests of asphalt mortar were used to obtain model parameters and to study the influencing factors of fatigue performance. Three major factors were taken into consideration and varied during the test process, including the test temperature (5°C and 10°C), loading strain (700 $\mu\varepsilon$ and 1000 $\mu\varepsilon$), and the asphalt content of mortar (4.3%, 4.8%, and 5.5%). Hereafter, in this study, the asphalt content of 4.3%, 4.8%, and 5.5% was referred to as low, moderate, and high contents, respectively. One fatigue test of asphalt mixture under the condition of 5°C, 700 $\mu\varepsilon$ and moderate asphalt content was conducted. The test data were used to verify the FE model of asphalt mixture. The tests were conducted strictly according to the Chinese standard [27], and the bending stiffness modulus was calculated as shown in equations (1)–(3). A load with haversine amplitude is applied to simulate the strain-control loading condition (10 Hz). The fatigue life was defined as the load cycles when the bending stiffness modulus decreased to 50% compared to the initial one:

$$\sigma_t = \frac{(L \times P)}{(w \times h^2)}, \tag{1}$$

where σ_t is the maximum tensile stress of the centre bottom of the test beam, Pa; L is the length of the test beam, m; P is

TABLE 1: Base properties of asphalt binder.

Index	Results	Requirements
Penetration (25°C, 100 g, 5s)/0.1 mm	71	60–80
Ductility (10°C)/cm	31.3	≥20
Softening point (°C)	48.2	≥46
Flash point (°C)	287	260
Density (g/cm³)	1.031	—

TABLE 2: Base properties of mineral powder.

Index		Results	Requirements
Granularity (%)	<0.6 mm	100	100
	<0.15 mm	96.5	90–100
	<0.075 mm	85.3	75–100

peak load during test, N; w is the width of the test beam, m; and h is the height of the test beam, m.

$$\varepsilon_t = \frac{(12 \times \delta \times h)}{(3 \times L^2 - 4 \times a^2)}, \tag{2}$$

where ε_t is the maximum tensile strain of the centre bottom of the test beam, m/m; δ is the maximum deflection of the beam centre, m; and a is the spacing distance between two adjacent chucks.

The bending stiffness modulus was defined as follows:

$$S = \frac{\sigma_t}{\varepsilon_t}, \tag{3}$$

where S is the bending stiffness modulus, Pa; σ_t is the maximum tensile stress of the centre bottom of the test beam, Pa; and ε_t is the maximum tensile strain of the centre bottom of the test beam, m/m.

The damage of the centre bottom of the test beam at the Nth load cycle was calculated as follows:

$$D_N = 1 - \frac{S_N}{S_0}, \tag{4}$$

where D_N is the damage at the Nth load cycle; S_N is the bending stiffness modulus at the Nth load cycle, Pa; and S_0 is the initial bending stiffness modulus, which is the bending stiffness modulus at the 50th load cycle according to the Chinese specification [27].

3.3. Fatigue Damage Model Development for Asphalt Mortar.

As a design index according to the Chinese standard [26], the requirement of the tensile strain in the layer bottom is much more significant to ensure fatigue resistance. Thus, only focused on the tensile damage, which dominated within the pavement structure under the cyclic load, the tensile damage evolution model was proposed to characterize the mixture performance. Since the bending stiffness modulus (S) obtained from the tests can only represent the structure performance, it is a macroindex characterizing the whole beam rather than materials. Thus, the fatigue damage model was developed further. Based on the principle of the continuum mechanics, when all the external load, boundary conditions, and structure size are

determined, the stress-strain state of point C can be obtained obviously. By establishing the mathematic relation of damage between the beam structure and centre bottom point, the damage evolution model of a tiny point can be obtained. And this point damage model is regarded as the micromaterial properties under the cyclic load regardless of the macrostructures. The details of the fatigue damage model are shown in Figure 2.

D_c, ε_c, and σ_c are defined as the damage, strain, and stress of point C after N times cyclic load, while the D, ε, and σ are the damage, strain, and stress of random points in the cross section of the beam after N times cyclic load. The constitutive model and fatigue damage evolution model are developed as an initial form as shown in equations (5) and (6).

The constitutive model:

$$\sigma = (1 - D)E\varepsilon. \tag{5}$$

The fatigue damage evolution model:

$$\frac{dD}{dN} = A[\varepsilon(1 - D)]^p, \tag{6}$$

where E is the elastic modulus and A and p are the coefficients to be determined, related to the material properties.

When derived the evolution formulas for mortar, the mortar was assumed as an isotropic continuum medium. So the four-point bending fatigue test for mortar could be regarded as a plane stress problem. The stress and strain vertical to the cross section were selected for formula derivation then. For random points in the cross section of the beam, equilibrium condition should be reached at any time, and then equations (8) and (9) can be obtained as follows:

$$\int_0^1 \sigma w h d\lambda = 0, \tag{7}$$

$$\int_0^1 \sigma w h^2 \lambda d\lambda = M, \tag{8}$$

$$\lambda = \frac{y}{h}, \tag{9}$$

where λ is the relative position of random points along the Y-coordinates direction; y is the Y-coordinates of the random points; M is the bending moment of the cross section of the beam, N·m; and the other parameters are the same as the formers.

Since the cyclic load, more microcracks and microholes will appear during the process especially in the centre bottom area of the beam. This causes damage and will make the neutral axis move upwards gradually. Thus, λ_N is proposed and defined as the new position of the neutral axis after N times cyclic load as follows:

$$\lambda_n = \frac{y_n}{h}, \tag{10}$$

where λ_n is the relative position of neutral axis along the Y-coordinates direction after N times cyclic load; y_n is the

TABLE 3: Base properties of aggregates.

Coarse part					
Size (mm)	16–13.2	13.2–9.5		9.5–4.75	4.75–2.36
Density (g/cm³)	2.771	2.791		2.688	2.735
Fine part					
Size (mm)	2.36–1.18	1.18–0.6	0.6–0.3	0.3–0.15	0.15–0.075	<0.075
Density (g/cm³)	2.715	2.718	2.716	2.717	2.720	2.752

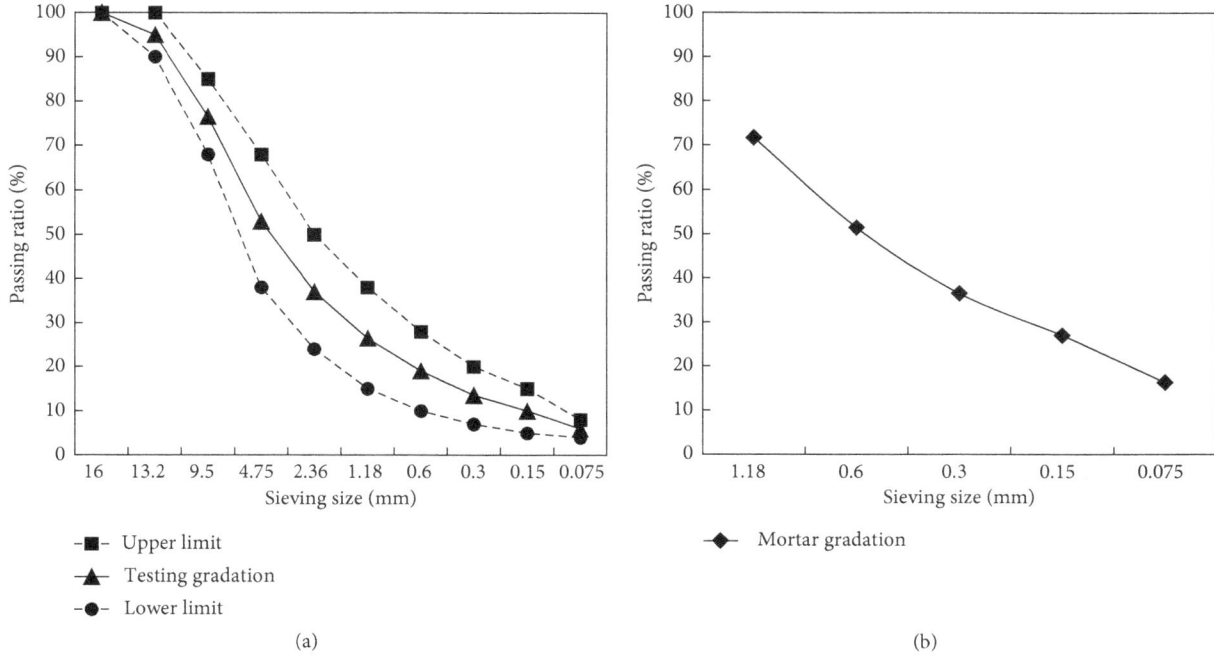

FIGURE 1: Testing gradations: (a) gradations for the asphalt mixtures; (b) gradations for the asphalt mortar.

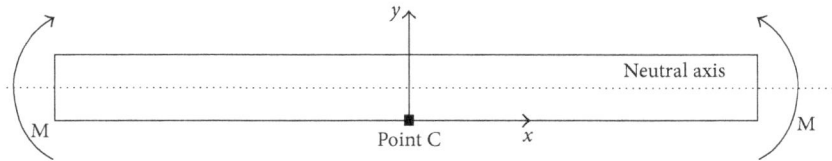

FIGURE 2: Fatigue damage of the point C.

Y- coordinates of the neutral axis after N times cyclic load, m; and the other parameters are the same as the formers.

Substituting equation (5) into equations (8) and (9), we obtain

$$\int_0^{\lambda_n} whE(1-D)\varepsilon d\lambda + \int_{\lambda_n}^1 whE\varepsilon d\lambda = 0,$$
$$\int_0^{\lambda_n} wh^2 E(1-D)\varepsilon\lambda d\lambda + \int_{\lambda_n}^1 wh^2 E\varepsilon\lambda d\lambda = M, \quad (11)$$

where the parameters are the same as the formers.

In the strain control mode, the strain of the centre bottom of the beam remains the same; thus, we obtain

$$\varepsilon = \frac{y}{\rho}, \quad (12)$$

where ρ is the curvature of the beam.

Defining $\eta = (\lambda_n - \lambda)/\lambda_n$ and substituting it into equation (12), the strain of random points can be obtained as follows:

$$\varepsilon = \varepsilon_c \eta, \quad (13)$$

$$d\lambda = -\lambda_n d\eta. \quad (14)$$

Substituting equations (13) and (14) into (9), we obtain

$$\int_0^1 \eta D d\eta = \frac{1}{\lambda_n} - \frac{1}{2\lambda_n^2}, \quad (15)$$

$$\int_0^1 \eta \left[\int_0^N \left(\frac{dD}{dN} \right) dN \right] d\eta = \frac{1}{\lambda_n} - \frac{1}{2\lambda_n^2}. \quad (16)$$

Taking equation (6) into (16), we obtain

$$\int_0^N A[\varepsilon(1-D)]^p dN = 2\left(\frac{1}{\lambda_n} - \frac{1}{2\lambda_n^2}\right). \quad (17)$$

To the point C, $\lambda = 0$, combined with the equation (6), we obtain

$$\int_0^N A[\varepsilon_c(1-D_c)]^p dN = D_c. \quad (18)$$

The damage of the point C can be determined through equations (17) and (18) as follows:

$$D_c = 2\left(\frac{1}{\lambda_n} - \frac{1}{2\lambda_n^2}\right). \quad (19)$$

The unknown parameter λ_n can be solved through equation (19) as follows:

$$\lambda_n = \frac{1}{1 + \sqrt{1 - D_c}}. \quad (20)$$

Apply the integral operation on both sides of equation (6), the fatigue damage evolution model for each point can be developed as shown in equation (21), and the Point C is shown in equation (22):

$$D = 1 - \left[1 - (-p + 1)A\varepsilon^p N\right]^{(1/(-p+1))}, \quad (21)$$

$$D_c = 1 - \left[1 - (-p + 1)A\varepsilon_c^p N\right]^{(1/(-p+1))}. \quad (22)$$

3.4. FEM Modelling of Asphalt Mixture. Virtual models were developed based on the FEM software named ABAQUS. Two model forms were taken into consideration. One is the virtual asphalt mortar and another is the asphalt mixtures. The virtual four-point bending fatigue tests for mortar were carried out to verify the FEM implementation of the proposed fatigue damage model. After that, the precise damage model of the mortar was imported into the virtual mixtures, and the virtual four-point bending fatigue tests for mixture were conducted. The simulation results were compared with the laboratory results of asphalt mixture to verify the mesostructure FE model. More simulations were conducted to analyze the effect of air void on damage evolution.

The coarse aggregates were regarded as pure elastomer without any damage evolution. When conducted the virtual tests, Young's modulus and Poisson's ratio of the coarse aggregates were set as 5.55 GPa and 0.15, respectively. The virtual specimens were developed with a 377 mm length and 50 mm height in two dimensions, which were identical to the realistic. The final virtual specimens for asphalt mixture are shown in Figure 3(a). Free meshing algorithm with quad-dominated element shape was used. And the type is CPS8R (an 8-node biquadratic plane stress quadrilateral, reduced integration). The green, white, and red areas represent aggregates, asphalt mortar, and air voids, respectively. Asphalt mortar was considered as an isotropic material, and its FE model is shown in Figure 3(b). Structured meshing

algorithm with quad element shape was used. And the element type is also CPS8R.

As shown in Figure 3(c), the position of four virtual clamps was the same as the real test. Clamp C was fixed in all degrees of freedom. A vertical displacement boundary condition with haversine amplitude was applied to clamp A to simulate the strain-control loading condition (10 Hz, $700\,\mu\varepsilon$, and $1000\,\mu\varepsilon$). The horizontal displacement and rotation of clamp A were constrained. The clamps were connected to the sample using surface-to-surface contact with geometric properties. The clamps were considered as pure elastomer with Young's modulus far greater than that of the sample.

The virtual mixtures including the asphalt mortar, irregular shape voids, and coarse aggregates were also developed in ABAQUS. Prior to FEM modelling, pre-processing for the random composition generations was completed based on Matlab. A user-defined routine was coded in Matlab to help develop the irregular shapes of particles and finally form the total mixtures as shown in Figures 4 and 5. As shown in Figure 4, polar coordinates were utilized in the irregular shapes plotting. Several groups of coordinate values were determined randomly within a designed range firstly and then a circumscribing polygon was plotted to form the final shapes. When developed the mixtures, particles of different grades were generated in turns based on their sizes. As shown in Figure 5, the rectangle was plotted as a virtual container, and then particles of 16 mm size were plotted firstly followed by 13.2 mm, 9.5 mm, 4.75 mm, etc. After the particle generations were completed, the voids were plotted in the same way. As shown, three kinds of mixtures were developed randomly. The blue lines represented the coarse aggregates while the red ones are the voids. The percent air voids in bituminous mixtures (VV) were 0%, 4%, and 8%, and the percent voids in mineral aggregate in bituminous mixtures (VCA) were 32.3%, 32.1%, and 30.7%, respectively. It should be noted that when plotted the irregular shapes of particles and voids, a judgment routine was conducted in Matlab to avoid the shape overlaps. If there existed a particle or void already, the next particle or void generation would not be permitted here until another available location was found. After the mixture images were generated successfully finally, they were imported to ABAQUS for structure and mesh generations as shown in Figure 6.

3.5. Model Parameter Determination of the Asphalt Mortar. Since the damage evolution of the asphalt mortar dominates within the whole mixtures, its model should be further specified by determining the related parameters. As illustrated in equation (22), the fatigue damage variable is related to two material parameters (A and p). A nonlinear minimization algorithm using differential evolution method is performed on the target error function F as

$$\min F(A, p) = \frac{1}{N}\sqrt{\sum_{i=1}^N \left(1 - \frac{D_{c,i}}{D_{t,i}}\right)^2}, \quad (23)$$

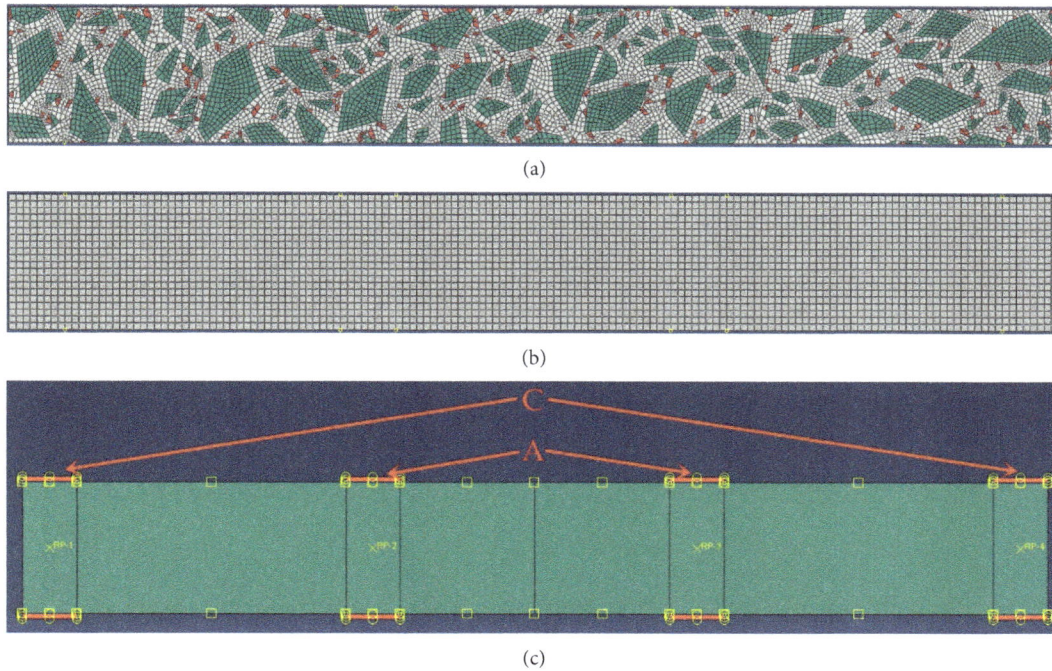

FIGURE 3: FEM modelling: (a) asphalt mixture sample; (b) asphalt mortar; (c) clamps.

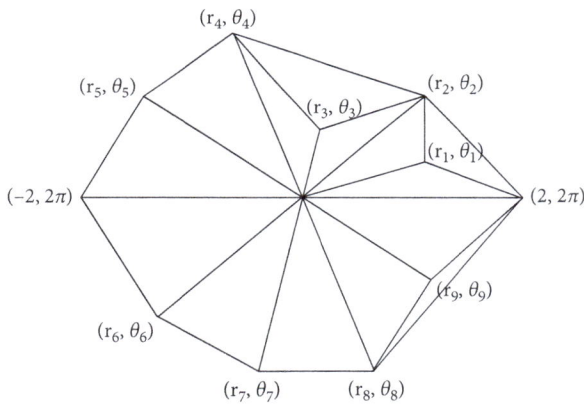

FIGURE 4: Irregular shape modelling for particles.

where $D_{c,i}$ is the calculated damage and $D_{t,i}$ is the damage obtained from laboratory results using equation (4).

By fitting the laboratory results, these undetermined parameters could be confirmed. Figures 7 and 8 are the fitting results of the proposed fatigue damage evolution models. Different conditions including the temperature, loading strain, and asphalt content were taken into considerations. The values of A and p applied for these conditions are all calibrated and summarized in Table 4. As shown in Figures 7 and 8, it is concluded that the proposed damage model could well characterize the fatigue performance of the asphalt mortar in the first two stages. In the third stage, the damage evolved rapidly converting the inner microcracks into macrofracture. In this stage, the macrocracks dominated, and therefore the modulus of the structure decreased sharply. Thus, focusing on the damage evolution only without the macrofracture mechanism, the fitting lines could only characterize the first two stages. And it has been verified in Figures 7 and 8 that the proposed damage evolution models could well predict the performance of asphalt mortar before fracture.

4. Results and Discussion

4.1. Analysis of the Laboratory Fatigue Tests. The results of the four-point bending fatigue tests in laboratory are summarized in Figures 9 and 10. Figure 9 illustrates the influences of the temperatures and strain levels on the fatigue performance of asphalt mortar. As shown in Figure 9, when compared the curves of 5°C, 700 $\mu\varepsilon$ with the 5°C, 1000 $\mu\varepsilon$ in Figures 9(a)–9(c), it is concluded that the strain level has a negative effect on the changes of the bending stiffness modulus (S). When comes to the curves of 5°C, 1000 $\mu\varepsilon$ and 10°C, 1000 $\mu\varepsilon$, it is found that as the temperature increases, the asphalt mortar will have better fatigue resistances. Moreover, the strain level plays a much more important role than temperature in affecting the fatigue performance of asphalt mortar which can be obviously summarized from Figure 9. The fatigue lives of 5°C, 700 $\mu\varepsilon$ are much more than the 10°C, 1000 $\mu\varepsilon$ despite the asphalt content. It is believed that the load strain decreased by 300 $\mu\varepsilon$ has a more positive effect than increasing the temperature by 5°C. Thus, avoiding the heavy traffic is significant to maintain the pavement permanent performance. As for Figure 9(c), different trends could be also found that the bending stiffness modulus (S) of the 5°C, 1000 $\mu\varepsilon$ was a litter larger than the other two at high asphalt contents. This is because of the lowest temperature and largest strain level which are the worst conditions for the mortar. When the asphalt content is high, the lower

(a)

(b)

(c)

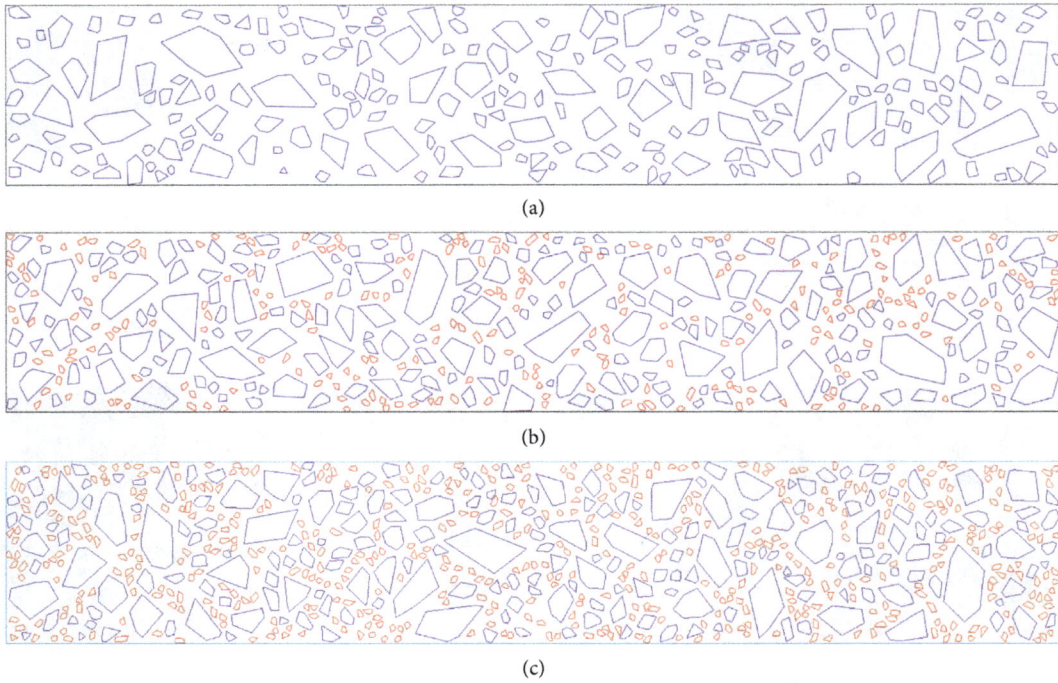

FIGURE 5: Irregular shape modelling for mixtures: (a) VCA = 32.3%, VV = 0%; (b) VCA = 32.1%, VV = 4%; (c) VCA = 30.7%, VV = 8%.

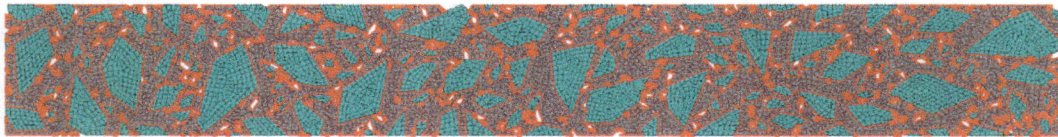

FIGURE 6: Final FEM modelling for the asphalt mixtures.

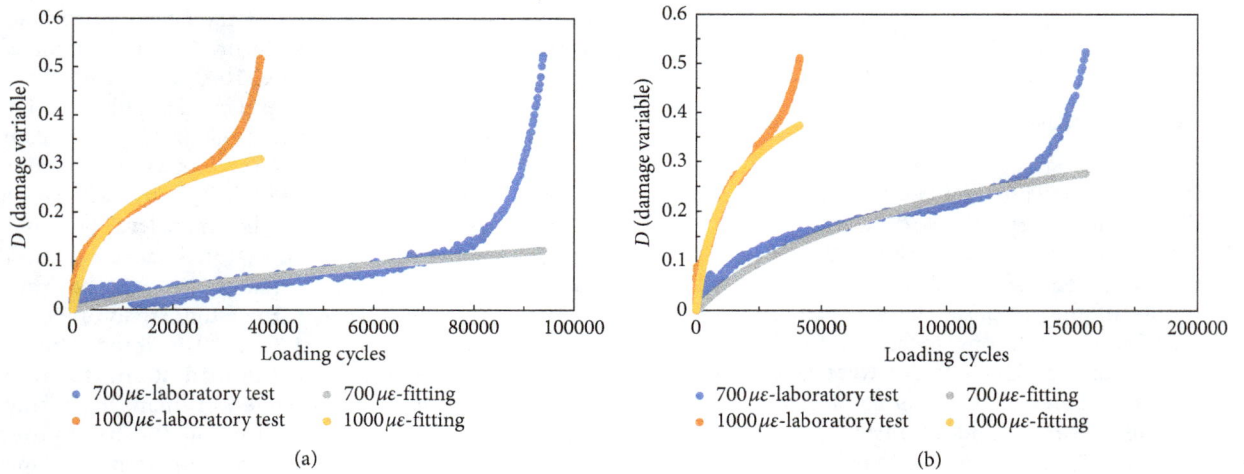

(a)

(b)

FIGURE 7: Continued.

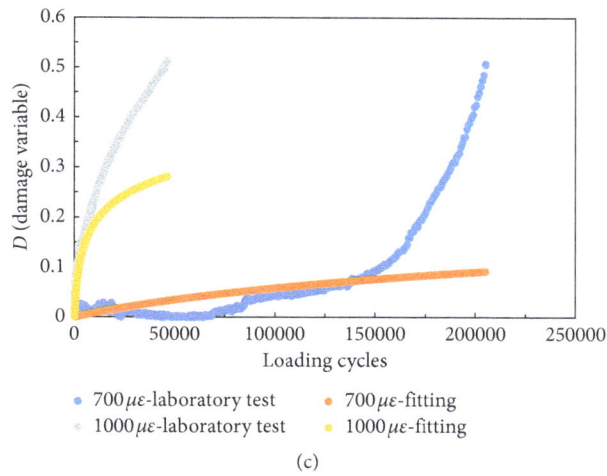

FIGURE 7: Fatigue performance prediction of the asphalt mortar at 5°C: (a) low asphalt content; (b) moderate asphalt content; (c) high asphalt content.

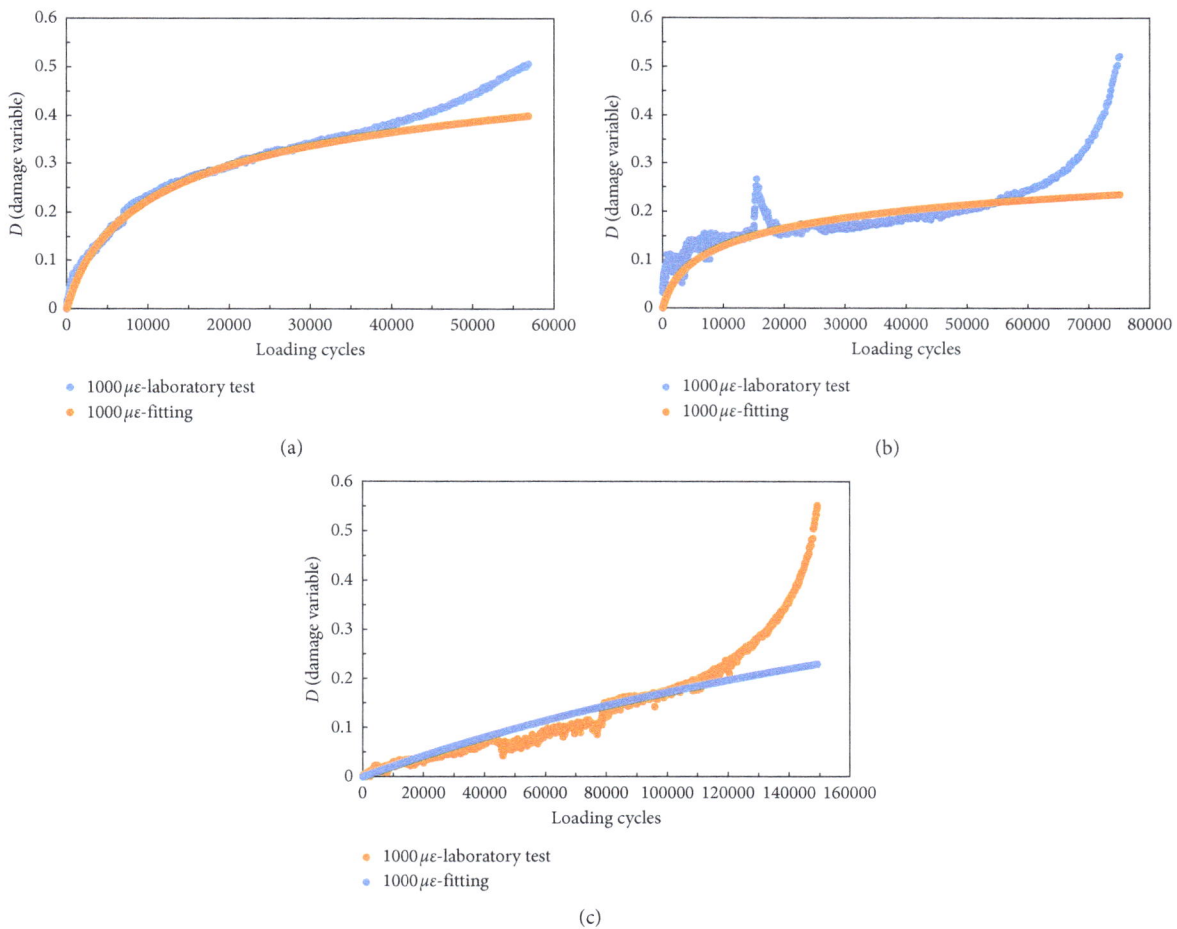

FIGURE 8: Fatigue performance prediction of the asphalt mortar at 10°C: (a) low asphalt content; (b) moderate asphalt content; (c) high asphalt content.

temperature means the beam structure is more similar to the elastic solid. So, the performance of the mortar at the high asphalt content was slightly different from the low and moderate ones. With the worst conditions (highest asphalt content at the lowest temperature), the mortar would become stiffer than others. So, the bending stiffness

TABLE 4: Fitting parameters of the proposed damage evolution model under different conditions.

Asphalt content (%)	Temperature (°C)	Loading strain (με)	A	p
Low	5	700	6.1e − 5	9
	5	1000	6.1e − 5	9
	10	1000	6e − 5	7
Moderate	5	700	4.7e − 5	6.1
	5	1000	4.7e − 5	6.1
	10	1000	4.6e − 5	15.7
High	5	700	9.3e − 5	13
	5	1000	9.3e − 5	13
	10	1000	2.3e − 6	3

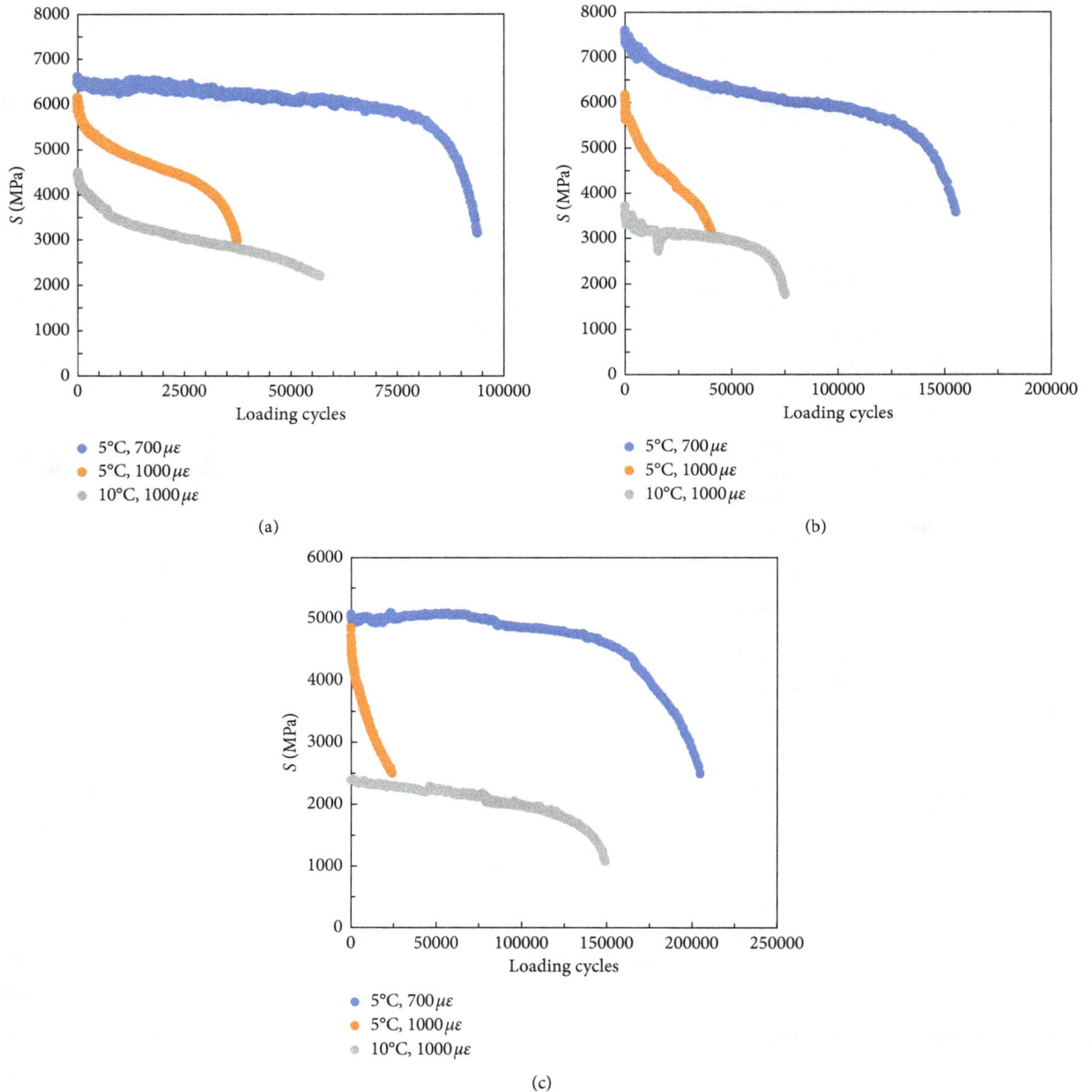

(a)

(b)

(c)

FIGURE 9: Influences of the temperatures and strain levels on the fatigue performance of asphalt mortar: (a) low asphalt content; (b) moderate asphalt content; (c) high asphalt content.

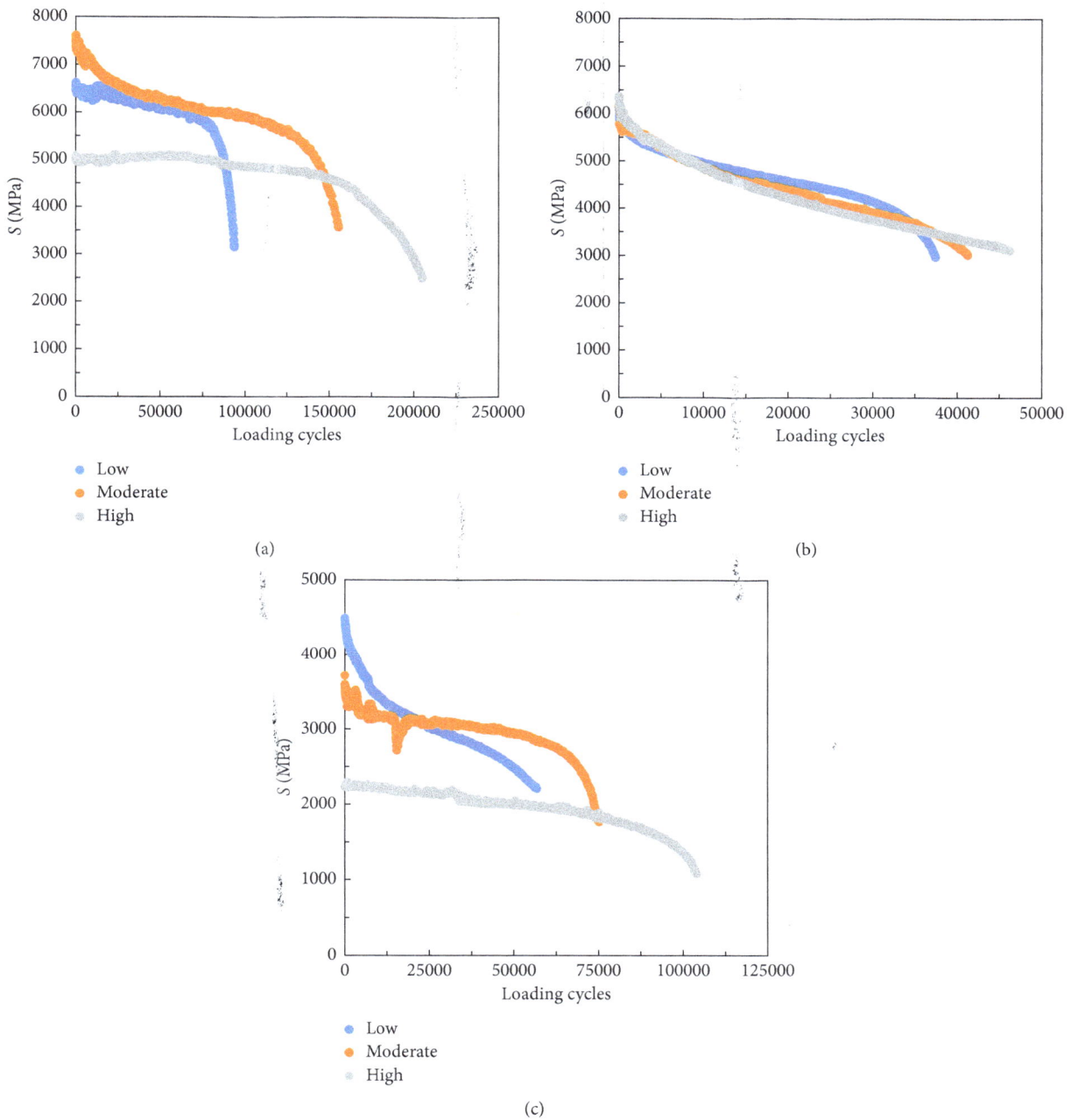

FIGURE 10: Influences of the asphalt content on the fatigue performance of asphalt mortar: (a) 5°C, 700 $\mu\varepsilon$; (b) 5°C, 1000 $\mu\varepsilon$; (c) 10°C, 1000 $\mu\varepsilon$.

modulus (S) of the 5°C, 1000 $\mu\varepsilon$ was a litter larger than the other two at the beginning. Under the largest strain level of 1000 $\mu\varepsilon$, it would decline sharply at the fastest speed which could be seen in Figure 9(c).

Figure 10 shows the influences of the asphalt content on the fatigue performance of asphalt mortar. As shown, the asphalt content has a positive effect on the fatigue performance of the asphalt mortar despite the loading conditions. Compared to the asphalt mixtures, although there is no coarse aggregate skeleton, the curve trends of the mortar also go through three stages known as starting stage, stable stage, and unstable stage. It is a rough verification of the

assumption that the damage evolution mainly processed within the mortar and the aggregates only form the skeleton in mixtures. From the perspective of mechanical calculation, these coarse aggregates can be regarded as boundary conditions directly.

4.2. Verification of the FEM Implementation of the Fatigue Damage Model. The proposed fatigue damage model was implemented into the ABAQUS software via UMAT user subroutine. The FE model of the asphalt mortar was constructed as described in Section 3.4. In the real test,

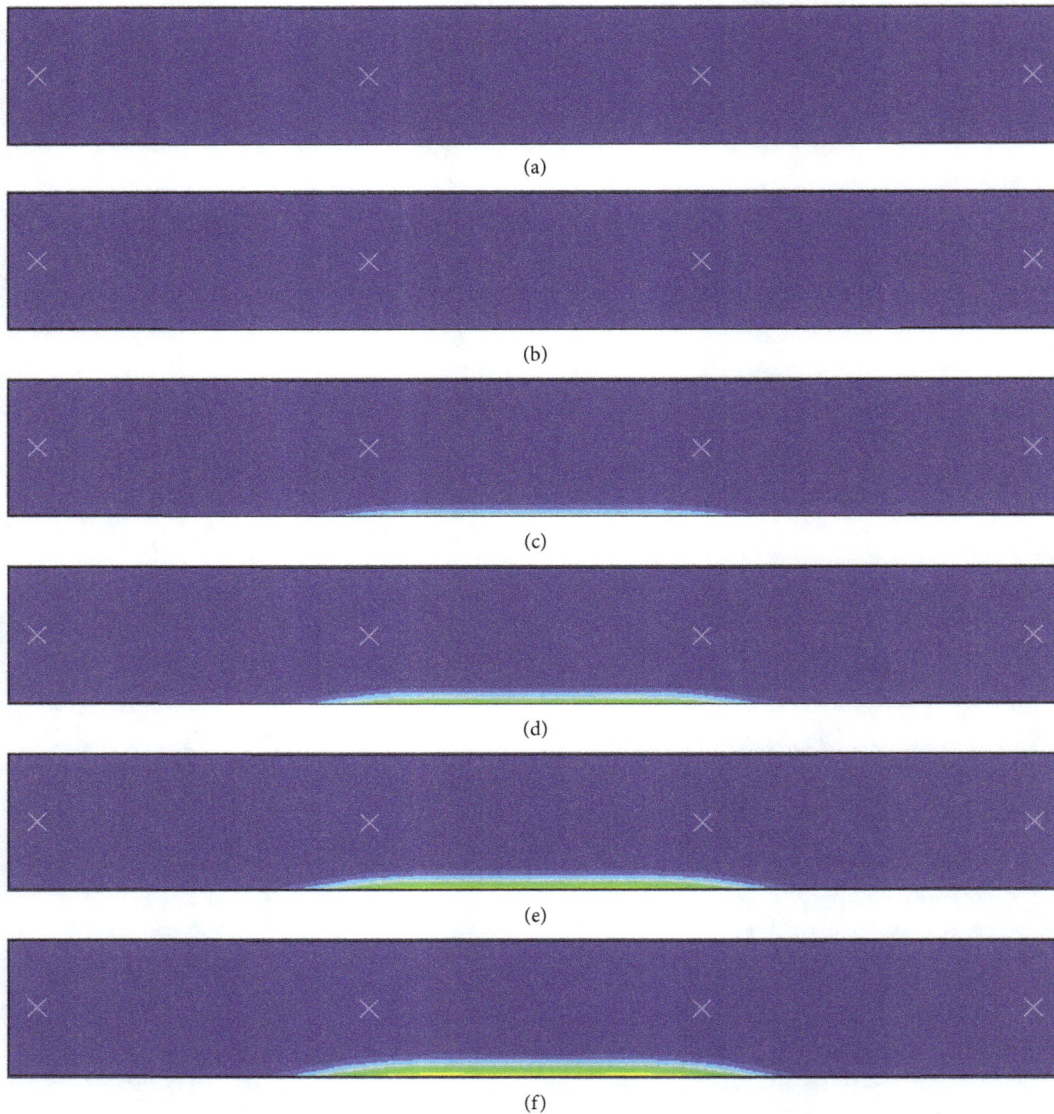

FIGURE 11: Damage evolution of asphalt mortar with load cycles. (a) Load cycles $N = 200$, $D_{max} = 2.041e - 4$; (b) load cycles $N = 2000$, $D_{max} = 5.492e - 3$; (c) load cycles $N = 20000$, $D_{max} = 5.278e - 2$; (c) load cycles $N = 40000$, $D_{max} = 8.840e - 2$; (e) load cycles $N = 60000$, $D_{max} = 1.137e - 1$; (f) load cycles $N = 77600$, $D_{max} = 1.422e - 1$.

maximum deflection and peak load of each cycle were recorded. Correspondingly, vertical displacement of the beam centre and the contact force between clamp A and the sample were recorded in the simulations. Additionally, the damage variable in the UMAT subroutine was also outputted.

The four-point bending beam fatigue tests of the asphalt mortar specimens with different asphalt contents, different temperatures, and different loading strains as described previously were simulated. The simulation results of the sample with moderate asphalt content under test condition of 5°C and 700 $\mu\varepsilon$ are shown in Figure 11.

It can be seen from Figure 11 that in the whole fatigue evolution process of the asphalt mortar sample, the damaged area slightly increased, showing an upward expansion trend. The main damage area was concentrated between clamp A

on the bottom of the sample, and the damage variables were substantially the same at the same height. This is because the middle part of the sample was in a purely bending state and the tensile strain of the sample was the same at the same height. In other parts of the sample, the specimen was subjected to compressive stress, or the tensile strain generated was very small, which was negligible compared with the area where the damage had occurred. The damage area was relatively small, which accounted for about 1/30 of the area of the sample.

The load cycles of the simulations were recorded, and the calculated damage variables at the end of simulations were compared with those of the tests. The results are summarized in Table 5. It indicated that the FE implementation can accurately characterize the mechanical behavior of the proposed fatigue model.

TABLE 5: Comparison between simulated damage variable and measured damage variable under different test conditions.

Asphalt content (%)	Temperature (°C)	Loading strain ($\mu\varepsilon$)	Simulated damage	Measured damage
	5	700	0.13	0.132
4.3	5	1000	0.31	0.307
	10	1000	0.42	0.39
	5	700	0.23	0.282
4.8	5	1000	0.38	0.357
	10	1000	0.23	0.24
	5	700	0.09	0.098
5.5	5	1000	0.18	0.173
	10	1000	0.27	0.22

4.3. Fatigue Performance Predictions of the Asphalt Mixture. The fatigue damage evolution of the asphalt mixtures was analyzed further based on the verified FE implementation of the proposed fatigue model.

A four-point bending fatigue test for asphalt mixture was conducted under the condition of 5°C, 700 $\mu\varepsilon$ and moderate asphalt content. The void ratio of the sample is 4%. A virtual test with the same void ratio was constructed using random packing method described in Section 3.4. A fatigue test was simulated under the same test condition. The peak load and maximum deflection were outputted, and the bending stiffness modulus was calculated using equations (1)–(3). As shown in Figure 12, the proposed FE model can well characterize the fatigue performance of the asphalt mixture in the first two stages.

Three virtual mixture samples with 0% VV, 4% VV, and 8% VV were constructed. Fatigue test simulations were conducted. Figure 13 shows the details of the virtual simulations at the 100000th loading cycle. As shown, the voids content plays a negative role in the fatigue performance of the asphalt mixtures. The maximum damage variables are 0.2042, 0.2232, and 0.2755 for the 0% VV, 4% VV, and 8% VV, respectively. Since there are no voids in Figure 13(a), the maximum damage occurred in the particles' edge. Differently, when comes to Figure 13(b) and 13(c), it is found that the boundary of the voids tend to cause damage more easily. The maximum damage always arose from the void edges. Comparing the simulation results of Figures 13(a)–13(c), it is concluded that as the void content increased, the damage evolved more sharply and the damage tended to arose from the void edges due to a stress concentration.

The reduction of the bending stiffness modulus of three virtual fatigue tests is shown in Figure 14. It can be seen that as the void ratio increased, the decay rate of the stiffness modulus increased. The difference between curves of samples with 4% VV and 8% VV was much larger than that between curves of samples with 0% VV and 4% VV. It indicated that the damage accumulation increased nonlinearly when the void ratio increased linearly. The void ratio had a significant influence on the fatigue performance of asphalt mixture.

5. Conclusions

This paper proposed a fatigue damage evolution model of the asphalt mortar based on the four-point fatigue

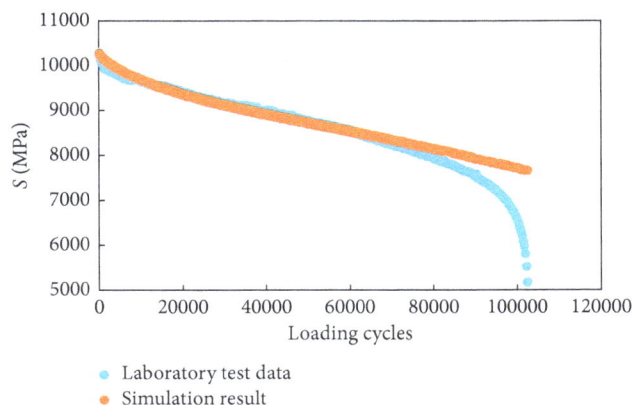

FIGURE 12: Comparison between simulated damage variable and measured damage variable of asphalt mixture.

bending tests. By fitting the laboratory results, the proposed model was verified, and the related parameters were determined under different loading conditions. By importing the damage model into FEM simulation, the major influences on the fatigue damage of the asphalt mixtures were analyzed further. The main conclusions drawn from the study are as follows.

The load strain has a negative effect on the bending stiffness modulus of asphalt mortar while the temperature and asphalt content have a positive one. When the asphalt content is fixed, decreasing the load strain is more effective than increasing the temperatures in improving the fatigue life of the mortar. Since the environmental temperature is uncontrollable during the pavement usage, avoiding the heavy traffic is an effective means to maintain the pavement permanent performance.

The proposed damage model was established based on the continuum mechanics. It characterized the performance of the tiny point within the whole structure which can represent the micromaterial property rather than the macrostructure. By fitting the model curves with the realistic, it has been validated that the proposed methods can well characterize the fatigue life of the asphalt mortar in the first two stages.

The damage evolution curves of the mortar are similar with the asphalt mixtures. It is a rough verification of the assumption that the damage evolution mainly processed within the mortar and the coarse aggregates only form the

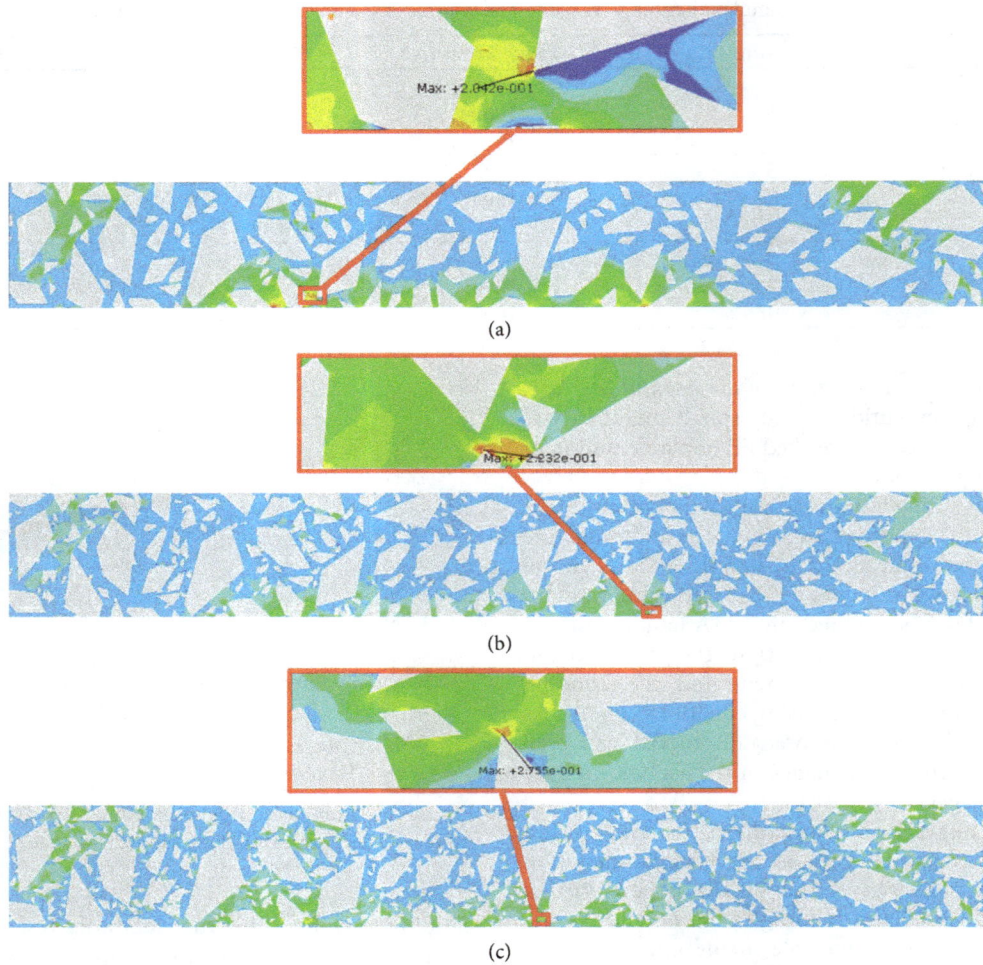

FIGURE 13: Influences of the voids on the fatigue damage of asphalt mixtures: (a) 0% VV; (b) 4% VV; (c) 8% VV.

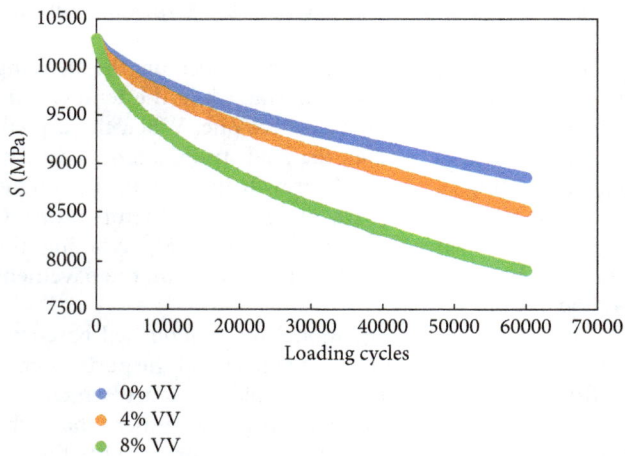

FIGURE 14: Bending stiffness modulus of virtual asphalt mixture samples with different air void ratios.

contents increased, the damage grew more sharply. By comparing the simulation results of virtual samples with different void ratios, it indicates that the damage accumulation increased nonlinearly when the void ratio increased linearly. The void ratio had a significant negative influence on the fatigue performance of asphalt mixture.

Conflicts of Interest

The authors declare that they have no conflicts of interest regarding the publication of this article.

Acknowledgments

The study was financially supported by the Science Foundations of Nanjing Institute of Technology (YKJ201423), National Natural Science Foundation of China (51878164),

skeleton in mixtures serving as boundary conditions. Through the FEM simulation, it shows that the damage always arose from the void edges firstly, and as the void

Jiangsu Natural Science Foundation (BK20161421), Scientific Research Foundation of Graduate School of Southeast University (YBJJ 1843), and Graduate Innovation Project of Jiangsu Province (KYCX18_0147).

References

[1] E. Gasthauer, M. Mazé, J. P. Marchand, and J. Amouroux, "Characterization of asphalt fume composition by GC/MS and effect of temperature," *Fuel*, vol. 87, no. 7, pp. 1428–1434, 2008.

[2] W. Wang, P. Song, and Y. Miao, "Research summary of the composition characteristics and performance of mineral aggregate of asphalt mixture," *Municipal Engineering Technology*, vol. 26, no. 8, pp. 13–16, 2014.

[3] T. Ma, Y. Zhong, T. Tang, and X. Huang, "Design and evaluation of heat-resistant asphalt mixture for permafrost regions," *International Journal of Civil Engineering*, vol. 14, no. 5, pp. 339–346, 2016.

[4] G. D. Airey, A. C. Collop, S. E. Zoorob, and R. C. Elliott, "The influence of aggregate, filler and bitumen on asphalt mixture moisture damage," *Construction and Building Materials*, vol. 22, no. 9, pp. 2015–2024, 2008.

[5] T. Ma, X. Ding, D. Zhang, X. Huang, and J. Chen, "Experimental study of recycled asphalt concrete modified by high-modulus agent," *Construction and Building Materials*, vol. 128, pp. 128–135, 2016.

[6] Y. Tan, H. Zhang, H. Xu, Y. Wang, and X. Yao, "Effect of fine aggregate form, angularity and texture on the viscoelastic properties of asphalt mortar," *Asphalt Pavements*, vol. 1-2, no. 1, pp. 637–648, 2014.

[7] M. R. M. Hasan, M. O. Hamzah, M. V. D. Ven, and J. Voskuilen, "Disruption of air voids continuity based on permeability loss due to mortar creep," *Construction and Building Materials*, vol. 116, pp. 347–354, 2016.

[8] B. G. Wang, L. I. Ping, Z. Q. Zhang, and H. W. Sun, "Influence of mineral powder on aging properties of asphalt mortar," *Journal of Chang'an University (Natural Science Edition)*, vol. 4, pp. 6–9, 2007.

[9] F. Wang, Z. Liu, and S. Hu, "Temperature stability of compressive strength of cement asphalt mortar," *ACI Materials Journal*, vol. 107, no. 1, pp. 27–30, 2010.

[10] M. F. Woldekidan, M. Huurman, and A. C. Pronk, "Linear and nonlinear viscoelastic analysis of bituminous mortar," *Transportation Research Record: Journal of the Transportation Research Board*, vol. 2370, no. 1, pp. 53–62, 2013.

[11] S. G. Hu, T. Wang, F. Z. Wang, Z. C. Liu, T. Gao, and J. Z. Zou, "Freezing and thawing resistance of cement asphalt mortar," *Key Engineering Materials*, vol. 400-402, pp. 163–167, 2009.

[12] P. Cundall, "A computer model for simulating progressive large scale movement in block rock systems," in *Proceedings of International Symposium. Fracture, ISRM*, pp. 11–8, Nancy, France, September 1971.

[13] S. Kobayashi, S. I. Oh, T. Altan, and A. Chaudhary, "Metal forming and the finite-element method," *Journal of Materials Shaping Technology*, vol. 8, no. 1, p. 65, 1990.

[14] J. Zhang, J. Peng, J. Zhang, and Y. Yao, "Characterization of stress and moisture-dependent resilient behaviour for compacted clays in South China," *Road Materials and Pavement Design*, pp. 1–14, 2018.

[15] J. H. Zhang, J. Li, Y. S. Yao, J. Zheng, and F. Gu, "Geometric anisotropy modeling and shear behavior evaluation of graded crushed rocks," *Construction and Building Materials*, vol. 183, pp. 346–355, 2018.

[16] S. Y. Zhu, Q. Fu, C. B. Cai, and P. D. Spanos, "Damage evolution and dynamic response of cement asphalt mortar layer of slab track under vehicle dynamic load," *Science China Technological Sciences*, vol. 57, no. 10, pp. 1883–1894, 2014.

[17] P. Wang, H. Xu, R. Chen, J. Xu, and X. Zeng, "Effect of cement asphalt mortar debonding on dynamic properties of CRTS II slab ballastless track," *Advances in Materials Science and Engineering*, vol. 2014, Article ID 769248, 6 pages, 2014.

[18] X. Ding, T. Ma, and W. Gao, "Morphological characterization and mechanical analysis for coarse aggregate skeleton of asphalt mixture based on discrete-element modeling," *Construction and Building Materials*, vol. 154, pp. 1048–1061, 2017.

[19] W. P. Sun, Y. R. Kim, and R. A. Schapery, "A viscoelastic continuum damage model and its application to uniaxial behavior of asphalt concrete," *Mechanics of Materials*, vol. 24, no. 4, pp. 241–255, 1996.

[20] H. J. Lee and Y. R. Kim, "Viscoelastic continuum damage model of asphalt concrete with healing," *Journal of Engineering Mechanics*, vol. 124, no. 11, pp. 1224–1232, 1998.

[21] J. L. Zheng and S. T. Lu, "Nonlinear fatigue damage model for asphalt mixtures," *China Journal of Highway and Transport*, vol. 22, no. 5, pp. 21–28, 2009.

[22] Y. R. Kim, C. Baek, B. S. Underwood, V. Subramanian, M. N. Guddati, and K. Lee, "Application of viscoelastic continuum damage model based finite element analysis to predict the fatigue performance of asphalt pavements," *KSCE Journal of Civil Engineering*, vol. 12, no. 2, pp. 109–120, 2008.

[23] B. Underwood, C. Baek, and Y. Kim, "Simplified viscoelastic continuum damage model as platform for asphalt concrete fatigue analysis," *Transportation Research Record Journal of the Transportation Research Board*, vol. 2296, no. 1, pp. 36–45, 2015.

[24] T. Boukharouba, M. Elboujdaini, and G. Pluvinage, *Damage and Fracture Mechanics: Failure Analysis of Engineering Materials and Structures*, Springer Science + Business Media, Berlin, Germany, 2009.

[25] J. E. Kliewer, H. Zeng, and T. S. Vinson, "Aging and low-temperature cracking of asphalt concrete mixture," *Journal of Cold Regions Engineering*, vol. 10, no. 3, pp. 134–148, 1996.

[26] Ministry of Transport of the People's Republic of China, *Technical Specification for Construction of Highway Asphalt Pavements*, China Communications Press, Beijing, China, 2004.

[27] Ministry of Transport of the People's Republic of China, *Standard Test Methods of Bitumen and Bituminous Mixtures for Highway Engineering*, China Communications Press, Beijing, China, 2011.

6

Properties of Direct Coal Liquefaction Residue Modified Asphalt Mixture

Jie Ji,[1] Hui Yao,[2] Di Wang,[3] Zhi Suo,[3] Luhou Liu,[4] and Zhanping You[5]

[1]*School of Civil Engineering and Transportation, Beijing University of Civil Engineering and Architecture and Beijing Urban Transportation Infrastructure Engineering Technology Research Center, Beijing 100044, China*
[2]*Department of Civil and Environmental Engineering, Michigan Technological University and School of Traffic and Transportation, Changsha University of Science and Technology, 1400 Townsend Drive, Houghton, MI 49931, USA*
[3]*Beijing Urban Transportation Infrastructure Engineering Technology Research Center and Beijing Collaborative Innovation Center for Metropolitan Transportation, Beijing 100044, China*
[4]*School of Civil Engineering and Transportation, Beijing University of Civil Engineering and Architecture and Beijing Cooperative Innovation Research Center on Energy Saving and Emission Reduction, Beijing 100044, China*
[5]*Department of Civil and Environmental Engineering, Michigan Technological University, 1400 Townsend Drive, Houghton, MI 49931, USA*

Correspondence should be addressed to Hui Yao; huiyao@mtu.edu

Academic Editor: Antonio Riveiro

The objectives of this paper are to use Direct Coal Liquefaction Residue (DLCR) to modify the asphalt binders and mixtures and to evaluate the performance of modified asphalt mixtures. The dynamic modulus and phase angle of DCLR and DCLR-composite modified asphalt mixture were analyzed, and the viscoelastic properties of these modified asphalt mixtures were compared to the base asphalt binder SK-90 and Styrene-Butadiene-Styrene (SBS) modified asphalt mixtures. The master curves of the asphalt mixtures were shown, and dynamic and viscoelastic behaviors of asphalt mixtures were described using the Christensen-Anderson-Marasteanu (CAM) model. The test results show that the dynamic moduli of DCLR and DCLR-composite asphalt mixtures are higher than those of the SK-90 and SBS modified asphalt mixtures. Based on the viscoelastic parameters of CAM models of the asphalt mixtures, the high- and low-temperature performance of DLCR and DCLR-composite modified asphalt mixtures are obviously better than the SK-90 and SBS modified asphalt mixtures. In addition, the DCLR and DCLR-composite modified asphalt mixtures are more insensitive to the frequency compared to SK-90 and SBS modified asphalt mixtures.

1. Introduction

Currently, the design method of asphalt pavement in China is based on the static and elastic layer system models [1]. The stress and strain of each layer can be calculated using the static modulus of asphalt mixtures. However, the static modulus is used to represent the property of each layer which is inaccuracy since the loading on the pavement is dynamic. It is necessary to use the dynamic modulus to calculate the mechanical property of each layer. The dynamic modulus was used in the United States in 1980. The National Cooperative Highway Research Program- (NCHRP-) 465 report indicates that the dynamic modulus of asphalt mixtures can be used

to evaluate the permanent deformation [2]. The dynamic modulus test can also be used to evaluate the service quality of the subgrade and pavement [3]. The NCHRP-702 and 580 reports present the standard and accuracy of the dynamic modulus test, and the reports suggest that the high- and low-temperature performance of asphalt mixtures can be predicted using the master curve of the dynamic modulus [4, 5]. The NCHRP-629 report also reveals that the durability of asphalt pavement can be evaluated using the dynamic modulus of asphalt mixture [6]. The permanent deformation and cracking growth pattern of asphalt mixtures were studied using the dynamic modulus test, and the initiation of cracks started with the change in the phase angle of asphalt mixtures, and the permanent deformation and fracture model were proposed [7]. The relationship between the rutting and dynamic modulus was also established [8]. The dynamic moduli of the modified asphalt mixtures were fitted by master curves [9]. The dynamic modulus of asphalt mixtures can be investigated using the actual stress and strain response of pavement [10]. The dynamic modulus is a basic parameter for the design of asphalt pavement in many countries, and it has been widely accepted [11, 12]. In the specifications of American Association of State Highway and Transportation Officials (AASHTO), *the design guideline for new pavement and regenerated pavement 2002* and *the guideline for asphalt pavement mechanics-empirical design method* (2002) was put forward by the NCHRP program [13, 14], and the dynamic modulus of asphalt mixtures can be considered one of the essential parameters in pavement design. It is possible to track the dynamic and viscoelastic behaviors of asphalt mixtures over a full temperature range through the master curve of asphalt mixtures [15]. The master curve of the dynamic modulus was plotted using the Christensen-Anderson-Marasteanu (CAM) model, and the viscoelastic properties of asphalt mixture were characterized [16, 17]. In addition, the performance of asphalt mixtures can be influenced by the properties of asphalt binders and aggregates. Many modifiers were used to modify and improve the performance of asphalt binders including polymer, waste materials, and by-products, and different surface treatments were also used to enhance the adhesion between aggregates and binders. Currently, the Direct Coal Liquefaction Residue (DCLR) is the main byproduct produced in the process of the direct coal liquefaction, which accounts for 30% of the total amount of raw coals [18]. The DCLR contains 30%–50% heavy oil and asphaltene materials [19, 20], and it has a potential to be developed for a modifier. The DCLR is mainly used as a fuel for heating, which not only causes serious environmental pollution but also reduces its economic value and leads to a waste of valuable resources.

At the beginning of the last century, researchers began to study the properties and applications of DCLR including the main structure and pyrolysis characteristics [21]. The DCLR modified asphalt binder was prepared using blending ESSO-70 asphalt binder and DCLR, and the optimum DCLR content was 7%–21% [22]. If the DCLR content was 5%, the properties of the DCLR modified asphalt binder met the technical standard of asphalt binders (Penetration grade number 50) [23]. The high-temperature performance of asphalt binders was improved [24], but low-temperature performance was reduced by the addition of DCLR [25]. The surface energies of the DCLR modified asphalt binders were calculated by Wilhelmy plate method and the microscopic properties were examined [26, 27]. The characteristics of the DCLR modified asphalt binder were researched by means of Thermogravimetric Analysis-Fourier Transform Infrared Spectroscopy (TG-FTIR), Fourier transform infrared spectroscopy (FTIR), and Fluorescence Optical Microscopy (FOM). The heavy oil can enhance the ductility and penetration of DCLR modified asphalt binders while the asphaltenes and preasphaltenes increase the softening point of DCLR modified asphalt binders [28]. The modified asphalt binder was prepared by the addition of the tetrahydrofuran soluble fraction (THFS)

and the benzaldehyde was used as the cross-linking agent. The conditions during the preparation of the modified asphalt binders were studied, such as mixing temperature, ratios of THFS, and cross-linking agent. The results show that the DCLR modified asphalt binder has better properties with the utilization of the cross-linking agent [28]. The mesophase pitches were prepared by the hydrogenation and polycondensation from the Shenhua DCLR. The element analysis and FTIR were used to investigate the composition and structure of DCLR modified asphalt binders. The effects of tetrahydrogen naphthalene and reaction temperatures were studied, as well as the morphologies of mesophase pitches [29].

Based on the discussions of the DCLR materials, it can be seen that DCLR can be used as an asphalt modifier to improve high-temperature properties of asphalt binders, while it may have a negative impact on low-temperature performance. Therefore, it is meaningful to use DCLR to conduct research on the dynamic modulus or properties under consideration of the environmental issue and economic value. The master curve was plotted to understand the dynamic modulus of asphalt mixtures. The Christensen-Anderson-Marasteanu (CAM) model was also used to study the viscoelastic behaviors of DCLR and DCLR-composite modified asphalt mixtures.

2. Objectives and Test Methods

2.1. Objectives and Experimental Plan. The objectives of this project are to use the DCLR to modify the asphalt binders and mixtures and to evaluate the performance of modified asphalt mixtures. The experimental plan of this study includes the following: (1) prepare three types of modified asphalt binders based on the SK-90 base asphalt binder, including Styrene-Butadiene-Styrene (SBS) modified asphalt binder, DCLR modified asphalt binder, and DCLR-composite modified asphalt binder; (2) design four asphalt mixtures based on the gradation of AC-20 (AC: Asphalt Concrete), which were SK-90 asphalt mixture, SBS modified asphalt mixture, DCLR modified asphalt mixtures, and DCLR-composite modified asphalt mixtures; (3) obtain the dynamic moduli of the asphalt mixtures and analyze viscoelastic properties of the asphalt mixtures; (4) establish a CAM model of the asphalt mixtures.

2.2. Test Methods. In accordance with the *Test Methods of Asphalt and Asphalt Mixtures for Highway Engineering (JTGE20-2011)*, the properties of the SK-90 base asphalt binder and three types of modified asphalt binders were tested based on the penetration and the Strategic Highway Research Program (SHRP) Performance Grade (PG) systems. According to the *Test Methods of Aggregate for Highway Engineering (JTG E42-2005)*, the properties of aggregates were measured. The specimens were prepared for the compaction test according to the T 0738-2011 of *Test Methods of Asphalt and Asphalt Mixtures for Highway Engineering (JTGE20-2011)*. The test specimens are formed by the compaction apparatus with dimensions of 450 mm in length, 150 mm in width, and 170 mm in height. The specimens were core drilled into

TABLE 1: Physical properties of DCLR, SBS, and rubber powders.

(a)

DCLR	Apparent gravity/(g/cm^3)	Density/(g/cm^3)	Water content/%	25°C Penetration/(0.1 mm)	Softening point/°C
Test results	1.12	1.23	0.60	2.0	170.0

(b)

SBS	Block ratio	Tensile strength/MPa	Elongation at break/%	Percentage of liquid volume/%
Test results	40/60	≥12	≥650	0

(c)

Rubber powders	Density/(g/cm^3)	Water content/%	Mental content/%	Fiber content/%
Test results	1.13	≥0.65	≥0.07	≥0.11

TABLE 2: Properties of the asphalt binders (penetration system).

Items	SK-90 asphalt binder	SBS modified asphalt binder	DCLR modified asphalt binder	DCLR-composite modified asphalt binder
25°C penetration/(0.1 mm)	81.0	61.2	35.1	33.4
Softening point/°C	51.0	65.4	59.2	77.5
10°C ductility/cm	51.8	68.2/32.3 (5°C)	5.7	12.2
After Rolling Thin Film Oven (RTFO) test				
Mass loss/%	+0.1	−0.2	+0.2	−0.1
Penetration ratio/%	64.1	64.2	69.3	79.5
10°C ductility/cm	8.0	39.6/21.8 (5°C)	4.2	9.7

cylindrical specimens with a diameter of 100 mm and height of 150 mm after cooling to room temperature for 24 h. The dynamic modulus test was carried out under the control of Universal Testing Machine UTM-25 with the sinusoidal load stress. The test temperatures are 5°C, 15°C, 35°C, and 50°C and the test frequencies are 25 Hz, 10 Hz, 5 Hz, 1 Hz, and 0.1 Hz. In addition, the dynamic moduli of asphalt mixtures were measured without the confinement.

3. Test Materials

3.1. Modifier. The DCLR was produced from China Shenhua Coal to Liquid and Chemical Co., Ltd. The Styrene-Butadiene-Styrene (SBS) was purchased from Sinopec Yanshan Petrochemical Co., Ltd., and rubber powders were bought from Antai Rubber Co., Ltd. In accordance with *Test Methods of Asphalt and Asphalt Mixtures for Highway Engineering (JTGE20-2011)*, the physical properties of test materials were measured and are listed in Table 1.

3.2. Asphalt Binders. The SK-90 asphalt binder was used as the base asphalt binder, which was produced from South Korea. 3.4% SBS was added in the asphalt binder by mass of SK-90 asphalt binder, and the SBS modified asphalt binder was formed. The DCLR-composite material contains 10% DCLR, 2% SBS, and 15% rubber powder by mass of SK-90 asphalt binder. These materials were added to the SK-90 asphalt binder, and the DCLR-composite modified asphalt binder was prepared at a temperature of around 135°C.

According to *Test Methods of Asphalt and Asphalt Mixtures for Highway Engineering (JTGE20-2011)*, properties of the asphalt binders were measured and are shown in Table 2. In addition, the Dynamic Shear Rheometer (DSR) and Bending Beam Rheometer (BBR) tests were employed to evaluate the performance of asphalt binders under the Rolling Thin Film Oven (RTFO) and Pressure Aging Vessel (PAV) aging conditions. The DSR and BBR results of the asphalt binders are shown in Table 3.

Table 2 shows the test results of asphalt binders including the penetration, softening point, and ductility. The SK-90 asphalt binder had a high penetration at 25°C, and the penetrations of modified asphalt binders decreased after the modification by SBS, DCLR, and DCLR-composite. It indicates that the modified asphalt binders become hard after the modification. The softening points of the modified binders increased after modification compared to the SK-90 base asphalt binder, and the DCLR-composite modified asphalt binder improved the most. It is likely that high-temperature performance of the modified asphalt binders was enhanced after modification. It can be expected that the ductility of the modified binders decreased greatly after modification. The SBS improved the ductility of modified asphalt binder. The low-temperature performance of the binders possibly degraded after modification for the DCLR modified binder. The ductility of the modified asphalt binders improved after RTFO compared to the SK-90 binder. Table 3 shows the test results of different asphalt binders after different aging conditions based on the Superpave PG system. The PG grade

TABLE 3: Properties of the asphalt binders (SHRP PG system).

Stages	Temperature/°C	SK-90 asphalt binder	SBS modified asphalt binder	DCLR modified binder	DCLR-composite modified asphalt binder	Superpave spec.
Unaged ($G^*/\sin\delta$ (kPa))	58	2.18	8.46	6.56	37.31	≥1.1
	64	0.96	6.10	2.75	20.24	
	70	—	2.98	1.2	11.09	
	76	—	1.23	0.61	6.39	
	82	—	0.84	—	3.86	
	88	—	—	—	0.66	
RTFO ($G^*/\sin\delta$ (kPa))	58	4.62	17.19	21.670	49.44	≥2.2
	64	1.97	9.29	8.862	27.13	
	70	—	4.74	3.821	15.82	
	76	—	2.5	1.73	7.54	
	82	—	1.35	—	4.54	
	88	—	—	—	2.09	
PAV ($G^* \cdot \sin\delta$ (kPa))	25	1958	2274	4658	2863	≤5000
	22	3014	3826	6266	3920	
	19	4555	5215	—	5516	
	16	6681	—	—	—	
PAV (stiffness (MPa))	−6	82.63	95.45	86.083	39.65	≤300
	−12	184.03	198.65	220.91	81.52	
	−18	306.23	211.32	325.20	164.39	
	−24	—	316.49	—	300.21	
PAV (m-value)	−6	0.34	0.34	0.33	0.42	≥0.3
	−12	0.32	0.31	0.27	0.34	
	−18	0.29	0.30	—	0.30	
	−24	—	0.28	—	0.29	
PG		58-22	76-22	70-16	82-28	

Note: G^*: complex shear modulus; δ: phase angle; $G^*/\sin\delta$: rutting factor; $G^* \cdot \sin\delta$: fatigue factor; RTFO: Rolling Thin Film Oven; PAV: Pressure Aging Vessel; and PG: Performance Grade.

of SBS modified asphalt binder improved from 58-22 to 76-22. The PG of DCLR modified asphalt binder was from 58-22 to 70-16, and the PG of the DCLR-composite asphalt binder was from 58-22 to 82-28. This indicates that the high-temperature performance of the modified asphalt binders improved, and the low-temperature performance of DCLR-composite modified asphalt binder was also enhanced compared to the base asphalt binder.

3.3. Aggregates. The limestone was used as the aggregate material in this study, which included 9.5–20 mm coarse aggregate, 4.75–9.5 mm coarse aggregate, and 0–4.75 mm fine aggregate. The limestone powder was used as a mineral powder. The properties of aggregates were measured in accordance with Test Methods of Aggregate for Highway Engineering (JTG E42-2005) and are shown in Tables 4 and 5.

Tables 4 and 5 show the specific gravity, wear loss, angularity, and sand equivalent of the coarse and fine aggregates, as well as the gravity, water content, hydrophilic coefficient, and plasticity index. The results of aggregates meet the

requirements of standards, and the aggregates can be used to make the asphalt mixture samples. The same aggregate was used for the mixture in this project.

3.4. Asphalt Mixture. The AC-20C (AC: Asphalt Concrete) asphalt mixture was adopted, and the gradation of the asphalt mixture is presented in Figure 1. The asphalt mixture was mixed and compacted based on the Marshall and Superpave systems. The mixing and compaction temperatures of modified asphalt mixtures were based on the temperatures of the base asphalt mixture. The mixing temperature of DCLR and SBS modified asphalt mixtures is around 160°C, and the compaction temperature is around 155°C, as well as the SK-90 asphalt mixture. The mixing and compaction temperatures of DCLR-composite modified asphalt mixtures are 175°C and 170°C, respectively. Table 6 shows the volume indexes of the asphalt mixtures at the optimum asphalt content. The dynamic stability is used to access the resistance to permanent deformation and the tensile strength ratio (TSR) is used to evaluate the moisture damage of asphalt mixtures. The

TABLE 4: Properties of coarse and fine aggregates.

(a)

Properties of coarse aggregate	4.75–9.5 mm	9.5–20 mm	Spec.
Apparent specific gravity/(g/cm^3)	2.80	2.85	≥2.60
Gross volume relative density/(g/cm^3)	2.71	2.76	—
Wear loss in Los Angeles/%	—	17.8	≤28
Washing < 0.075 m particle content/%	0.1	0.2	≤1

(b)

Properties of fine aggregate	Test results	Spec.
Apparent specific gravity/(g/cm^3)	2.78	≥2.60
Bulk relative specific gravity/(g/cm^3)	2.68	—
Angularity/s	43.2	≥30
Sand equivalent/%	65.0	≥60

TABLE 5: Properties of mineral powder.

Items	Test results	Spec.
Apparent specific gravity/(g/cm^3)	2.73	≥2.5
Water content/%	0.52	≤1
Size range		
<0.075 mm	100	100
<0.15 mm	99.75	90–100
<0.6 mm	88.56	75–100
Hydrophilic coefficient	0.71	<1
Plasticity index	2.8	<4

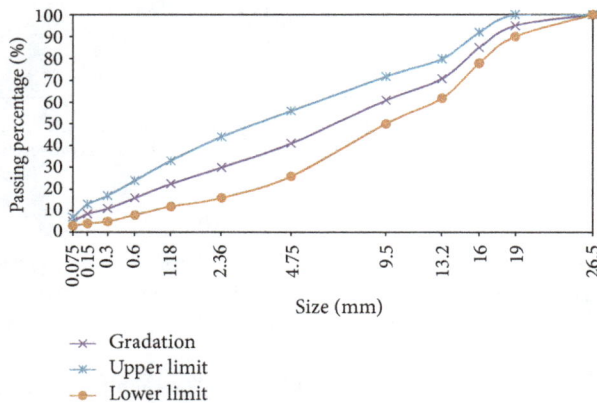

FIGURE 1: Gradation of aggregates in AC-20 asphalt mixture.

properties of the asphalt mixtures are shown in Table 7 after dynamic stability and tensile strength ratio tests.

The compaction parameters and volumetric properties of asphalt mixtures were displayed, and the dynamic stability and tensile strength ratio were tested. The dynamic stability and TSR of the modified asphalt mixtures increased compared to the base mixture, and the dynamic stability and TSR of DCLR-composite modified asphalt mixture were higher than those of other mixtures. This implies that the DCLR-composite modified asphalt mixture has better overall performance (high-temperature performance and moisture

susceptibility) compared to other mixtures. It is possible that the rubber powder, SBS, and DCLR in the DCLR-composite enhance the high-temperature performance and moisture resistance. The DCLR and SBS both can improve the resistance to rutting in asphalt mixtures, and it is deduced that the rubber powder could be effective to the enhancement of moisture resistance in asphalt mixtures.

4. Results and Discussions

4.1. Dynamic Modulus Test. The dynamic modulus of the asphalt mixture was tested by Universal Testing Machine (UTM-25). The test conditions of the dynamic modulus are that test temperatures are 5°C, 15°C, 35°C, and 50°C, and frequencies are 25 Hz, 10 Hz, 5 Hz, 1 Hz, and 0.1 Hz. The dynamic modulus and phase angle were collected during testing. The dynamic modulus is the ratio of stress to strain under different conditions, and the phase angle indicates the viscous part of asphalt mixtures. Dynamic modulus (E^*) and phase angle (δ) of four asphalt mixtures are shown in Figures 2–5.

The dynamic moduli of DCLR and DCLR-composite modified asphalt mixtures were 17008 MPa and 16723 MPa at 15°C and 10 Hz, respectively. Based on the definition of high-modulus asphalt mixture in France or China, the modulus of the mixture should be higher than 14000 MPa at 15°C and 10 Hz [30–32]. DCLR and DCLR-composite modified asphalt mixtures met the technical standards of high-modulus asphalt mixtures. The dynamic moduli of the asphalt mixtures declined with the increase in temperature and increased with the increase in frequency. At the same temperature, the higher the frequency is, the higher the dynamic modulus of the asphalt mixture is. At a high temperature and low frequency, the dynamic modulus of asphalt mixture is minimum. Therefore, in the summer, it is potential that permanent deformations occur in the slow lanes and parking lots, such as rutting at high temperatures. Furthermore, the dynamic moduli of the asphalt mixtures increased rapidly in the range of 0.1 HZ–5 Hz and increased slowly and approached stability when the frequency exceeded 5 Hz.

The phase angle of asphalt mixtures is a key parameter to characterize the viscoelastic property. The smaller the phase

TABLE 6: Volumetric indexes of the asphalt mixtures.

Items	SK-90 asphalt mixture	SBS modified asphalt mixture	DCLR modified asphalt mixture	DCLR-composite modified asphalt mixture
Bulk relative specific gravity/(g/cm^3)	2.422	2.560	2.503	2.547
Maximum theoretical specific gravity/(g/cm^3)	2.596	2.615	2.632	2.623
VV/%	4.4	4.5	4.3	4.6
VMA/%	13.3	13.3	13.2	13.3
VFA/%	68.1	65.2	67.4	66.5
OAC/%	4.2	4.2	4.2	4.3

Note: VV: volume of air voids; VMA: volume of voids in mineral aggregate; VFA: volume of voids filled with asphalt; and OAC: optimum asphalt content.

TABLE 7: Performance of the asphalt mixtures.

Types	Dynamic stability/(times/mm)	Failure strain/$\mu\varepsilon$	Residual stability/%	TSR/%
SK-90 asphalt mixture	943.88	2683	80.05	76.38
SBS modified asphalt mixture	2452.38	2798	84.40	84.15
DCLR modified asphalt mixture	2604.86	1552	83.64	83.78
DCLR-composite modified asphalt mixture	9867.65	3070	100.05	86.61

Note: DCLR: Direct Coal Liquefaction Residue; DCLR-composite: 2% SBS, 15% rubber powder and 10% DCLR by mass of SK-90 asphalt binder; and TSR: tensile strength ratio.

angle is, the more elastic the asphalt mixture is. The larger the phase angle is, the more viscous the asphalt mixture is. When the temperature was lower than 35°C, the phase angle of the asphalt mixtures decreased with the rise in frequency and declines at the range of 0.1 Hz–5 Hz. When the temperature is higher than 35°C, the phase angle of the asphalt mixtures increased with the increase of the frequency at the range of 0.1 Hz–1 Hz and stay stability after 1 Hz. This indicates that the asphalt mixture is more elastic at a low temperature and high frequency; and asphalt mixture is more viscous at a high temperature and low frequency. Furthermore, it is found that the influence of the temperature on the viscoelasticity of asphalt mixtures is more than that of the frequency.

When the frequency was constant, the phase angle of asphalt mixture increased with the increase of temperature, and it indicates that asphalt mixture is more viscous at high temperatures. Asphalt mixtures demonstrated a more viscous state under high temperatures and low frequencies. The comparison of the dynamic modulus and phase angle of the asphalt mixtures at different temperatures under a loading frequency of 10 Hz is demonstrated in Figure 6, since this frequency is equivalent to a vehicle speed of 65–70 km/h [33].

The dynamic moduli of DCLR and DCLR-composite modified asphalt mixtures were higher than those of SK-90 and SBS modified asphalt mixture at different temperatures at 10 Hz. The dynamic moduli of the asphalt mixtures declined with the increase of temperature. The dynamic moduli of the asphalt mixtures declined slowly at 0°C–35°C, while the dynamic moduli of the asphalt mixtures dropped fast and finally approached the same level at 35°C–50°C. This indicates that the deformation resistance of asphalt mixture gradually declined with the increase of temperature. The DCLR and

DCLR-composite modified asphalt mixtures had a good resistance to deformation at high temperatures compared to SK-90 and SBS modified asphalt mixtures. The phase angles of DCLR and DCLR-composite modified asphalt mixture were lower than those of SK-90 and SBS modified asphalt mixture at different temperatures. The phase angle of the asphalt mixtures increased with the rise in temperature. The phase angle of the asphalt mixtures increased rapidly at 5°C–15°C and 35°C–50°C, while at 15°C–35°C, the phase angle of the asphalt mixtures increased slowly. This shows that the viscous part of the asphalt mixtures became strong with the rise in temperature. Compared to SK-90 and SBS modified asphalt mixtures, the DCLR and DCLR-composite modified asphalt mixture had a good elastic property. The phase angle of DCLR-composite modified asphalt mixture was smaller than that of DCLR modified asphalt mixture at a high temperature, which was due to the addition of SBS and rubber powders. It means that it is more elastic at high temperatures compared to the DCLR modified asphalt mixture.

4.2. Master Curve of the Dynamic Modulus. According to the time-temperature equivalent principle of viscoelastic materials, the dynamic modulus curve at different temperatures and frequencies can be composed into a smooth curve (master curve) at a reference temperature through a shift. The master curve can be used to predict the viscoelastic properties of asphalt mixtures at a low frequency or high frequency that is difficult to reach in the lab. The master curves of dynamic modulus of the asphalt mixtures were plotted in Figure 7 based on [34]

$$\log\left(E^*\right) = \delta + \frac{\alpha}{1 + e^{\beta + \lambda \lg \omega_{\text{red}}}}, \tag{1}$$

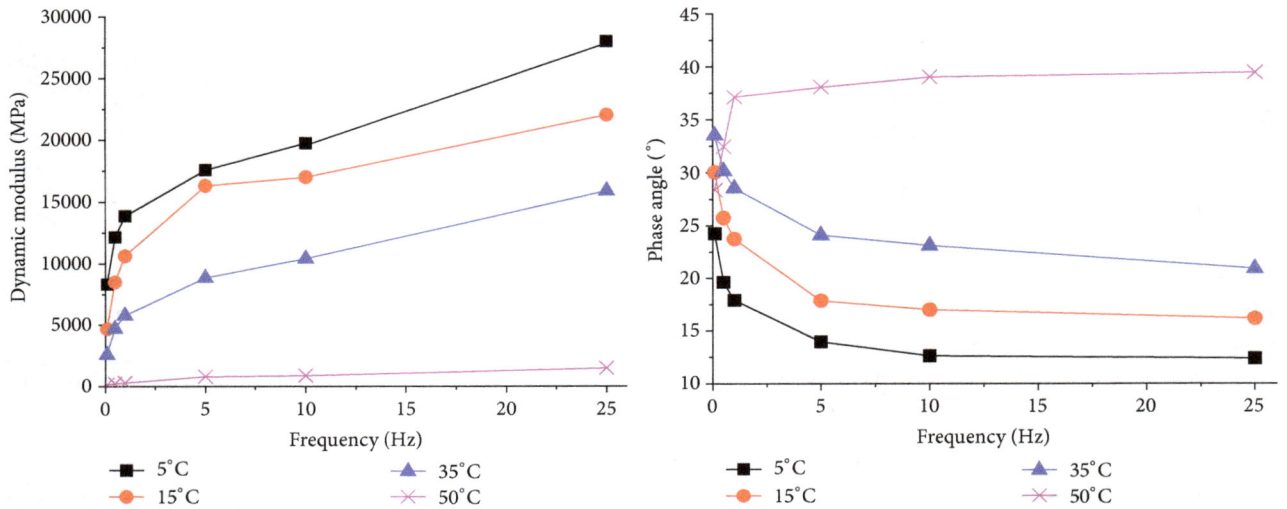

FIGURE 2: The dynamic modulus and phase angle of DCLR modified asphalt mixture.

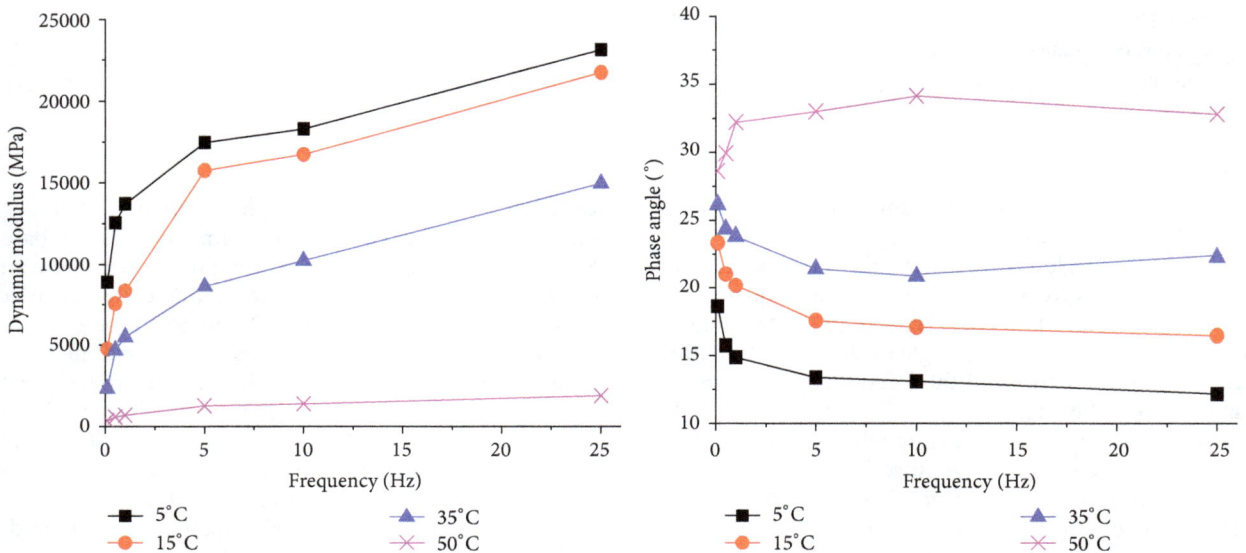

FIGURE 3: The dynamic modulus and phase angle of DCLR-composite modified asphalt mixture.

where E^* is the dynamic modulus of asphalt mixtures; ω_{red} is the reduced frequency under the reference temperature; λ, α, β, and ω are the regression coefficients.

It can be seen from Figure 7 that the master curve tended to change slowly and approach an asymptote. The DCLR-composite modified asphalt mixture had the highest dynamic modulus value. The order is followed by the DCLR, SBS, and SK-90 asphalt mixtures. The dynamic moduli of DCLR and DCLR-composite modified asphalt mixtures were closer and higher than those of the SK-90 and SBS modified asphalt mixtures when the loading frequency was higher than 0.1 Hz. The dynamic modulus of the asphalt mixtures increased after the addition of DCLR and DCLR-composite, and this indicates that this addition improves the resistance to permanent deformation in the asphalt mixtures at high temperatures. It is

likely that the DCLR and DCLR-composite modified asphalt mixtures could be used for the parking lots or slow lanes due to the effective prevention of permanent deformations.

4.3. Christensen-Anderson-Marasteanu (CAM) Model. On the basis of the Christensen-Anderson (CA) model, the Christensen-Anderson-Marasteanu (CAM) model was further developed. The CAM model has a clearly physical meaning [35] compared to the CA model. This paper used the CAM model to study the viscoelastic behaviors of DCLR and DCLR-composite modified asphalt mixtures. The CAM model mainly consists of four equations: the complex modulus master curve, the storage modulus master curve, the phase angle, and the temperature-displacement factor.

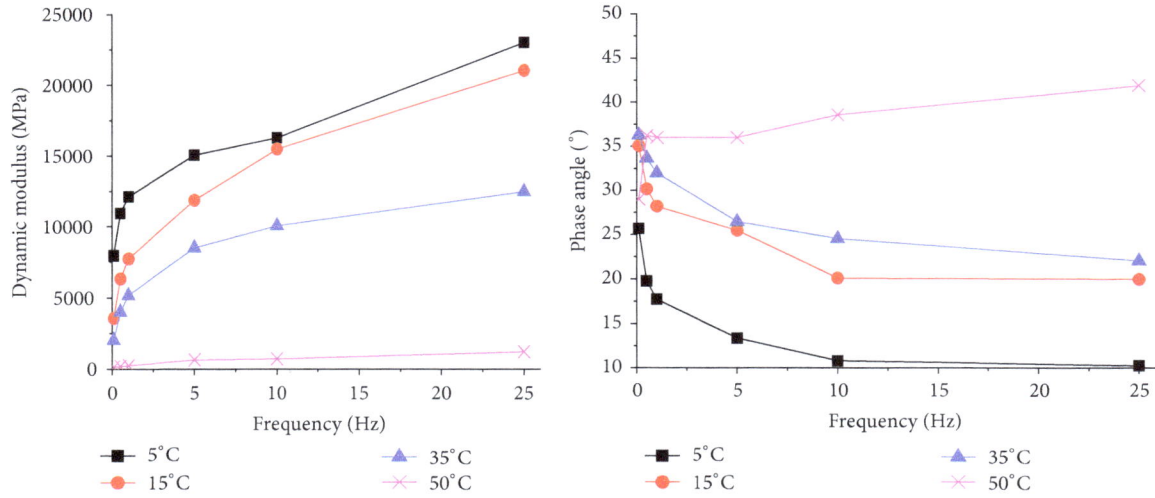

FIGURE 4: The dynamic modulus and phase angle of SBS modified asphalt.

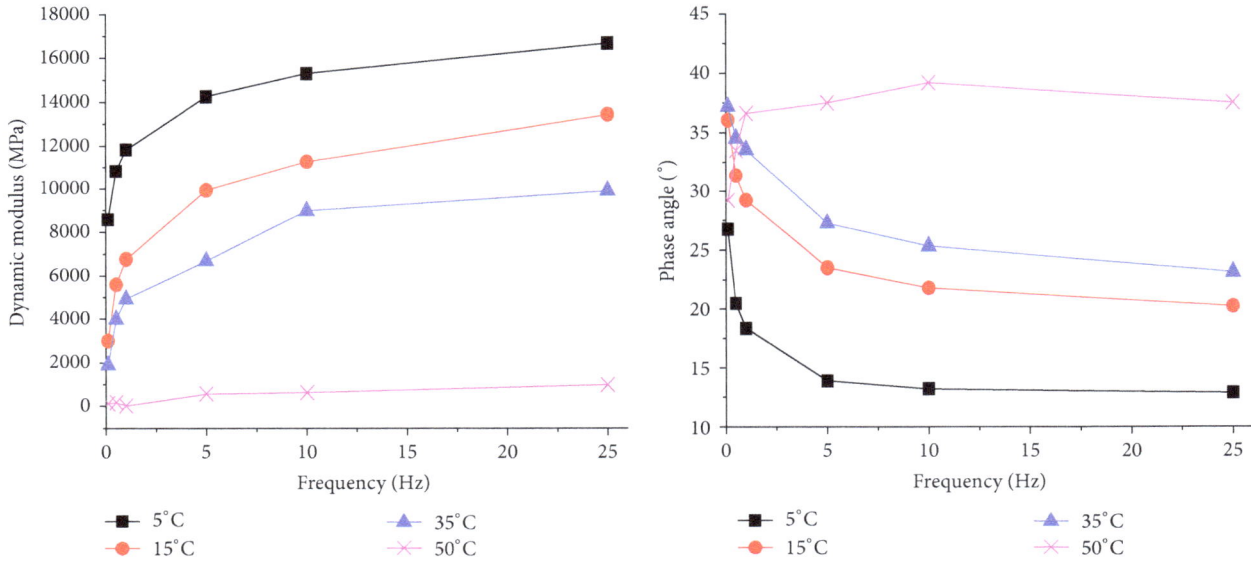

FIGURE 5: The dynamic modulus and phase angle of SK-90 asphalt mixture.

4.3.1. Master Curve for the Complex Modulus. The equation for describing the complex modulus master curve of asphalt mixtures in the CAM model is shown in formula (2). The starting frequency when the master curve enters into the high temperature or low frequency limit state is defined as the second limit frequency. The range between the two limit frequencies is called the rheological region. In this range, the rheological properties of asphalt mixture were affected by frequency and temperature, and the phase change of asphalt mixture mainly occurred in this region. The regions outside the two limit frequencies are called the low frequency steady state zone and the high frequency steady state zone. In these zones, the rheological properties of asphalt mixture were not affected by the frequency or temperature. The modulus corresponding to the limit frequency of low frequency steady state is called the complex modulus in equilibrium state G_e^*,

and the modulus corresponding to the limit frequency of high frequency steady state is called the complex modulus in the glass state G_g^*. In addition, the turning point, of which asphalt mixtures transition from a low frequency steady state to a rheological region, it is called the low frequency turning point f_c. The changing point, of which asphalt mixtures transition from rheological to a high frequency steady state, is defined as the high frequency turning point f_c'. The intercept of G_e^* and G_g^* in logarithmic coordinates is denoted R (see (3)), which relates to morphological parameters m and k. A high R value indicates that the change from the elastic behavior to the viscous behavior is easier.

$$G^* = G_e^* + \frac{G_g^* - G_e^*}{\left[1 + (f_c/f')^k\right]^{m_e/k}}, \qquad (2)$$

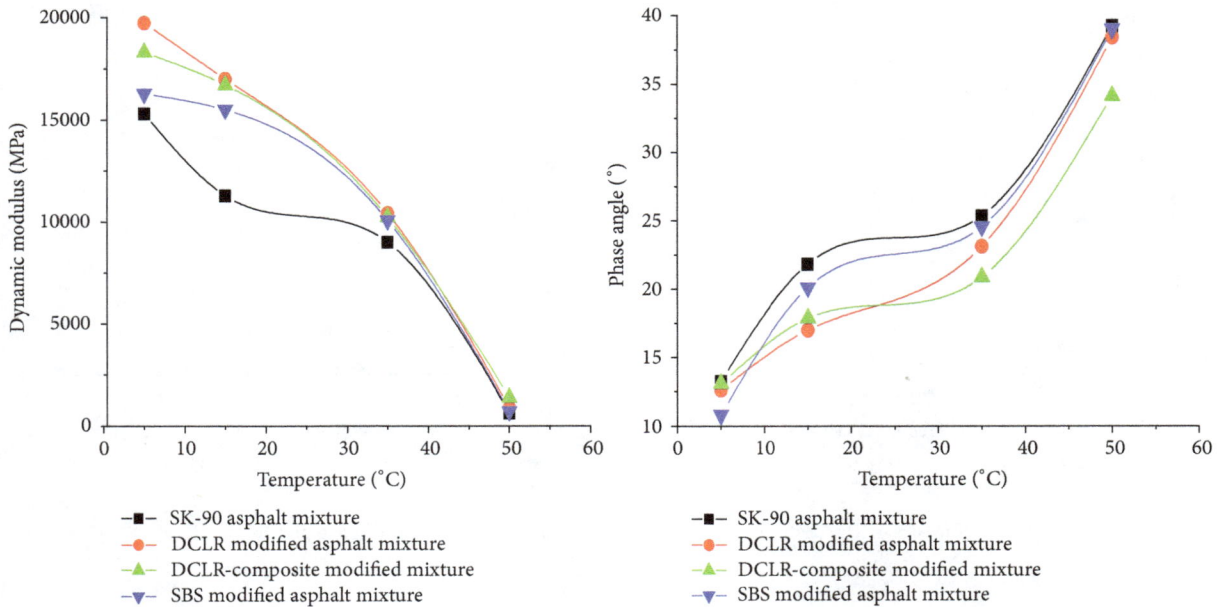

FIGURE 6: Comparison of dynamic modulus and phase angle between the asphalt mixtures at 10 Hz.

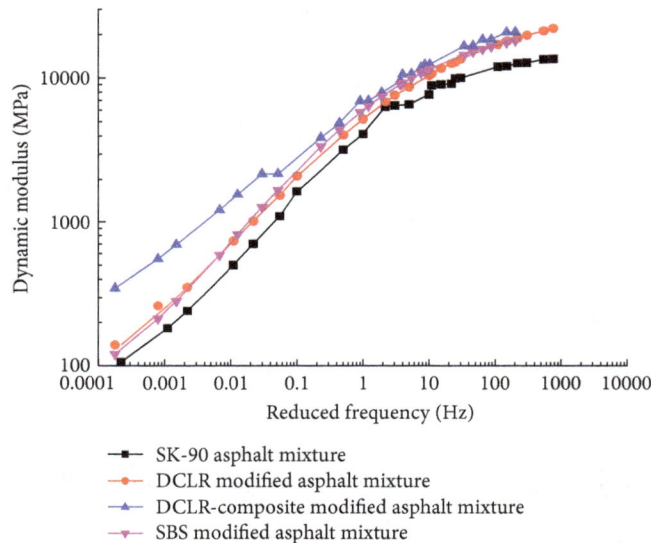

FIGURE 7: The dynamic modulus master curves of the four asphalt mixtures.

$$R = \log \frac{2^{m_c/k}}{1 + \left(2^{m_c/k} - 1\right)\left(G_e^*/G_g^*\right)}, \tag{3}$$

where G^* is the dynamic modulus; G_e^* is the complex modulus of the equilibrium state; G_g^* is the complex modulus of the glass state; f_c is the elastic limit threshold, which is the critical frequency for asphalt mixture transitioning from the viscous flow zone into the rheological zone; m_e and k are the dimensionless morphological parameters.

4.3.2. Dynamic and Viscoelastic Properties of Asphalt Mixtures.
Based on the curve fitting by the CAM model, the viscoelastic parameters of the CAM model are shown in Figure 8. Different parameters relate to different properties in asphalt mixtures, and the results and discussions are shown as follows.

The parameter G_e^* describes the resistance to rutting in asphalt mixtures at high temperatures. The parameter G_e^* results of DCLR and DCLR-composite modified asphalt mixtures were much larger than those of SK-90 and SBS modified asphalt mixture. This shows that the addition of DCLR and DCLR-composite can significantly improve the rutting resistance of asphalt mixtures at high temperatures. The parameters G_g^* and f_c depict the resistance to permanent deformation in asphalt mixtures at low temperatures. The coefficients

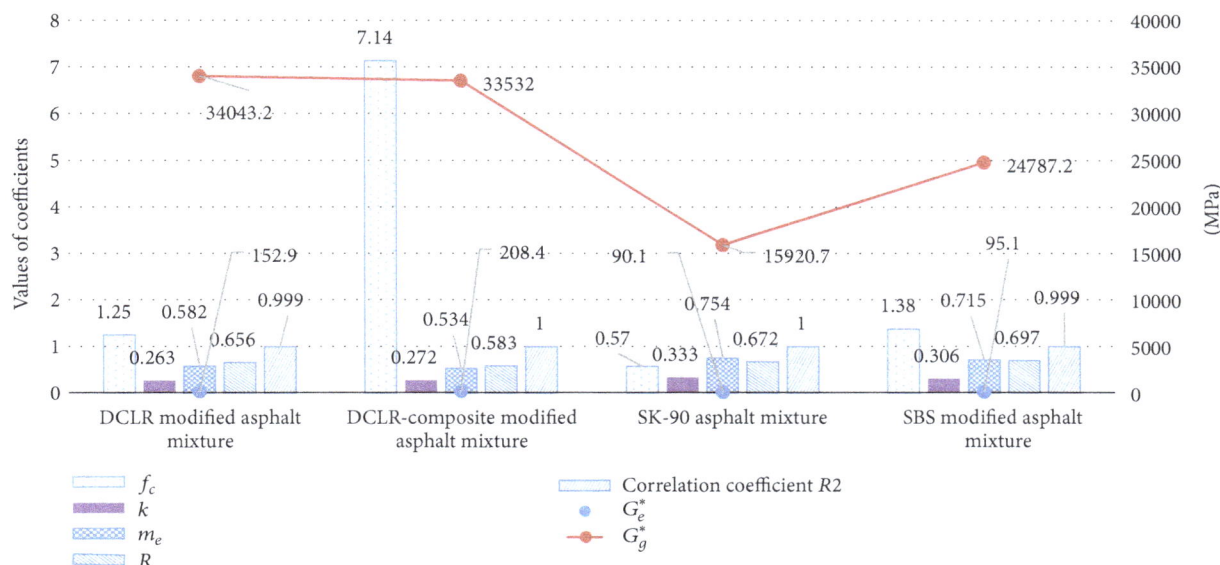

FIGURE 8: Viscoelastic parameters of the CAM model.

G_g^* and f_c of DCLR and composite DCLR modified asphalt mixtures were higher than those of SK-90 and SBS modified asphalt mixture. It indicates that the DCLR and DCLR-composite modified asphalt mixtures have a better resistance to deformation at low temperatures or high frequencies. Small m_e and R represent a less sensitivity to the frequency since m_e and R denote the sensitivity of asphalt mixtures to the frequency. The SBS modified asphalt mixture has an easy transition from the elastic part to the viscous part compared to other mixtures. Based on the results of the curve fitting, the DCLR and DCLR-composite modified asphalt mixtures had a lower sensitivity to the frequency. The correlation degrees of CAM model fitting of the asphalt mixtures were all above 0.999, which proves that the CAM model can characterize the viscoelastic behavior of the asphalt mixtures.

5. Conclusions

The DCLR and DCLR-composite were used to modify the base asphalt binder, and the properties of modified asphalt mixtures were analyzed compared to SK-90 and SBS modified asphalt mixtures. The viscoelastic properties of asphalt mixtures were studied using the CAM model, and the following conclusions can be drawn.

(1) The DCLR and DCLR-composite modified asphalt mixtures had higher dynamic moduli and smaller phase angles than those of the SK-90 and SBS modified asphalt mixtures. This indicates that the DCLR and DCLR-composite modified asphalt mixtures are more elastic compared to the SK-90 and SBS mixtures.

(2) The dynamic modulus of the DCLR-composite modified asphalt mixture was higher than those of other

mixtures when the frequency was smaller than 0.1 Hz. Otherwise, the dynamic moduli of the DCLR and DCLR-composite modified asphalt mixtures were close to each other and higher than those of SK-90 and SBS modified asphalt mixtures. The high- and low-temperature performance of the DCLR-composite modified asphalt mixture were better than the other mixtures, as well as the DCLR modified asphalt mixtures.

(3) The utilization of the CAM model helps analyze the viscoelastic properties of asphalt mixtures, and a good fit and correlation are observed. The resistance to permanent deformation in asphalt mixtures was enhanced by the addition of the DCLR and DCLR-composite in the base asphalt mixture. The DCLR and DCLR-composite modified asphalt mixtures exhibited a less sensitivity to the frequency based on the parameter results of the CAM model, such as coefficients m_e and R.

Competing Interests

The authors declare that they have no competing interests.

Acknowledgments

This project is supported by the Importation and Development of High-Caliber Talents Project of Beijing Municipal Institutions (Grant no. PXM2013-014210-000165) and the National Natural Science Foundation of China (51478028). The authors wish to express their gratitude to Mr. Jinqi Gao for his assistance in the laboratory work.

References

[1] H. S. R. I. o. M. o. Communications, *Standard Test Methods for Asphalt and Asphalt Mixtures for Highway Engineering (JTG E20-2011)*, People's Communication Press, Beijing, China, 2011.

[2] M. W. Witzcak, *Simple Performance Test for Superpave Mix Design*, vol. 465, Transportation Research Board, 2002.

[3] A. Loizos, G. Boukovalas, and A. Karlaftis, "Dynamic stiffness modulus for pavement subgrade evaluation," *Journal of Transportation Engineering*, vol. 129, no. 4, pp. 434–443, 2003.

[4] R. F. Bonaquist, *Precision of the Dynamic Modulus and Flow Number Tests Conducted with the Asphalt Mixture Performance Tester*, vol. 702, Transportation Research Board, 2011.

[5] M. W. Witczak, *Specification Criteria for Simple Performance Tests for Rutting*, vol. 1, Transportation Research Board, 2007.

[6] R. F. Bonaquist, *Ruggedness Testing of the Dynamic Modulus and Flow Number Tests with the Simple Performance Tester*, 2008.

[7] Y. Zhang, R. Luo, and R. L. Lytton, "Characterizing permanent deformation and fracture of asphalt mixtures by using compressive dynamic modulus tests," *Journal of Materials in Civil Engineering*, vol. 24, no. 7, pp. 898–906, 2012.

[8] A. K. Apeagyei, "Rutting as a function of dynamic modulus and gradation," *Journal of Materials in Civil Engineering*, vol. 23, no. 9, pp. 1302–1310, 2011.

[9] H. Zhu, L. Sun, J. Yang, Z. Chen, and W. Gu, "Developing master curves and predicting dynamic modulus of polymer-modified asphalt mixtures," *Journal of Materials in Civil Engineering*, vol. 23, no. 2, pp. 131–137, 2011.

[10] J.-C. Wei, S.-P. Cui, and J.-B. Hu, "Research on dynamic modulus of asphalt mixtures," *Journal of Building Materials*, vol. 11, no. 6, pp. 657–661, 2008.

[11] T. R. Clyne, M. O. Marasteanu, X. Li et al., *Dynamic and Resilient Modulus of Mn/DOT Asphalt Mixtures*, Minnesota Department of Transportation: Department of Civil Engineering, University of Minnesota, 2003.

[12] T. Pellinen and M. Witczak, "Stress dependent master curve construction for dynamic (complex) modulus (with discussion)," *Journal of the Association of Asphalt Paving Technologists*, vol. 71, pp. 281–309, 2002.

[13] M. Witczak, D. Andrei, and W. Houston, "Development of the 2002 guide for the design of new and rehabilitated pavement structures," Inter Team Technical Report (Seasonal 1), Resilient Modulus as Function of Soil Moisture—Summary of Predictive Models, College of Engineering and Applied Sciences, Department of Civil and Environmental Engineering, Arizona State University, 2000.

[14] J. P. Hallin, "Development of the 2002 guide for the design of new and rehabilitated pavement structures: phase II," Report for National Cooperative Highway Research Program, Transportation Research Board, National Research Council, 2004.

[15] E. Chailleux, G. Ramond, C. Such, and C. De La Roche, "A mathematical-based master-curve construction method applied to complex modulus of bituminous materials," *Road Materials and Pavement Design*, vol. 7, no. s1, pp. 75–92, 2006.

[16] X. Zhang, F. Chi, L. Wang, and J. Shi, "Study on viscoelastic performance of asphalt mixture based on CAM model," *Journal of Southeast University*, vol. 24, no. 4, pp. 498–502, 2008.

[17] Y. Yin, *Research on Dynamic Viscoelastic Characteristics and Shear Modulus Predicting Methods for Aspgalt Mixtures Based on Dynamic Mechanical Analysis (DMA) Means*, South China University of Technology, Guangzhou, China, 2010.

[18] C. Wu and K. Xie, *Direct Coal Liquefaction*, Chemical Industry Press, Beijing, China, 2010.

[19] W. Zhang, J. Jin, H. Yu, and L. Tian, "Coal direct hydrogenation liquefaction process," *Clean Coal Technology*, vol. 7, no. 3, pp. 31–33, 2001.

[20] X. Gu, "Properties and utilization of coal direct liquefaction residue," *Clean Coal Technology*, vol. 18, no. 3, pp. 24–31, 2012.

[21] S. Khare and M. Dell'Amico, "An overview of conversion of residues from coal liquefaction processes," *The Canadian Journal of Chemical Engineering*, vol. 91, no. 10, pp. 1660–1670, 2013.

[22] L. He, *Study on the preparation and performance of asphalt modified by coal liquefaction residue [M.S. thesis]*, Highway School, Chang'an University, Xi'an, China, 2013.

[23] Y. Zhang, *Exploratory Study on Direct Coal Liquefaction Residue Modified Asphalt*, Engineering, Northwest University, Xi'an, China, 2012.

[24] Y.-S. Zhao and J. Ji, "Study on the performance of direct coal liquefaction residue modified mixture asphalt," in *Challenges and Advances in Sustainable Transportation Systems: Plan, Design, Build, Manage, and Maintain*, p. 319, ASCE, 2014.

[25] J. Ji, Y.-S. Zhao, and S. F. Xu, "Study on properties of the blends with direct coal liquefaction residue and asphalt," *Applied Mechanics and Materials*, vol. 488-489, pp. 316–321, 2014.

[26] D.-R. Zhang, R. Luo, Y. Chen, S.-Z. Zhang, and Y. Sheng, "Performance analysis of DCLR-modified asphalt based on surface free energy," *China Journal of Highway and Transport*, vol. 29, no. 1, pp. 22–28, 2016.

[27] Q. Zhang, H. Pei-wen, and B. Zheng-yu, "Research on preparation and adhesion of emulsified asphalt modified with waterborne epoxy resin," *Journal of Highway and Transportation Research and Development*, no. 9, pp. 9–14, 2015.

[28] J. Chen, M. Sun, X.-M. Dai et al., "Asphalt modification with direct coal liquefaction residue based on benzaldehyde crosslinking agent," *Journal of Fuel Chemistry and Technology*, no. 9, pp. 1052–1060, 2015.

[29] Z. Liu, H. Song, Z. Ma et al., "Preparation of mesophase pitch from the residue of coal liquefaction by solvent hydrogenation," *Carbon Techniques*, no. 2, pp. 18–22, 2016.

[30] J.-F. Corte, "Development and uses of hard-grade asphalt and of high-modulus asphalt mixes in France," *Transportation Research Circular*, vol. 503, pp. 12–31, 2001.

[31] T. C. R. I. o. L. Province, *Technical Specification Construction of High Modulus Asphalt Mixture (DB21/T 1754-2009)*, People's Communication Press, Beijing, China, 2009.

[32] X. Li, *Mechanical Engineering and Control Systems: Proceedings of the 2015 International Conference on Mechanical Engineering and Control Systems (MECS2015)*, World Scientific Publishing, 2016.

[33] G. A. Tannoury and R. C. E. University of Nevada, *Laboratory Evaluation of Hot Mix Asphalt Mixtures for Nevada's Intersections—Phase II*, University of Nevada, Reno, Nev, USA, 2007.

[34] H. Yao, Z. You, L. Li, S. W. Goh, and C. Dedene, "Evaluation of the master curves for complex shear modulus for nano-modified asphalt binders," in *Proceedings of the 12th COTA International Conference of Transportation Professionals (CICTP '12)*, pp. 3399–3414, August 2012.

[35] M. Zeng, H. U. Bahia, H. Zhai, M. R. Anderson, and P. Turner, "Rheological modeling of modified asphalt binders and mixtures (with discussion)," *Journal of the Association of Asphalt Paving Technologists*, vol. 70, pp. 403–441, 2001.

7

Evaluating RLWT Rutting Test of Asphalt Mixtures Based on Industrial Computerized Tomography

Wenliang Wu[ID]**, Zhi Li**[ID]**, Xiaoning Zhang**[ID]**, and Minghui Li**[ID]

School of Civil Engineering and Transportation, South China University of Technology, Guangzhou 510640, China

Correspondence should be addressed to Zhi Li; 41548915@qq.com

Academic Editor: Giorgio Pia

To eliminate the effects of image's light and shade difference when separating and distinguishing the material composition, a method is put forward, namely, ring-type and partitions threshold segmentation. It means setting up different segment threshold for different areas of the same image and then combining these different areas into one image. Furthermore, by analyzing the CT image before and after the RLWT rutting test for the drilling specimen and Marshall specimen and taking the volume of air voids and the angle (alpha) between max main axis and X axis, the differences of two kinds of specimens' macrotest results were discussed from internal structure distribution. Here, we show that there are differences between macrotest results of two kinds of specimens because of internal air voids and aggregate distribution, which should be considered for compliance testing.

1. Introduction

With the increasing traffic volume and the effects of heavy-duty axle loading, pavement rutting has become one of major distresses occurring in asphalt pavements. It can influence the safety and the driving comfort directly when the rutting depth reaches a certain limit [1]. A variety of methods, such as French Pavement Rutting Tester (PRT), Hamburg Wheel-Tracking Device (WTD), Asphalt Pavement Analyzer (APA), and Rotary Loaded Wheel Tester (RLWT), have been used for estimating the rutting performance of the asphalt concrete pavement in the world. The Rotary Loaded Wheel Tester, developed in the US in the late 1990s, has been paid more and more attention because of its convenience and ease of use. It can evaluate the rutting performance by measuring the mixture sample and pavement coring specimens whose diameter is 100 mm or 150 mm. The further study on quality control conformance testing about applying RLWT to asphalt pavement has begun [2, 3].

As a unique nondestructive evaluation technique, industrial computerized tomography (CT) can inspect the reaction of material interior structures dynamically and quantificationally and provide a new method to researching interior structures. The computer-aided design of asphalt mixtures based on industrial CT, a prospective technology, is predicted as "the next generation of infrastructure materials research" and "analysis modeling and design method based on system." Masad used CT to research asphalt mixture very early, and he paid more attention to the volumetric properties, contact relationship, and numerical simulation [4, 5]. Wang et al. have done some research on 3D reconstruction technology and performed a large number of numerical simulation test based on 3D reconstruction technology [6, 7]. The author's team researched the CT technology for asphalt mixture and made certain progress [8]. The main challenge of CT technology is that the brightness of CT image varies according to the distance with central axis because of the low ray power of industrial CT machine. And the CT image cannot give obvious peak, and this challenge leads to some problems in distinguishing different materials accurately.

Based on the foregoing background, this paper developed a new method to segment threshold, namely, ring-type and partitions threshold segmentation. This means setting up different segment thresholds for different areas of the same image and then combining these different areas into one image. Furthermore, by analyzing the CT image before and after the RLWT rutting test for pavement core and indoor specimen and taking the volume of air voids and the angle alpha between max main axis and X axis, the differences of two kinds of specimens' macrotest results were discussed from internal structure distribution.

2. Ring-Type and Partitions Threshold Segmentation

Threshold segmentation is one way to distinguish the voids, asphalt mastic, and aggregates through segmenting images, whose accuracy largely depend on the selection of threshold T1 and T2 [9]. Many researchers take the 2-mode method to analyze the images. This way can generate threshold by applying point correlation technique, assuming that the image between the two peaks is the same medium and preliminarily determining the threshold T1 and T2 and then ascertaining the T1 and T2 by trial and error. However, after carefully comparing the slice images of different positions scanned by CT, it can be found that the gray level information of each image is not consistent completely, and the gray value of the same kind of material scanned in different position is not same. The gray level information of single image cannot generally represent the whole sample distributing disciplinarian. In addition, from the original image scanned by CT (Figure 1), it can be found that the surrounding of the image is darker than that in the center. Figure 2 shows the image which is consisted by the gray value 189–255 of pixel points. And, it can be seen from Figure 2 that the aggregate around the image can be roughly distinguished, but the one in the center is not clear. The threshold segmentation uses gray value to distinguish the material. Therefore, it is very necessary to adjust image so that the same material can be in the same gray level. Ring-type and partitions threshold segmentation helps to solve this problem.

In view of nonuniformity that the gray value of the same material in the CT image gradually increases from the center to the edge, the CT image is divided into different ring-type images. The specific steps are as follows: (1) dividing the whole image into a series of overlapping by 50% ring-type subimage (the center image is round.); (2) using Otsu's method [10] for each subimage to calculate the gray threshold of the aggregate and the background; (3) finally, combining the sub-image processed in step 2.

The following is an example about the application of the ring-type and partitions threshold segmentation when the number of the ring is five.

(1) The first step is dividing the whole image into five subimages and dealing with image uniformity processing separately. The result is shown in Figure 3.

(2) Then establishing the gray histogram of each subimage (as is shown in Figure 4). Table 1 shows gray value statistics of the subimage.

Finally, the appropriate image segmentation algorithm is selected (Otsu's method is selected in this article), and each subimage is segmented. Then, the aggregate particles are

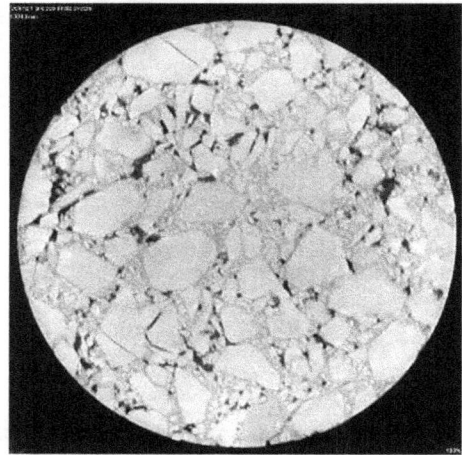

FIGURE 1: Original image of CT scanning.

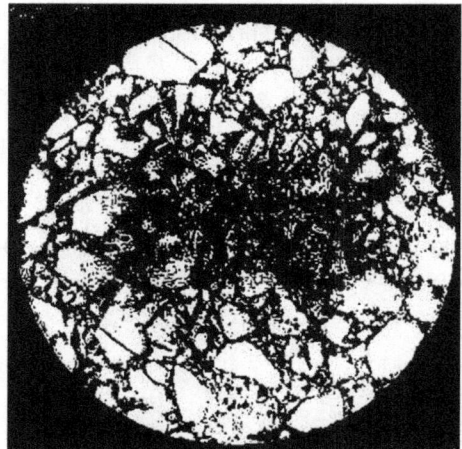

FIGURE 2: Image of gray level from 189 to 245.

isolated and the subimages are combined, which only contains aggregate particles. Figure 5 shows the final aggregate image.

This article has attempted other normal methods (fuzzy c-mean method, GMM-EM method, and otsu method) to analyze the CT image, as shown in Figures 6–8. From Figures 5 and 8, it is shown that using ring-type and partitions threshold segmentation method is much better to extracting aggregates than segmenting the image directly. Furthermore, a normal method requires human interaction, and its efficiency is low with the specimen's radius growing. Ring-type and partitions threshold segmentation is a method of variable threshold processing and can solve the problem of uneven background since low-energy X-ray.

The ring-type and partitions threshold segmentation method is suitable for situation that the gray of the same material is changing gradually from the center to the edge in CT image. For poor quality images, the quality of image segmentation can be improved by appropriately increasing the number of rings.

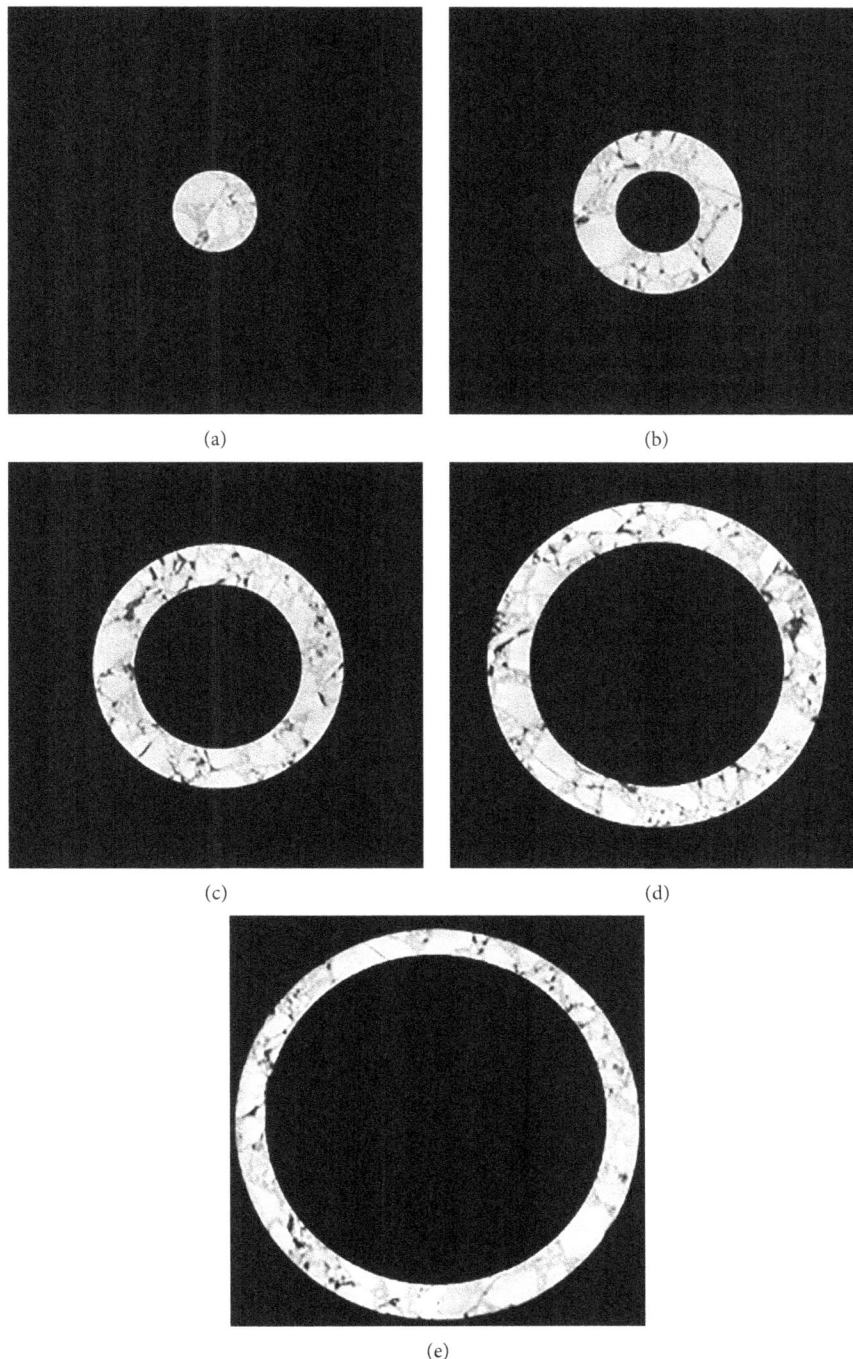

FIGURE 3: Subimage after segmentation: (a) central disk, (b) second ring, (c) third ring, (d) fourth ring, and (e) fifth ring.

3. RLWT Rutting Test

RLWT is a one direction device with no horizontal displacement, as shown in Figure 9. Figure 10 shows the stress pattern of the specimen. There are ten small rubber wheels arranged at the edge of a big rotary wheel. When the shaft spins once, every rubber wheel will load the specimen once. In this test, a 125 Newton force and 0.7 MPa contact pressure is applied to the surface of the specimen through free-spinning rubber wheel. Most of the sample will be tested over 16,000 load application. Besides, it will test and record rut depth and the corresponding loading time automatically. And the maximum rut depth of the test is 6.35 mm, and the maximum loading number is 60,000.

4. Summary and Discussion

Drilling three core samples is performed in the middle layer of AC-20 asphalt pavement. Table 2 presents the gradation of

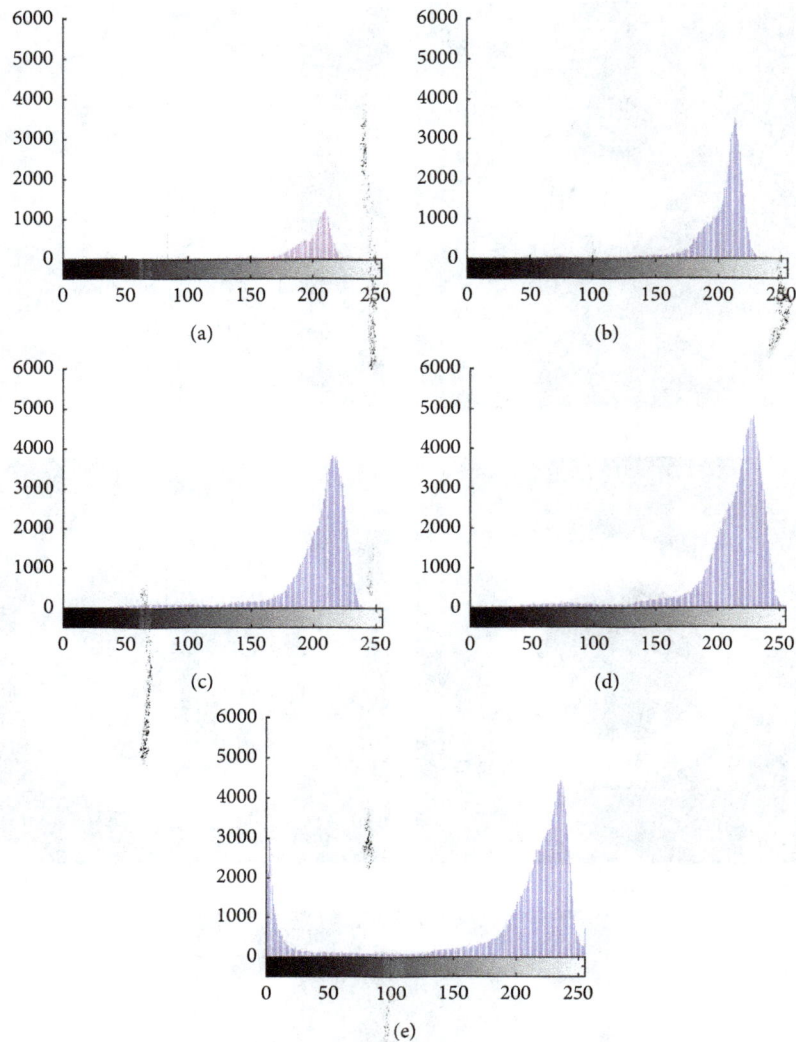

FIGURE 4: Gray histogram of the subimage: (a) central disk, (b) second ring, (c) third ring, (d) fourth ring, and (e) fifth ring.

TABLE 1: Statistical table of gray level.

Subimage	Statistical starting point	Statistical end point	Peak value	Average	Standard deviation
Central disk	158	233	211	202	6.86
Second ring	128	240	214	204	7.78
Third ring	121	244	217	206	9.26
Fourth ring	116	255	230	215	9.63
Fifth ring	114	255	236	218	10

AC-20 asphalt mixtures. Scanning the two kinds of the sample before and after the RLWT and processing the CT image.

4.1. *Void Ratio Distribution Characteristics.* The specimen's study area is chosen after scanning, eliminating the effect of external drop shadow, and determining the void segment threshold value T1 by trial method. The air void is assumed to be the part of the specimen, whose gray value between 0 and

T1, as shown in Figure 11. Furthermore, the air voids content can be measured by the image method through calculating the ratio of the volume of void against the total volume of the specimen. Table 3 presents the specimen's air voids.

From Table 2, it can be found that the range of void is very wide. And the air void contents of the in-place sample's variability are much greater than those of the Marshall specimen. The air void content measured by the image method is less than the one measured by the standard method, and the latter is about 1.2 times of the former. The

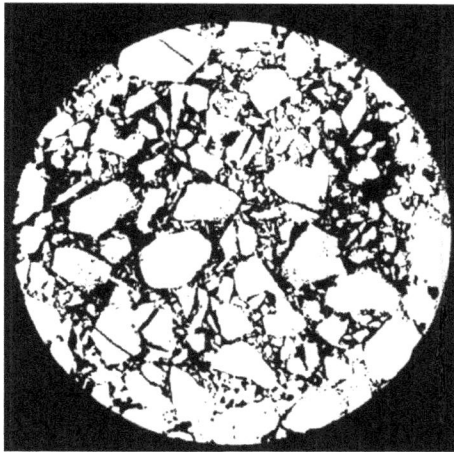

FIGURE 5: Aggregate image after segmentation.

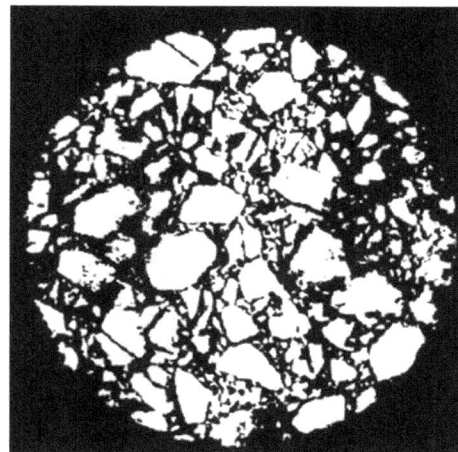

FIGURE 6: Fuzzy c-mean method.

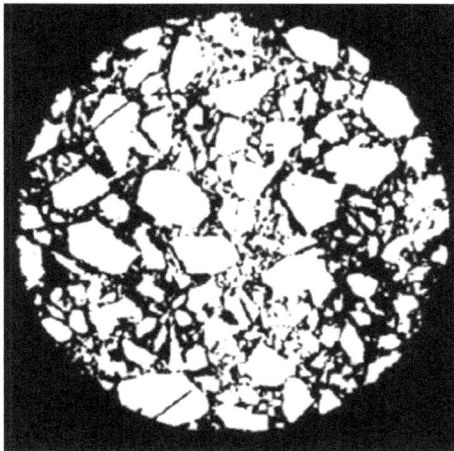

FIGURE 7: GMM-EM method.

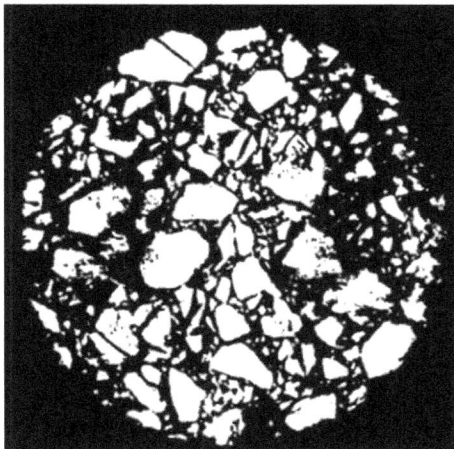

FIGURE 8: Otsu method.

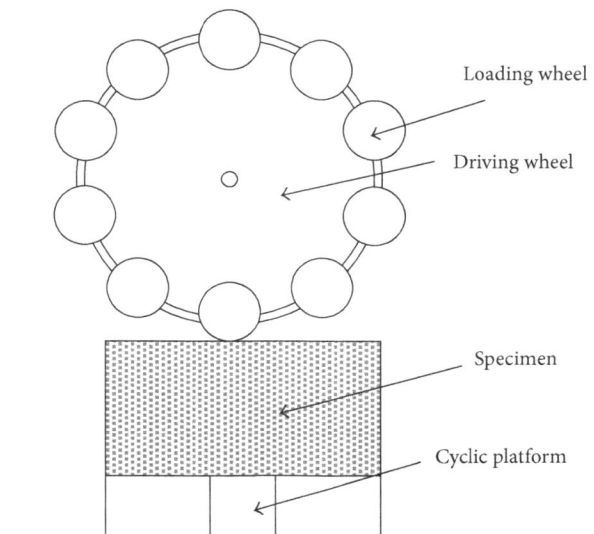

FIGURE 9: Loading sketch of RLWT.

the open pore is processed as part of background in the image processing. After analyzing upper 3 mm air void of the sample, it can be found that the upper air void of the in-place sample is 5.5%, but the sample shaped in laboratory is only 4.09%. The difference of the upper air void directly led to the sample's performance difference. When the pavement compaction degree is low, also means the initial void is high, the aggregate particles cannot interlock very well with each other, and the contact point is unstable. Besides, the compaction rutting might occur under the dynamic loading because of lacking bearing capacity.

4.2. Aggregate Particles Distribution Characteristics. For the line connecting two points on the edge of aggregate particles, there is a unique longest line called particle's main axis, and the main axis intersects the edge of the aggregate particles at the endpoints. The direction of the particle axis, represented by alpha ($0 \leq$ alpha $\leq 90°$), is the angle between the particle axis and the X axis. The physical meaning of it is shown in Figure 12 [11].

reason is that some small pores cannot be identified because of the relationship of the image quality and the program application adaptability. Another important reason is that

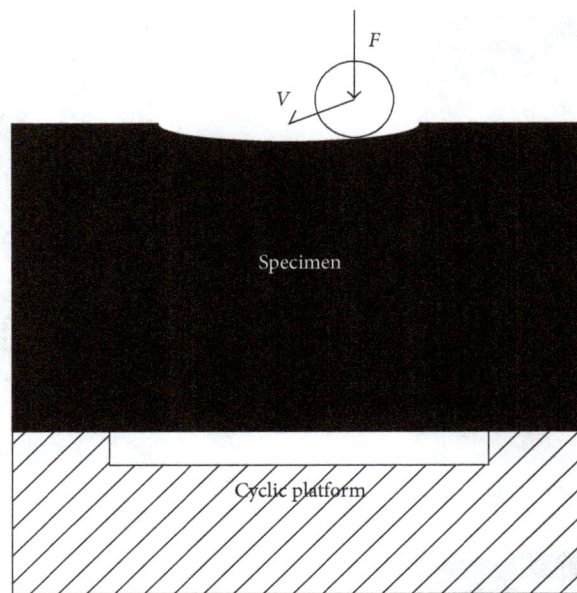

FIGURE 10: Stress pattern of specimen.

TABLE 2: Gradation of AC-20 asphalt mixtures.

Gradation (mm)	Percent passing												Asphalt-aggregate ratio (%)
	26.5	19	16	13.2	9.5	4.75	2.36	1.18	0.6	0.3	0.15	0.075	
AC-20	100	94.9	84.8	73.3	54.4	34.8	22.1	18.5	14.5	11.3	9.0	5.4	4.2

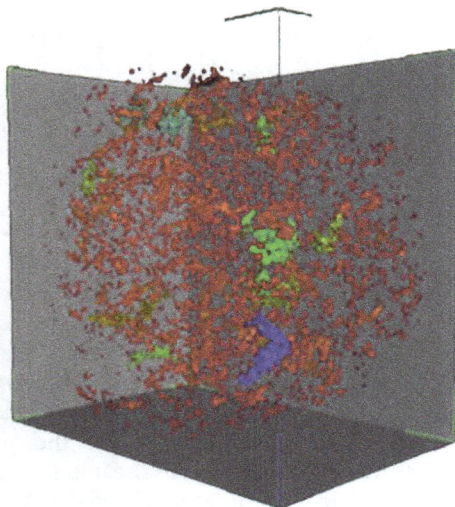

FIGURE 11: 3D image of air voids.

TABLE 3: Statistical table of air void ratio.

	Item	Drilling specimen	Marshall specimen
Image method	Threshold T_1	40	40
	Air void content of upper 3 mm specimen (%)	5.50	4.09
	Air void content (%)	3.7	3.0
	Maximum air void content (%)	6.55	6.2
	Minimum air void content (%)	2.5	1.6
	Variable coefficient	0.33	0.28
	Standard air void content (%) (submerged weight in water method)	4.71	3.53

According to the research of Chen and Liao [12], there is a good correlation between the direction of the particle axis and rutting resistance performance of mixture. Therefore, dividing the coarse aggregate into three types is based on the direction of particle axis: (1) the aggregate whose alpha is in 0°~30° is regraded as stable aggregate because of low center of gravity, (2) the aggregate whose alpha is in 30°~60° may become unstable because of higher center of gravity, and (3) the aggregate whose alpha is in 31°~60° is unstable. It is easy for angle change to appear and lead to pavement rutting.

While researching the distribution characteristics of aggregate particles, the CT image of the samples needs to be obtained first, followed by applying the ring-type and partitions threshold segmentation method (the CT image is shown in Figure 13(a)). Finally, testing the specimens 16,000 times and scanning the specimens at the same location should be performed. The aggregate particles image is shown in Figure 13(b). Figure 14 shows the relation between deformation and loading times. The alpha is counted before and after the rutting test in these 9 images. And the statistical method is used to analyze the change of alpha before and after the rutting test. The result is shown in Table 4.

It can be found from Figure 11 that the deformation of the drilling specimen is significantly larger than laboratory

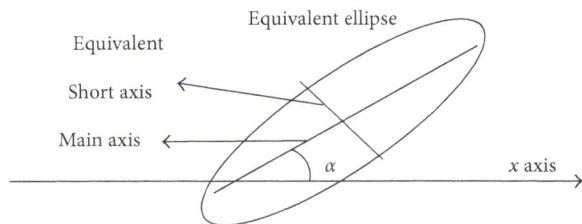

FIGURE 12: Particles' orientation of main axis.

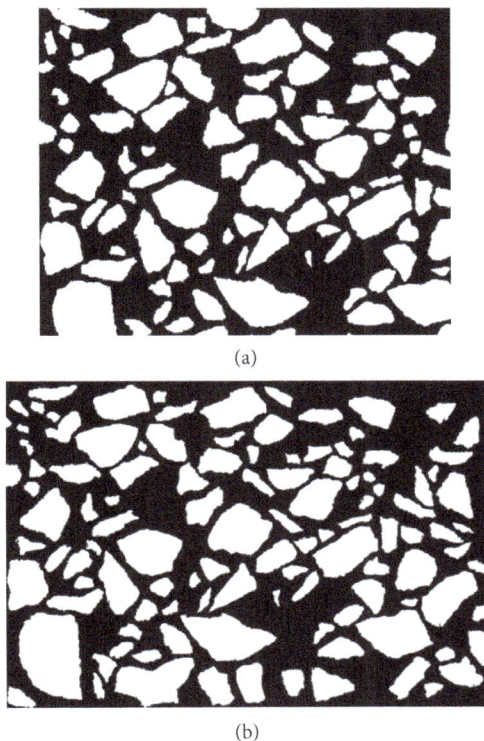

(a)

(b)

FIGURE 13: (a) Aggregate particles before the rutting test; (b) aggregate particles after the rutting test.

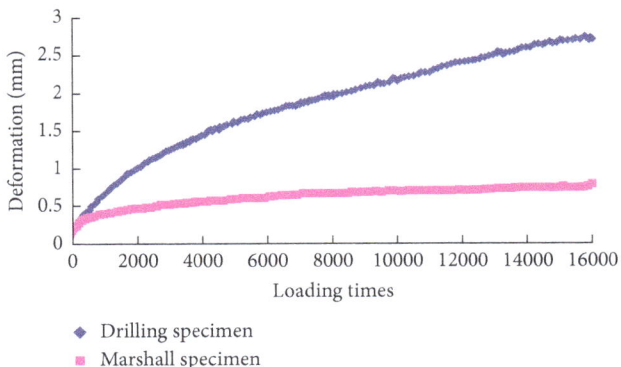

FIGURE 14: Relation between deformation and loading times.

TABLE 4: Statistical table of alpha.

Specimen type	Aggregate size	The proportion of alpha (%)			Average angle (°)
		0°~30°	31°~60°	61°~90°	
Drilling specimen before the rutting test	2.36~4.75 mm	46	34	20	35.83
	4.75~9.5 mm	69	16	15	29.60
	9.5~20 mm	62	19	19	27.08
Drilling specimen after the rutting test	2.36~4.75 mm	51	30	19	37.79
	4.75~9.5 mm	66	24	10	28.86
	9.5~20 mm	67	22	11	25.00
Marshall specimen before the rutting test	2.36~4.75 mm	44	28	28	39.87
	4.75~9.5 mm	54	29	17	39.59
	9.5~20 mm	47	30	23	43.98
Marshall specimen after the rutting test	2.36~4.75 mm	47	26	27	40.61
	4.75~9.5 mm	41	34	25	38.92
	9.5~20 mm	33	50	17	40.01

especially in the upper specimen. Therefore, drilling the core sample easily causes compactness rutting because of its higher porosity.

From Table 4, it can be found that the specimen is not only vertically compressed, but also affected by the shear stress because the base of rotary loaded wheel tester is suspended. And the shear effort is obvious, especially when the specimen is thin. The average alpha value of the drilling specimen and the laboratory Marshall specimen is 31°~60°. It indicates that both of the specimens under different shaping methods might convert from an unstable condition into stable condition. From the variance of alpha value before and after the rutting test, it can be found that the alpha of aggregate decreases gradually after the rutting test. The particle proportion of alpha in 61°~90° decrease significantly, while the one in 31°~60° increases evidently, and it changes a little in 0°~30° range. In other words, the specimen obtains higher stability after being tested by RLWT.

The variation of aggregate particles' alpha mainly occurred in the coarse aggregate (the size over 4.75 mm), especially the part of large particle (over 9.5 mm). But the alpha of fine aggregate particle changes insignificantly compared with the change of coarse aggregate particles. The change of large particles are mainly due to the conversion of 61°~90° aggregate particles into 31°~60° aggregate particles. The average alpha value of 2.36~4.75 mm aggregate particles increases after being tested by RLWT, while the average alpha value of coarse aggregate particles decreases.

Comparing the 0°~30° aggregate particles' proportion between the drilling specimen and the Marshall specimen in 0°~30° range, it can be found that the value of the drilling specimen is larger. Which means conversion ratio of the drilling specimen from unstable condition to stable condition is relatively small. When compared with the 31°~90° of large particle proportion, the proportion of the Marshall specimen is larger. In other words, the Marshall specimen has higher possibility of converting into an unstable condition. It also means that the large particle of the Marshall

Marshall specimen deformation. This is mainly due to the impact of the rolling method and construction factors. And it causes the larger air void contents of the drilling specimen,

specimen has larger rotation under loading. Comparing the 0°~30° aggregate particles' proportion between the drilling specimen and the Marshall specimen in 0°~30° range, it can be found that the value of the drilling specimen is larger. Which means conversion ratio of the drilling specimen from the unstable condition to stable condition is relatively small. When compared with the 31°~90° of large particle proportion, the proportion of the Marshall specimen is larger. In other words, the Marshall specimen has higher possibility of converting into the unstable condition. It also means that the large particle of the Marshall specimen has larger rotation under loading.

5. Conclusions

(1) In view of the situation that the gray of the same material changes gradually from the center to the edge in CT image, applying ring-type and partitions threshold segmentation method is better for the result and more accurate for particles extraction when compared with segmenting the whole image directly.

(2) The air void variability of the drilling core is higher than that of the Marshall specimen. The air void value measured by the image method is less than the one measured by the standard method. And the later is about 1.2 times of the former. The air void of drilling core within the 3 mm upper specimen is far more than that of the Marshall specimen. The difference of upper air void will make the difference of the specimen performance directly.

(3) After RLWT rutting test, it can be found that the alpha value of the drilling specimen and the Marshall specimen decreases, and the specimens convert from unstable condition into stable condition. But the change of each shift aggregate is different. What is more, the alpha change of the two kinds of specimens is different because of the shaping method.

(4) The two kinds of specimens, whose volumetric properties and mix proportions are same, have different internal material (void and aggregate particle) distribution, macroscopic volume parameters, and mechanics indexes because of the different shaping method. It should be paid more attention when we do compliance test of construction quality.

Conflicts of Interest

The authors declare that they have no conflicts of interest.

Acknowledgments

The work represented herein was the result of a team effort. The authors acknowledge the support of the National Natural Science Foundation of China (Grant nos. 51008131 and 51038004).

References

[1] X. N. Zhang, *Quality Control and Assure of Asphalt Pavement Construction*, China Communications Press, Beijing, China, 2009.

[2] H. Lu, L. L. Hu, and X. N. Zhang, "Influence of asphalt mixture interior factors on RLWT test," *Highway*, vol. 10, pp. 204–207, 2009.

[3] J. Xu, S. C. Huang, Y. C. Qin et al., "Mixes design for rut-filling micro-surfacing trials," *Petroleum Asphalt*, vol. 24, no. 2, pp. 25–29, 2010.

[4] E. Mahmoud and E. Masad, "Experimental methods for the evaluation of aggregate resistance to polishing, abrasion, and breakage," *Journal of Materials in Civil Engineering*, vol. 19, no. 1, pp. 977–985, 2007.

[5] E. Masad, A. Al Omari, and H. C. Chen, "Computations of permeability tensor coefficients and anisotropy of asphalt concrete based on microstructure simulation of fluid flow," *Computational Materials Science*, vol. 40, no. 4, pp. 449–459, 2007.

[6] L. B. Wang, X. Wang, L. Mohammad, and Y. Wang, "Application of mixture theory in the evaluation of mechanical properties of asphalt concrete," *Journal of Materials in Civil Engineering*, vol. 16, no. 2, pp. 167–174, 2004.

[7] L. B. Wang, J. D. Frost, L. Mohammad et al., *Three-Dimensional Aggregate Evaluation Using X-Ray Tomography Imaging*, Transportation Research Board (TRB), Washington, DC, USA, 2002.

[8] X. N. Zhang, "Basic methods of digital image technology for material structure of asphalt mixtures," *Journal of South China University of Technology*, vol. 40, no. 10, pp. 166–173, 2012.

[9] X. J. Li, J. F. Zhang, K. N. Liu et al., "Finite element modeling of geomaterial using digital image processing and computerized tomography identification," *Roch and Soil Mechanics*, vol. 27, no. 8, pp. 1331–1334, 2006.

[10] X. N. Zhang, Y. H. Duan, Z. Li et al., "Classification of asphalt mixture materials based on x-ray computed tomography," *Journal of South China University of Technology*, vol. 38, no. 10, pp. 120–124, 2011.

[11] Z. Li, *Analysis of Volume Constituent Characteristics for Asphalt Mixture Based on Digital Image Processing*, Harbin Institute of Technology, Harbin, China, 2002.

[12] J.-S. Chen and M.-C. Liao, "Evaluation of internal resistance in hot-mix asphalt (HMA) concrete," *Construction and Building Materials*, vol. 16, no. 6, pp. 313–319, 2002.

8

Studying Engineering Characteristics of Asphalt Binder and Mixture Modified by Nanosilica and Estimating Their Correlations

Moein Hasaninia[1] **and Farshad Haddadi**[2]

[1]*Department of Civil Engineering, Iran University of Science and Technology (IUST), Narmak, Tehran, Iran*
[2]*Department of Civil Engineering, Florida International University (FIU), Miami, FL, USA*

Correspondence should be addressed to Moein Hasaninia; moinhasani@civileng.iust.ac.ir

Academic Editor: Peter Majewski

The objective of this research was to investigate rutting and fatigue distresses in asphalt containing 2, 4, 6, and 8 percent of nanosilica (NC) and to find out the correlation between engineering properties of the modified binder and mixture asphalt. In order to study the effect of NC on the rutting and fatigue properties of modified binders, the multiple stress creep recovery (MSCR) and linear amplitude sweep (LAS) tests were carried out. The Marshall stability, dynamic creep, and four-point bending beam fatigue tests were used to evaluate performance characteristics of the mixtures. The binder and mixture tests all indicated an improvement of fatigue and rutting resistance using NC as a modifier. Furthermore, some statistical correlations between engineering properties were developed successfully.

1. Introduction

Binder plays an important role in preventing common distresses associated with asphalt mixture such as fatigue cracking and rutting. For instance, binder with enough adhesion and cohesion can significantly hinder the segregation and separation of aggregates from the pavement surface [1]. In order to improve the behavior of asphalt binder at different temperatures, many types of additives have been used. Among them, nanomaterials have been used by many researchers for their high surface area and capability of creating powerful networks in asphalt binder, culminating to an increase in the mixture's resistance to permanent deformation [2]. Performance characteristics of binders and asphalt mixtures have been affected to some extent due to the addition of nanoparticles such as nanoclay, nanolime, carbon nanofiber, and carbon nanotube [2, 3].

The effect of nanoclay as an asphalt binder additive on the mechanical properties of the asphalt mixture demonstrated a considerable increase on rutting resistance and resilient modulus of asphalt concrete (AC) samples. However, the additive has no considerable effect on low-temperature fatigue resistance of the modified sample [4].

Amirkhanian et al. studied the effects of carbon nanoparticles on the performance characteristics of asphalt binder. The viscosity, performance grade (PG), creep and creep recovery, and frequency sweep tests were carried out on modified binder. Experiment results indicate that the addition of the carbon nanoparticle was effective in increasing viscosity, failure temperature, complex modulus and elastic modulus, and as a result, rutting resistance of the binder [5].

Yao et al. used NC as a binder modifier. They added NC to an SBS-modified asphalt in 4 and 6 percent by the weight of the modified base binder. Experiment results showed that the value of viscosity at high temperatures decreased slightly; in fact, at low temperatures, modified binder with NC behaves similar to control binder samples; furthermore, NC improved binder antioxidation characteristics. The rutting and fatigue cracking performance of asphalt binder modified by NC was improved [6].

Among the advantages of NC are its functional features and low-cost production. NC is one of the new minerals which include potential useful features, such as huge surface area, good distribution, high absorption, high stability, and high percentage of purity.

Today, researchers are looking for binder tests that not only could demonstrate the mixture's performance-related characteristics of both modified and unmodified binder but also are easy and quick to conduct. Insufficiency of the performance grade (PG) binder specification as one of the common methods to evaluate binder performance, especially when it is modified or rejuvenated by additives, has been proved by many researchers [7, 8]. To address this issue and as a way to find a better performance-related test method, the LAS and MSCR tests were introduced to evaluate fatigue and rutting performance of the binder, respectively. The LAS test showed a good correlation with long-term pavement performance (LTPP) field fatigue cracking data [9]. Furthermore, unlike the existing SHRP test method, MSCR captures the nonlinear behavior of rutting phenomenon and correlates fairly well with field rutting data [7].

In this research study, the binder is modified by 2, 4, 6, and 8 percent of NC, and two important distresses of asphalt, rutting and fatigue, are evaluated through the LAS, MSCR, 4-point bending beam, and dynamic creep tests. Finally, some correlation between binder and mixture test results was developed successfully.

2. Materials

The asphalt binder used in this study was AC-60/70, provided by Pasargad Oil Company, Tehran. The characteristics of the binder are presented in Table 1. The required aggregates to produce the sample are taken from Asb-Cheran Mine located in Roudehen in the north of Tehran. Rock dust is used as the filler in the production of samples. The characteristics of the aggregates are presented in Table 2. The gradation of aggregates is according to AASHTO M323 and presented in Table 3 and Figure 1. NC used in this research study has a purity of more than 99%. The maximum diameter of the particles is 10 nm, and the surface area is 600 m^2/g. Its bulk density is less than 0.10 g/cm^3, and the true density is 2.4 g/cm^3.

3. Sample Preparation

NC is added to the asphalt binder by 2, 4, 6, and 8 percent of the original binder's weight. A high-shear mixing device is used to mix NC and binder with 4000 rpm for 2 hours at 135°C. The SEM images of the modified binder with 4 percent NC in three magnitudes are shown in Figure 2. Accordingly, particles' diameters are roughly between 50 and 150 nanometer.

The Marshall method was used to determine the stability, flow, and optimum binder of all asphalt samples (ASTM D2726 and ASTM D1559). Percentage of optimum binders obtained were 5.5, 5.3, 5.2, 5 and 4.9 at mixtures with 0%, 2%, 4%, 6% and 8% of NC content, respectively. The samples were compacted using a Gyratory compactor for the dynamic creep, indirect tensile strength, and resilient modulus tests. Samples used in fatigue tests were originally fabricated as slabs with dimensions of 5*30*40 cm using wheel track compactor. All the samples made at optimume binder and 4% air

TABLE 1: Physical properties measured of bitumen.

Parameter measured	Test method	Test value
Specific gravity at 25°C (g/cm^3)	AASHTO T228	1.01
Penetration at 25°C (0.1 mm)	AASHTO T49	60
Softening point (R&B) (°C)	AASHTO T53	56
Viscosity at 120°C (centistokes)	AASHTO T201	1055
Viscosity at 135°C (centistokes)	AASHTO T201	361
Viscosity at 160°C (centistokes)	AASHTO T201	170
Ductility at 25°C (cm)	AASHTO T51	>100

void. Then, they were sawn to the prismatic beams with dimensions of 38.5 mm × 63.5 mm × 50 mm, considering the AASHTO T321 standard [10].

The binder performance characteristic tests have been carried out on aged samples. Prior to the multiple stress creep recovery (MSCR) and linear amplitude sweep (LAS) tests, all the modified binder samples, as well as the 60/70 base binder, were aged in the rolling thin-film oven (RTFO) in order to represent a short-term aging condition.

4. Experimental Design

4.1. Multiple Stress Creep Recovery (MSCR). This test has been used to measure the percent of recovered strain (R) and unrecovered strain (j_{nr}) of asphalt binders. The elastic response of the binder under the shear stresses can be calculated by this test methodology. The aged samples in the RTFO process are used in this test method. In order to conduct the MSCR test, the dynamic shear rheometer (DSR) is used. The binder sample is put under a 0.1 kPa shear stress for a 1-second duration, followed by a 9-second rest period at the temperature of 60°C. This loading repeats for 10 cycles. Then, after the completion of the first ten cycles, a similar procedure will be applied to the sample with a stress level of 3.2 kPa. According to the ASTM D-7405-10a standard, at each 0.1 sec interval, the relevant output should be recorded [11].

4.2. Linear Amplitude Sweep (LAS) Test. This test was proposed by Johnson and Hintz to investigate the fatigue resistance of asphalt binders [12]. According to the AASHTO standard (AASHTO-TP 101-12-UL), the binder samples of 8 mm thickness are tested in the dynamic shear rheometer (DSR). All DSR tests are conducted on RTFO aged samples. The test is carried out under the strain-controlled mode with linearly increased load amplitudes from 0.1% to 30% strain in a total time of 310 seconds [13].

In the viscoelastic continuum damage (VECD) analysis, the binder fatigue performance parameter N_f can be calculated by

$$N_f = A_{35} \times \gamma^B, \tag{1}$$

where N_f is the number of cycles to failure, A_{35} is the damage intensity corresponding to 35 percent reduction of undamaged $|G^*|\sin\delta$, and B demonstrates the binder sensitivity to applied strain level. The parameters A_{35} and B are experimentally defined, and γ is the applied shear strain.

TABLE 2: Properties of used aggregates.

Properties	Method	Requirement	Values
Coarse aggregate			
Los Angeles abrasion (%)	AASHTO T96	30 max.	20
Water absorption (%)	AASHTO T85	5 max.	0.8
Bulk specific density (g/cm^3)	AASHTO T85	—	2.654
Flat and elongated (3–1) (%)	ASTM D4791	20 max.	12
Soundness (sodium sulfate) (%)	AASHTO T104	15 max.	4.8
Crushed content (one face) (%)	ASTM D5821	100 min.	100
Crushed content (two faces)	ASTM D5821	90 min.	100
Fine aggregate			
Water absorption (%)	AASHTO T84	—	1.4
Bulk specific density (g/cm^3)	AASHTO T84	—	2.617
Mineral filler			
Bulk specific density (g/cm^3)	AASHTO T84	—	2.702

TABLE 3: Gradation of the aggregates used in the study.

Sieve sizes	US	3/4″	1/2″	No. 4	No. 8	No. 50	No. 200
Sieve sizes	Metric	19	12.50	4.75	2.36	0.3	0.075
Passing (%)	HMA gradation	100	95	60	40	15	5

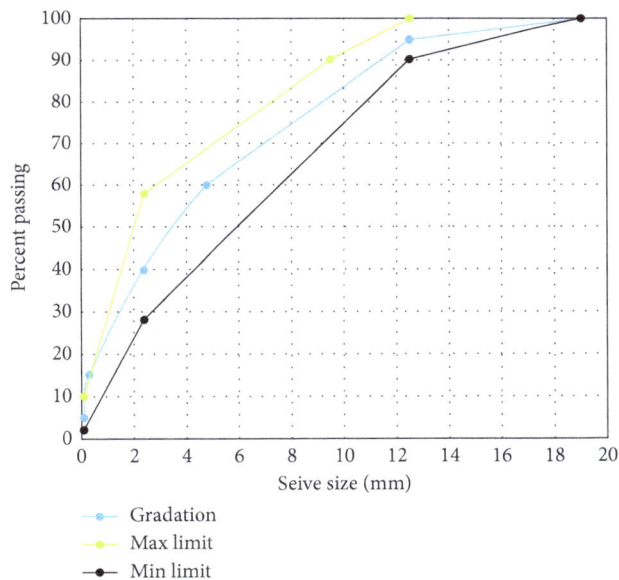

FIGURE 1: Gradation of designated aggregates.

4.3. Marshall Stability and Quotient.

The test was carried out according to ASTM D1559. Before the test, all samples were put in 60°C water for 30 minutes. Marshall stability is the peak resistance load obtained during a constant rate of the deformation load sequence, and Marshall flow is a measure of deformation of the asphalt mix determined during the stability test. The Marshall quotient equals the ratio of the Marshall stability to the value of the Marshall flow. The value of the Marshall quotient indicates the resistance of asphalt mixtures against permanent deformations and the value of rutting [14].

4.4. Resilient Modulus.

Resilient modulus is one of the important parameters in the pavement design procedure. The measurement of this parameter is in the form of pavement response under dynamic loads and the corresponding strains associated with them. The value of resilient modulus is measured based on the ASTM D4123 [15] standard. This test is conducted at temperatures of 5°C and 25°C, and the minimum numbers of loadings are 100. The value of resilient modulus (M_r) can be obtained by [16]

$$M_r = P \frac{(\mu + 0.27)}{(t\delta_h)}, \quad (2)$$

where P is the maximum dynamic load (N), μ is Poisson's ratio, t is the specimen length (mm), and δ_h is the horizontal recoverable deformation (mm).

4.5. Dynamic Creep Test.

In the present research, the dynamic creeping test has been used to evaluate the rutting property of the asphalt mixtures. The output creep curve of the test is made of three areas. In the present article, the flow number (F_n) parameter is used as representation of rutting resistance of asphalt mixtures, which is the number of cycles the creep curve enters from the second to the third phase [17].

4.6. Four-Point Beam Fatigue Test.

The fatigue life of the asphalt mixtures is evaluated by the 4-point bending beam test under AASHTO-T321 standard specifications. A constant sinusoidal loading was applied on beam specimens at constant strain levels of 600, 800, and 1000 microstrains until 50 percent reduction of initial stiffness.

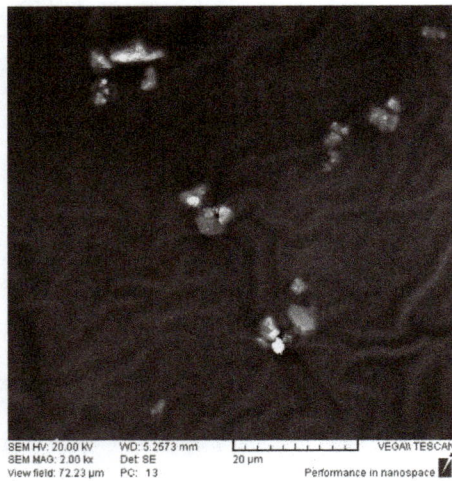

(a)

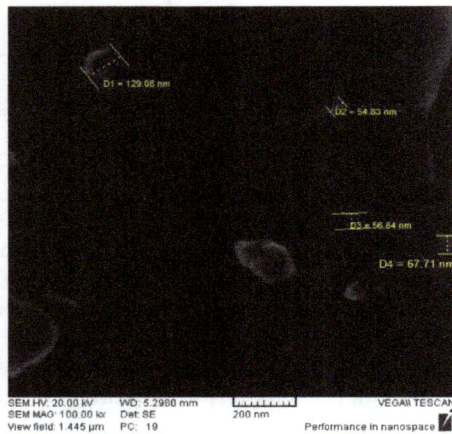

(b)

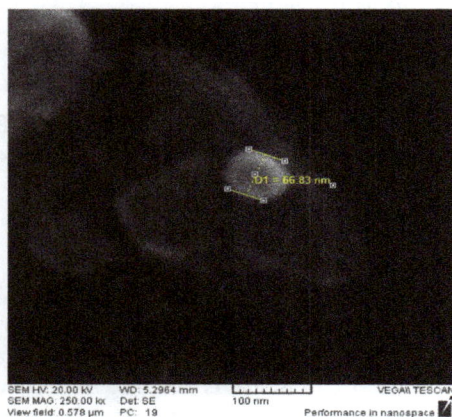

(c)

FIGURE 2: SEM images of NC with asphalt binder: (a) 20 μm, (b) 200 nm, and (c) 100 nm.

5. Results and Discussion

5.1. Binder Test

5.1.1. MSCR Test Results. The MSCR test encompasses different outputs. The accumulated strain versus time over

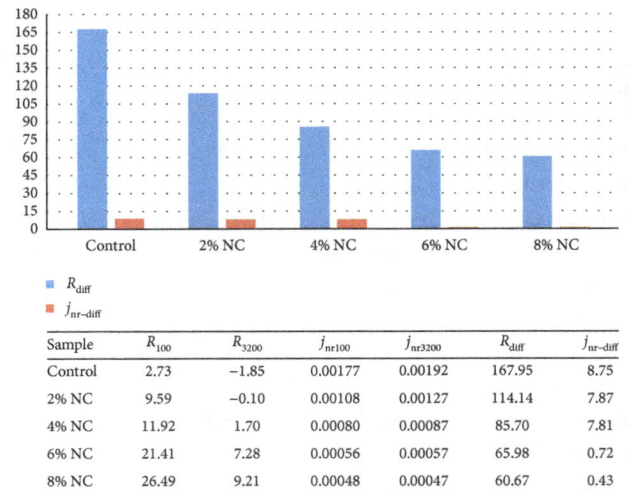

Sample	R_{100}	R_{3200}	j_{nr100}	j_{nr3200}	R_{diff}	$j_{nr-diff}$
Control	2.73	−1.85	0.00177	0.00192	167.95	8.75
2% NC	9.59	−0.10	0.00108	0.00127	114.14	7.87
4% NC	11.92	1.70	0.00080	0.00087	85.70	7.81
6% NC	21.41	7.28	0.00056	0.00057	65.98	0.72
8% NC	26.49	9.21	0.00048	0.00047	60.67	0.43

FIGURE 3: Summary of MSCR test results.

a 10-cycle period for the shear stress values of 100 Pa and 3200 Pa is shown in Figure 3. Results indicate a considerable decrease in accumulated strain for the NC-modified binders. An amount of 8% of NC decreases the accumulated strain of the base binder from 1.7% to 0.4% over a 10-cycle period under a 100 Pa shear stress. For the case of 3200 Pa applied shear load, the accumulated strain decreased from 62% to 14%. This decrease indicates a higher resistance of the modified binders against applied stresses and consequently decreases of binder deformation under cyclic loading.

The values of percent recovered strain (R) for base and modified binders are also shown in Figure 3. It can be seen that, by adding NC, there is a considerable increase in re-covered strain in the modified binders. Under a 100 Pa load, the recovered strain for base and modified binders with 8% of NC is 2.73 and 26.94, respectively. The recovered strain under 3200 Pa shear stress for the modified binder containing 8% of NC is 9.21, while the value for the base binder is less than zero. This negative R value is due to the fact that the binder has no recovered strain during the unloading process. Since any increase in percent recovered strain contributes to an increase in elastic response of the binder, adding NC causes an increase in recovered strain of the binder and improves the elastic response of it.

The permanent deformation characteristics of binders can be quantified by the j_{nr} parameter. A less value of j_{nr} indicates lower permanent deformation of the binder. The j_{nr} values under the 100 Pa and 3200 Pa shear loads for all base and modified binders are shown in Figure 3. It can be seen that adding NC can decrease the j_{nr} values, and as a result, the permanent deformation of the modified binders decreased. The stress sensitivity of the binder can be described by the $j_{nr-diff}$ and R_{diff} values. The less $j_{nr-diff}$ and R_{diff} values indicate less stress sensitivity, and the binder has similar behavior under the 100 Pa and 3200 Pa shear loads. The $j_{nr-diff}$ and R_{diff} values are also shown in Figure 3. The results indicate a considerable decrease in the $j_{nr-diff}$ and R_{diff} values. Therefore, adding NC can decrease the stress sensitivity of the binder.

TABLE 4: Linear amplitude sweep (LAS) test results.

Parameters	0% NC	2% NC	4% NC	6% NC	8% NC
A_{35}	33,200	49,334	57,990	134,765	146,928
B	−1.92	−2.09	−2.13	−2.35	−2.45

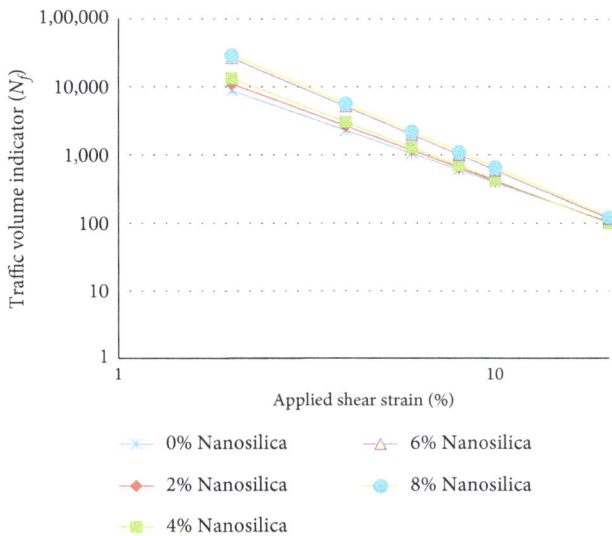

FIGURE 4: Fatigue models from the VECD analysis.

5.1.2. LAS Test Results. The results of linear amplitude sweep (LAS) tests are analyzed based on the theory of viscoelastic continuum damage. Utilizing this theory, A_{35} and B parameters should be calculated to assess the fatigue life of the asphalt binder. A higher value of A_{35} indicates better performance of the binder. Results showed that modified binders with NC had higher A_{35}. The A_{35} values for unmodified and modified binders are presented in Table 4.

The increase in the absolute value of B indicates that, by increasing the level of strain, the fatigue life will decrease in a higher rate. Moreover, any decrease in the absolute value of the parameter B indicates the reduction in the rate of fatigue life. The B values for unmodified and modified binders are presented in Table 4. Results show that adding NC can increase the binder sensitivity against loading level strain.

The trend of fatigue life of asphalt binders based on the VECD theory is depicted in Figure 4. Values in the Y-axis are the number of cycles to failure (N_f), which are indicative of traffic loads passing on the pavement, and the X-axis demonstrates applied shear strain. Results show that adding NC causes an increase in N_f values for lower shear strain levels. The N_f values for modified binders with 6 and 8 percent of NC are almost identical at low shear strain levels.

5.2. Mixture Performance Analysis

5.2.1. Marshall Stability, Flow, and Quotient Tests. Table 5 summarizes the engineering properties of the mixtures. The maximum increase of the Marshall stability, flow, and quotient is for the mixtures containing 8% of NC, which are improved by 31%, 12%, and 35%, respectively. The more the amount of Marshall quotient, the stronger the asphalt mixture is against permanent deformation [14].

5.2.2. Resilient Modulus Test. Resilient modulus (M_r) is an important factor in designing the pavement. Results of M_r show that, at 25°C temperature, the M_r value is 1.37 times greater than the base binder when the NC content is 8%, and at 5°C temperature, the M_r value is 1.24 greater than the unmodified binder when the NC content is 8%. Accordingly, NC improved fatigue resistance. Other research test results on polymer-modified asphalt showed that adding NC increases fatigue resistance at intermediate temperatures conducting the same test [18].

5.2.3. Dynamic Creep Test. Dynamic creep test results indicate that asphalt mixtures modified with NC have higher resistance against permanent deformation in comparison to control samples. Such an increase in NC can raise the flow number, which is a parameter of resistance against rutting. Increasing the amount of NC by 8% led to an almost 71% increase in the flow number. This finding is in a good agreement with other researchers [6, 18].

5.2.4. Four-Point Beam Fatigue Test. In the 4-point bending beam test, it was observed that, by increasing the percentage of NC, the variation trend of N_f in all three strain levels identically increased. Adding 8% of NC to the base binder increased the fatigue life of asphalt mixtures to the amount of 52%, 92%, and 65% under 600, 800, and 1000 microstrains, respectively.

Further details and discussions about engineering performance of NC-modified mixture are presented in another article by the same authors [19].

5.3. Regression Analysis. A series of linear regression models between binder test results, which are considered as independent, and mixture test results, which are considered as dependent, are developed. Linear regression is a statistical method that defines the relationship between two independent variables [20]. Summary of R-squared values and equations for each correlation is presented in Table 6. Independent values did not have strong correlation with each other, so it was possible to develop a reliable regression model and avoid the multicollinearity problem. Examples of two correlations of three independent variables are shown in Figure 5. On the other hand, the independent and dependent variables showed strong correlations with R-squared values close to 1. Furthermore, for each correlation, there is a fundamental relation between independent and dependent variables, which makes them comparable. For example, Nf and A35 correlation is investigated because both represent the fatigue performance; one for mixture and the other for binder. So, the same characteristics are correlated to see if it is possible to estimate one using the other. Also, MSCR test results are proved to have a good correlation with the rutting depth obtained from the field

TABLE 5: Summary of mixture test outcomes.

		Control	Mixture + 2% NC	Mixture + 4% NC	Mixture + 6% NC	Mixture + 8% NC
Average Marshall stability (KN)	—	10.23	11.30	12.10	12.73	13.40
Average Marshall flow (mm)	—	4.2	4.32	4.41	4.55	4.72
MQ (KN/mm)	—	2.44	2.62	2.74	2.80	2.84
N_f (4-point bending beam)	@600 $\mu\varepsilon$	106,092	120,477	141,428	151,623	158,685
	@800 $\mu\varepsilon$	76,738	85,541	113,843	139,781	147,428
	@1000 $\mu\varepsilon$	26,246	33,888	38,133	42,355	43,414
Resilient modulus	@25°C	3520.2	3707.5	3976.2	4420.6	4828.5
	@5°C	11,884.6	12,826.2	13,255.3	13,988.7	14,711.5

TABLE 6: Summary of correlation equations and regression coefficients.

Dependent variable (y)	Independent variable (x)	Equation	R^2
N_f @ 600	A_{35}	$y = 0.3763x + 103{,}883$	0.8097
N_f @ 800	A_{35}	$y = 0.5729x + 64{,}290$	0.9086
N_f @ 1000	A_{35}	$y = 0.1008x + 29{,}046$	0.8712
Dynamic creep	j_{nr100}	$y = -85{,}5191x + 3066.6$	0.9626
Dynamic creep	j_{nr3200}	$y = -760{,}429x + 3040$	0.9829
Marshall	j_{nr100}	$y = -309.13x + 2.978$	0.9915
Marshall	j_{nr3200}	$y = -272.86x + 2.9663$	0.9976
M_r at 5	R_{100}	$y = 114.28x + 11{,}676$	0.9816
M_r at 20	R_{3200}	$y = 55.483x + 3289.8$	0.9755

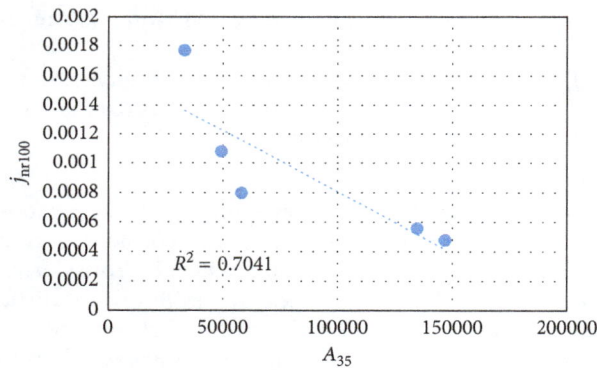

(a)

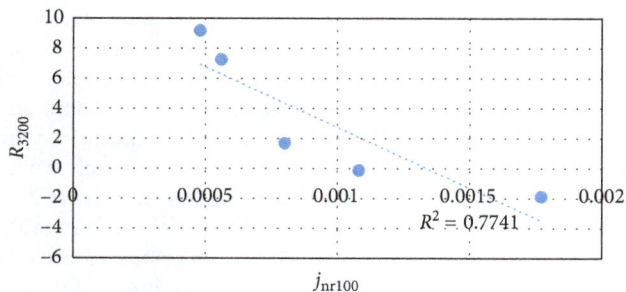

(b)

FIGURE 5: Correlation between independent vriables (a) A35 vs. Jnr100 and (b) Jnr100 vs. R3200.

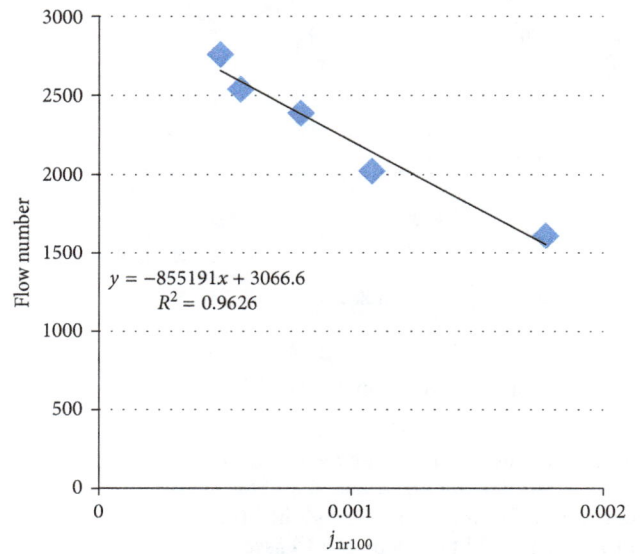

(a)

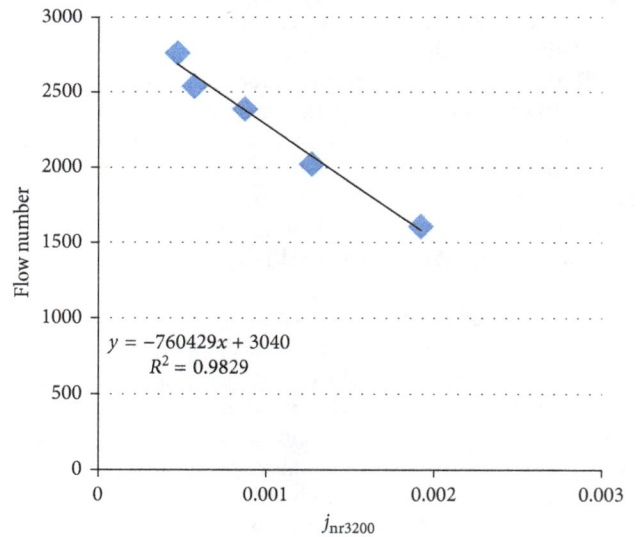

(b)

FIGURE 6: Flow number of mixture versus (a) j_{nr100} and (b) j_{nr3200}.

measurements [7]. Therefore, in this research, flow number, MQ, and M_r as parameters showing rutting susceptibility of mixture, are also correlated with MSCR test results.

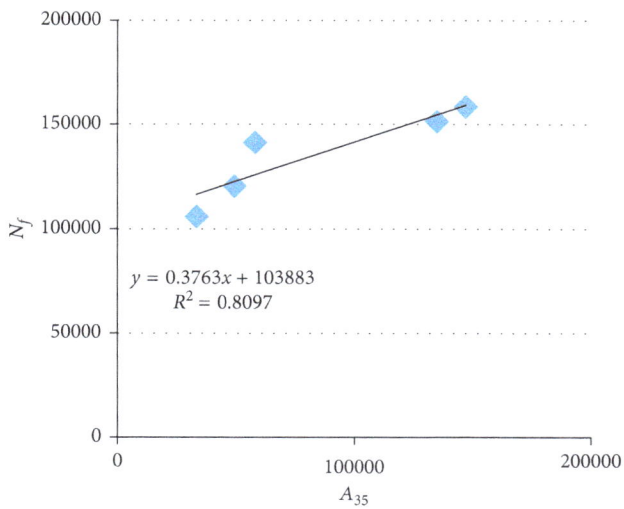

(a)

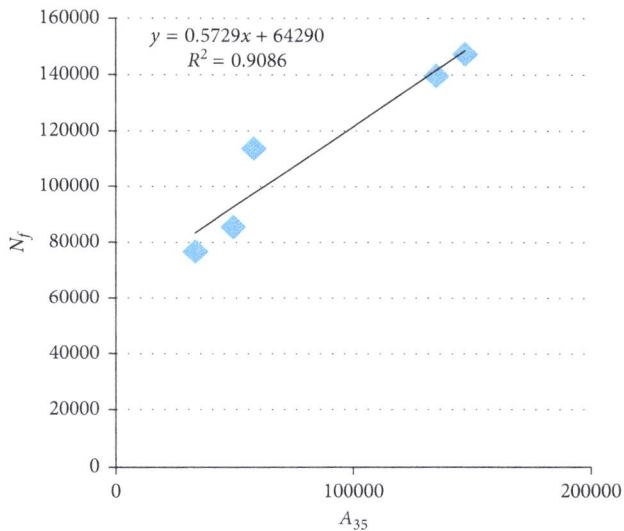

(b)

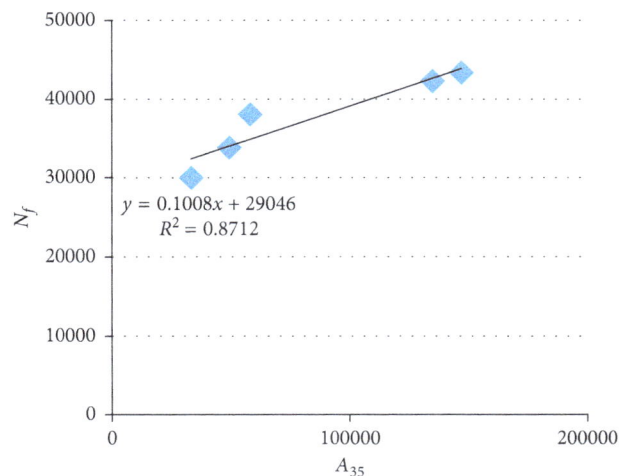

(c)

FIGURE 7: Fatigue life of mixture (N_f) at three strain levels: (a) 600 $\mu\varepsilon$, (b) 800 $\mu\varepsilon$, and (c) 1000 $\mu\varepsilon$ versus A_{35}.

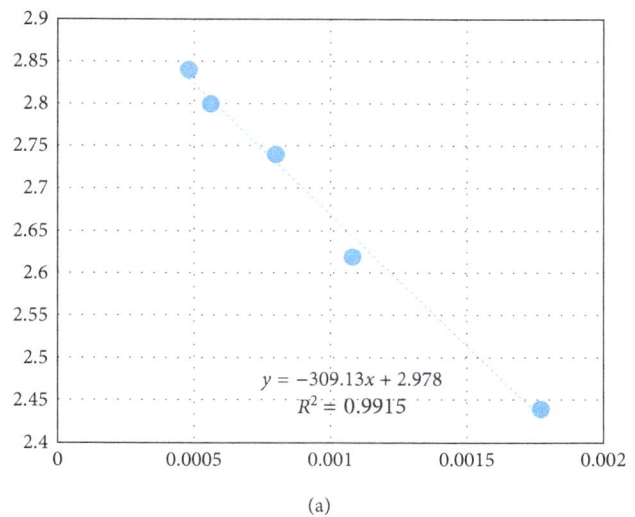

(a)

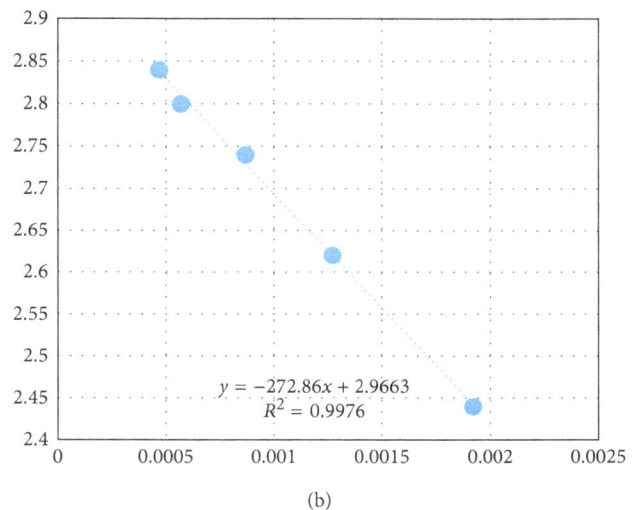

(b)

FIGURE 8: Marshall quotient versus (a) j_{nr100} and (b) j_{nr3200}.

5.4. Correlation between Flow Number and j_{nr}. Results of the flow number of mixture versus nonrecoverable compliance (j_{nr}) are depicted in Figure 6. It is observed that flow number and j_{nr} are inversely correlated, showing a good correlation (R^2 more than 0.9). This indicates that the MSCR binder test result has a close relation with the mixture's response, and using its data could give us a good estimation of the mixture's performance.

5.5. Correlation between A_{35} of LAS Test and Mixture N_f. Fatigue properties of the mixture are strongly correlated to those of binders. Therefore, modifying the binder could considerably alter fatigue behavior of the mixture, and the binder test could give us a good estimation of fatigue characteristics of the mixture [21]. In this research, results of the 4-point bending beam and LAS tests were studied in correlation with each other. According to Figure 7, the fatigue life obtained from the 4-point bending beam test is in a fairly well correlation with N_f values for the binder test.

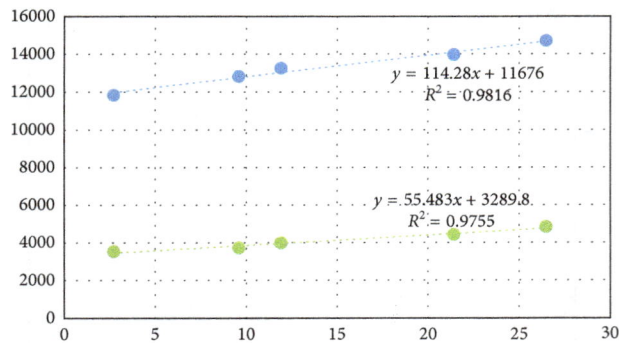

FIGURE 9: Resilient modulus at 5°C and 25°C versus R_{100}.

A_{35} parameter represents fatigue life of the asphalt binder subjected to a rate of 1% strain, which is the recommended value by the Superpave. Plotting A_{35} versus fatigue life of asphalt mixture shows a good relation between them. The best correlation was for the strain level of $800\,\mu\varepsilon$, and the lowest R^2 was for the strain level of $1000\,\mu\varepsilon$.

5.6. Correlation between Marshall Q and j_{nr}. The Marshall stiffness index (or Marshall quotient) represents the resistance of the material to shear stress and permanent deformation. Higher MQ means the mixture is stiffer and is more resistant against rutting. The permanent deformation characteristics of binders are quantified by the j_{nr} parameter. The lower value of j_{nr} indicates lower permanent deformation of the binder. Thus, the MQ correlation with unrecovered j_{nr} parameter makes sense as both are indicators of resistance against permanent deformation. As shown in Figure 8, it was observed that the MQ is inversely correlated with j_{nr}. This indicates that as the MQ increases, the unrecovered strain of the binder decreases, which means less susceptibility to rutting. R^2 is too close to 1 for five samples, which is a promising result so as to find the mixture performance using easier and faster binder tests.

5.7. Correlation between M_r *and* R_{100}. Resilient modulus is an indicator of a material's deflection behavior. In this research, resilient modulus at both 5°C and 25°C is correlated with R_{100} (recoverable deformation of the binder) obtained from the MSCR test. Results are shown in Figure 9. The correlation between resilient modulus at both 5°C and 25°C and R_{100}, had a regression coefficient more than 0.97.

6. Conclusion

(1) The results of the MSCR test indicate improvement of elastic properties of modified asphalt binders at a high temperature. They also show that NC not only increases rutting resistance in modified asphalt but also reduces its stress sensitivity. Results of the MSCR test are in line with the results obtained from the dynamic creep test, presenting a strong correlation between performance of the binder and mixture. Overall, it can be deduced that the binder

modified by NC contributes to resistance of the asphalt mixture against permanent deformations.

(2) The LAS test results indicate a considerable increase in the fatigue life of the NC-modified asphalt binder at low strain levels. On the other hand, the fatigue life of modified asphalt binders at high strain levels is a little less than that of the unmodified binders. Furthermore, the results of the four-point bending beam test demonstrate that adding NC up to 8 percent by binder content could considerably improve fatigue resistance, which confirms the results of the LAS test.

(3) According to the correlation of the flow number, MQ, and M_r with the results of the MSCR test, it could be concluded that the MSCR test results correlate very well with mixture test results related to rutting performance.

(4) According to a fairly well correlation between binder and mixture test outputs obtained in this research, it is promising to develop a phenomenological relation between characteristics of the binder and mixture using more extensive data and considering important parameters, in future research, so more time and materials are saved using binder tests instead of the mixture tests.

Conflicts of Interest

The authors declare that there are no conflicts of interest regarding the publication of this paper.

References

[1] M. Baqersad, A. Hamedi, M. Mohammadafzali, and H. Ali, "Asphalt mixture segregation detection: digital image processing approach," *Advances in Materials Science and Engineering*, vol. 2017, Article ID 9493408, 6 pages, 2017.

[2] C. Fang, R. Yu, S. Liu, and Y. Li, "Nanomaterials applied in asphalt modification: a review," *Journal of Materials Science & Technology*, vol. 29, no. 7, pp. 589–594, 2013.

[3] A. Kavussi and P. Barghabany, "Investigating fatigue behavior of nanoclay and nano hydrated lime modified bitumen using LAS test," *Journal of Materials in Civil Engineering*, vol. 28, no. 3, p. 04015136, 2015.

[4] M. Van de Ven, A. Molenaar, and J. Besamusca, "Nanoclay for binder modification of asphalt mixtures," in *Proceedings of the 7th International RILEM Symposium on Advanced Testing and Characterization of Bituminous Materials*, Rhodes, Greece, May 2009.

[5] A. N. Amirkhanian, F. Xiao, and S. N. Amirkhanian, "Characterization of unaged asphalt binder modified with carbon nano particles," *International Journal of Pavement Research and Technology*, vol. 4, no. 5, pp. 281–286, 2011.

[6] H. Yao, Z. You, L. Li et al., "Properties and chemical bonding of asphalt and asphalt mixtures modified with nanosilica," *Journal of Materials in Civil Engineering*, vol. 25, 2012.

[7] J. A. D'Angelo, "The relationship of the MSCR test to rutting," *Road Materials and Pavement Design*, vol. 10, no. 1, pp. 61–80, 2009.

[8] M. Mohammadafzali, H. Ali, J. Musselman, G. Sholar, S. Kim, and T. Nash, "Long-term aging of recycled asphalt binders:

a laboratory evaluation based on performance grade tests," in *Proceedings of the Airfield and Highway Pavements 2015*, pp. 617–627, Miami, FL, USA, June 2015.

[9] F. Zhou, W. Mogawer, H. Li, A. Andriescu, and A. Copeland, "Evaluation of fatigue tests for characterizing asphalt binders," *Journal of Materials in Civil Engineering*, vol. 25, no. 5, pp. 610–617, 2012.

[10] AASHTO-T 32, *Determining the fatigue life of compacted hot mix asphalt (HMA) subjected test repeated flexural bending*, American Association of State Highway & Transportation Officials, Washington, DC, USA, 2007.

[11] ASTM D7405-10, *Standard Test Method for Multiple Stress Creep and Recovery (MSCR) of Asphalt Binder Using a Dynamic Shear Rheometer*, ASTM International, West Conshohocken, PA, USA, 2010.

[12] C. M. Johnson, *Estimating Asphalt Binder Fatigue Resistance Using an Accelerated Test Method*, University of Wisconsin–Madison, Madison, WI, USA, 2010.

[13] C. Hintz, *Understanding Mechanisms Leading to Asphalt Binder Fatigue*, University of Wisconsin-Madison, Madison, WI, USA, 2012.

[14] A. Behnood and M. Ameri, "Experimental investigation of stone matrix asphalt mixtures containing steel slag," *Scientia Iranica*, vol. 19, no. 5, pp. 1214–1219, 2012.

[15] ASTM D4123-82, *Standard Test Method for Indirect Tension Test for Resilient Modulus of Bituminous Mixtures (Withdrawn 2003)*, ASTM International, West Conshohocken, PA, USA, 1995.

[16] Y. Huang, *Pavement Analysis and Design*, Peason Education, Inc., USA, 2004.

[17] K. E. Kaloush, M. W. Witczak, G. B. Way, A. Zborowski, M. Abojaradeh, and A. Sotil, *Performance Evaluation of Arizona Asphalt Rubber Mixtures Using Advanced Dynamic Material Characterization Tests. Final Report*, FNF Construction, Inc. and the Arizona Department of Transportation, Arizona State University, Tempe, AZ, USA, 2002.

[18] N. I. M. Yusoff, A. A. S. Breem, H. N. M. Alattug, A. Hamim, and J. Ahmad, "The effects of moisture susceptibility and ageing conditions on nano-silica/polymer-modified asphalt mixtures," *Construction and Building Materials*, vol. 72, pp. 139–147, 2014.

[19] M. Hasaninia and F. Haddadi, "The characteristics of hot mixed asphalt modified by nanosilica," *Petroleum Science and Technology*, vol. 35, no. 4, pp. 351–359, 2017.

[20] A. Massahi, H. Ali, F. Koohifar, and M. Mohammadafzali, "Analysis of pavement raveling using smartphone," in *Transportation Research Board 95th Annual Meeting (No. 16-6155)*, January 2016.

[21] M. Ameri, S. Nowbakht, M. Molayem, and M. H. Mirabimoghaddam, "A study on fatigue modeling of hot mix asphalt mixtures based on the viscoelastic continuum damage properties of asphalt binder," *Construction and Building Materials*, vol. 106, pp. 243–252, 2016.

9

Quantitative Analysis on Force Chain of Asphalt Mixture under Haversine Loading

Mingfeng Chang,[1] Pingming Huang,[2] Jianzhong Pei,[2] Jiupeng Zhang,[2] and Binhui Zheng[1]

[1]*School of Materials Science and Engineering, Chang'an University, Shaanxi, Xi'an 710061, China*
[2]*School of Highway, Chang'an University, Shaanxi, Xi'an 710064, China*

Correspondence should be addressed to Mingfeng Chang; mfchang99@126.com

Academic Editor: Meor Othman Hamzah

AC-13 asphalt mixture was taken as the research object to investigate the evolution and distribution laws of force chains. A digital specimen of AC-13 asphalt mixture was reconstructed using the discrete element method (DEM) to simulate the simple performance test (SPT). Next, the force chain information among aggregate particles was extracted to analyze the evolution, probability distribution, and angle distribution of force chains. The results indicate that the AC-13 mesoscopic model reconstructed using the DEM is feasible to simulate the mesoscopic mechanical properties of asphalt mixture by comparing the predicted results and laboratory test results. The spatial distributions of force chains are anisotropic. The probability distributions of normal force chains varying with the loading times are consistent. Furthermore, the probability distribution has the maximum value at the minimum f (the ratio of contact force to mean contact force); the peak value appears again at $f = 1.75$ and then gradually decreases and tends to be stable. In addition, the angle distributions of force chains mainly locate near 90° and 270°, and the proportions of strong force chains are slightly greater than 50%, but the maximum proportion is only 51.12%.

1. Introduction

Asphalt mixture is a multiphase composite material, which comprises aggregates, asphalt binder, and air voids. Aggregates that dominate by volume and mass in asphalt mixture are considered as the granular media, which are the main material to form a skeleton structure and resist external loading applied to the asphalt mixture. Therefore, the asphalt mixture exhibits the essential attribute of granular media in a certain sense. And there exists the interaction force among the contacting particles that forms the loading transmission paths in a granular system, so Bouchaud et al. proposed the concept of force chain after observing stress distribution, propagation, and arching within granular media [1, 2]. Currently, the research methods on force chains focus on the following areas: mechanical analysis, laboratory test, and numerical simulation.

Cates et al. argued that fragility was linked to the marginal stability of force chains within granular matter and described that the force chains propagated and fluctuated with scalar model and tensorial model in granular media [3–5]. Tordesillas et al. viewed the force chain buckling in a constrained granular medium from the structural mechanics, investigated the force chain evolution for cohesionless granular systems, and presented a regularized two-dimensional model for the force chain buckling [6–8]. Howell et al. carried out the experiments on a granular system consisting of photoelastic polymer disks subjected to shear and compressive loading and measured and calculated the probability distributions of normal and tangential force chains [9, 10]. Yi et al. studied the force chain transmission in a hexagonal-close-packed array granular system with and without point defect submitted to a concentrated loading using the aluminum-plastic board and carbon paper technique [11, 12]. Tordesillas et al. examined the relationship between force cycles and force chain in a dense granular material under quasistatic biaxial loading using a discrete element simulation and analyzed the spatial patterns of force chain buckling with multiscale characterization [13, 14]. Sun et al. proposed the criteria of strong force chain and angle and analyzed

TABLE 1: Aggregate properties.

Index	Unit	Technical requirements	Measured results
Crushed stone value	%	≤26	18.5
Los Angeles abrasion loss	%	≤28	16.2
Bulk density			
13.2 mm		—	2.785
9.5 mm	g/cm^3	—	2.798
4.75 mm		—	2.783
Water absorption	%	≤2.0	0.57
Silt content	%	≤1	0.9
Flat and elongated particle content	%	≤15	9.8

TABLE 2: Asphalt binder properties of SK 70[#].

Index	Unit	Technical requirements	Measured results
Penetration (25°C, 5 s, 100 g)	0.1 mm	60~80	72
Penetration index	—	−1.5~+1.0	−0.94
Ductility (15°C)	cm	≥100	>150
Softening point	°C	≥46	48.2
Solubility (trichloroethylene)	%	99.5	99.92
Paraffin content	%	2.2	1.72
Residue after thin film oven (163°C, 5 h)			
Mass loss	%	≤±0.8	0.1
Penetration ratio (25°C)	%	≥61	70.8
Ductility (15°C)	cm	≥15	15

the evolution of force chain structure morphology through simulating the uniaxial compression and biaxial test using the discrete element method [15, 16]. You et al. reconstructed the mesoscopic model of asphalt mixture and provided the force chain distributions within aggregates, within mastic, and between aggregates and mastic [17, 18].

It can be found that the researches on force chains concentrate on the granular material with regular morphology, single particle size, and smooth edges by the literature cited above. But few of them are involved in bond effect. However, the asphalt mixture comprises the aggregates with different morphology, geometric shapes, and broad particle size, and the asphalt binder with bond effect. The interaction between aggregates and asphalt binders changes with time and temperature that makes the anisotropic interaction more complex. Furthermore, its spatial and temporal evolution and distribution laws of force chains have not yet reported.

2. Determination of DEM Viscoelastic Mesoscopic Parameters

The SPT dynamic modulus test was developed to measure the dynamic moduli and phase angles of AC-13 asphalt mixture and mastic specimens. Burger's model parameters were calculated based on the test results, and the mesoscopic parameter values were determined by verifying the corresponding relationship between Burger's model parameters

and DEM mesoscopic parameters. Aggregates are limestone, asphalt binder is SK70[#], and their properties are listed in Tables 1 and 2. The asphalt mastic was obtained from AC-13 asphalt mixture's aggregate gradation by eliminating all the aggregates bigger than 2.36 mm except the asphalt binder. The gradations of AC-13 asphalt mixture and mastic are listed in Table 3. Optimum asphalt contents of AC-13 asphalt mixture and mastic are 4.49% and 11.92%, respectively.

The dynamic moduli and phase angles of asphalt mixture and mastic were measured at three temperatures of 5°C, 20°C, and 40°C and five frequencies of 1 Hz, 5 Hz, 10 Hz, 20 Hz, and 25 Hz. 5°C and 20 Hz were selected to analyze the mesoscopic responses of asphalt mixture and the characteristics of force chains. And the test results were used to calculate the mesoscopic parameters of asphalt mixture viscoelastic model. The mesoscopic parameters include stiffness within aggregates and adjacent aggregates, stiffness and viscosities within asphalt mastic, and stiffness and viscosities adjacent to asphalt mastic as well as stiffness and viscosities between aggregates and mastic.

3. Reconstruction of Digital Specimen and Verification of Mesoscopic Model

The asphalt mixture specimen prepared by gyratory compaction was drilled core to obtain a new specimen with diameter of 100 mm, which was cut along the diameter direction, and then a cross-sectional image (as shown in

TABLE 3: Gradations of AC-13 asphalt mixture and mastic.

Size (mm)	Percent passing of AC-13 (%)	Percent passing of asphalt mastic (%)
16.0	100	—
13.2	95.05	—
9.5	72.23	—
4.75	46.26	—
2.36	33.13	100
1.18	24.09	72.71
0.6	17.01	51.34
0.3	11.83	35.7
0.15	8.14	24.56
<0.075	5.09	15.35

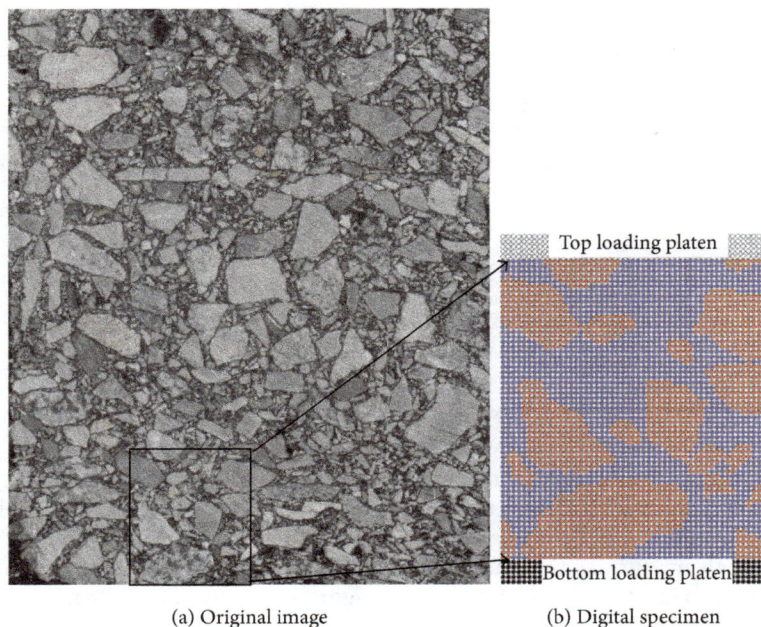

(a) Original image   (b) Digital specimen

FIGURE 1: Reconstruction of digital specimen.

Figure 1(a)) was received. The image was processed using the digital image processing technique (Image-Pro Plus 6.0, IPP 6.0), its pixel coordinates were extracted and the DEM was imported to generate a digital specimen of asphalt mixture combining Matlab. The whole digital specimen has 50,500 particles, and so many particles affect the computational efficiency. Consequently, this paper captured part of the digital specimen to conduct a numerical simulation, and the digital specimen with 3060 particles was demonstrated in Figure 1(b). The red parts represent the aggregates, the blue parts represent the asphalt mastic, and the upper and lower parts represent the loading platens. The Haversine loading is applied by the top loading platen and the bottom loading platen is served as a fixed base, which are consistent with the SPT laboratory test.

It is necessary to verify the developed DEM model's reasonability and the consistency between simulation and laboratory results. The frequency applied by the top loading platen was 20 Hz, and the time-step was 1.285×10^{-8} s/step.

The stresses and strains were extracted after running for 0.1 s (two periods), and the stress and strain curves were plotted in Figure 2. Dynamic moduli and phase angles from the DEM simulation were calculated by

$$|E^*| = \frac{\sigma_{max} - \sigma_{min}}{\varepsilon_{max} - \varepsilon_{min}},$$

$$\Phi = \frac{\Delta t}{T},$$

(1)

where σ_{max} and σ_{min} are the maximum and minimum values of stresses; ε_{max} and ε_{min} are the corresponding maximum and minimum strains; Δt is the time difference between the adjacent peak stress and strain; and T is loading period and 20 Hz corresponds to the loading period of 0.05 s.

The predicted dynamic modulus and phase angle were calculated according to the stresses and strains and (1). It is found that the difference of dynamic modulus between measured and predicted results is 0.784 GPa, and its error

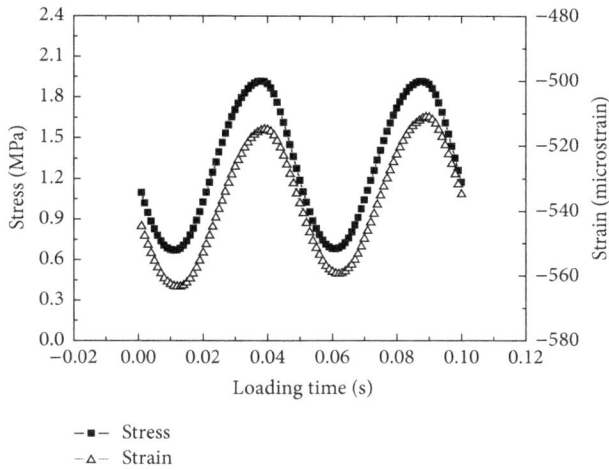

FIGURE 2: Applied stress and strain responses at 20 Hz.

is 2.9%. Besides, the difference of phase angle between measured and predicted results is 1.8°. Therefore, there are slight differences between measured and predicted dynamic moduli and phase angles. The reconstructed DEM model has the ability to analyze the mesoscopic responses of AC-13 asphalt mixture.

4. Results and Discussion

4.1. Force Chain Evolution.

There were produced mesoscopic responses in aggregates, in mastic, and at the interfaces of aggregates/mastic after applying the Haversine loading at 20 Hz for the AC-13 digital specimen. The anisotropic force chain networks directly reflected the mesoscopic responses within the asphalt mixture and the force chain evolution internal aggregates and internal mastic and at the interfaces of aggregates/mastic. The force chain networks were traced at four loading times (0.125×10^{-3} s, 0.25×10^{-3} s, 0.375×10^{-3} s, and 0.5×10^{-3} s, which corresponded to 9728 steps, 19455 steps, 29184 steps, and 38910 steps) and were described in Figure 3. The blue parts represent compressive force chain networks and the red parts represent tensile force chain networks.

Figure 3 shows that the spatial distributions of force chains are anisotropic under the Haversine loading. The thicker lines represent the greater force chains, which are mainly the compressive force chains in vertical direction and the main loading transmission paths. Moreover, the force chains that transfer the smaller loading in the other directions are the tensile force chains and these smaller force chain networks distribute widely. The maximum contact forces are 2.970×10^4 N, 3.332×10^4 N, 3.611×10^4 N, and 3.958×10^4 N at four loading times indicating that their acting forces among particles enhance as the loading time grows. This is attributed to the sensitivity of particles inside the specimen to the Haversine loading while it transfers from top to bottom. The interlocked action among particles strengthens, and the particle system adapts to the loading by means

of self-organized behaviors. In addition, the force chains within aggregates are all the compressive force chains (blue parts), and the tensile force chains (red parts) are all within mastic and at the interfaces of aggregate/mastic. The reason is that the unbroken aggregates form the skeleton structure subjected to and transferring the Haversine compressive forces, while the Haversine loading produces the stripping trend between asphalt and aggregates within mastic and at the interfaces of aggregate/mastic.

4.2. Probability Distribution of Force Chains.

It is difficult to excavate the quantitative distribution information of force chain networks due to their spatial and temporal heterogeneity. Currently, the macroscopic statistical characteristics of force chains are the main way to describe the mesoscopic information of force chain networks. Probability distribution of force chains $P(f)$ is conducted as a macroscopic quantitative indicator, where f is the ratio of contact force to mean contact force. The probability distributions of normal force chains is provided in Figure 4 at four loading times.

As can be seen in Figure 4, the probability distributions of normal force chains $P(f)$ have no significant differences at four loading times and the maximum probability distribution ($P(f) = 0.2495$) at 0.5×10^{-3} s is greater than the other three loading times. $P(f)$ decreases with the increase of f when $f \leq 0.75$; $P(f)$ is proportional to f when $0.75 < f \leq 1.75$, and the peak occurs at $f = 1.75$; $P(f)$ is inversely proportional to f when $f > 1.75$ and it remains unchanged basically when $f \geq 3.0$. The variation laws of $P(f)$ illustrate that the probability distributions of smaller normal compressive and tensile force chains are greater, and the probability distribution of larger normal compressive force chains is the largest at $f = 1.75$.

4.3. Angle Distribution of Force Chains.

Force chains possess great sensitivity and uncertainty due to their anisotropic spatial extension directions depending on the magnitude and time of external loading. The angles between the force chain direction and horizontal direction are served as the angle distribution quantitative indicators. A statistical analysis is employed to obtain the angle distribution characteristics of force chains in $0°$–$360°$ (as shown in Figure 5) by extracting the normal and shear force chains.

Figure 5 presents that these angle distributions of force chains mainly reside nearby $90°$ and $270°$; the angle distribution proportions in $0°$–$180°$ are significantly greater than those in $180°$–$360°$, which relate to the Haversine loading directions. The asphalt mixture specimen is applied at the vertical Haversine loading that mainly transmits along the vertical direction. But, the asphalt mixture as a composite material comprises aggregate particles with different particle sizes and geometric shapes, so that the angle distributions mainly distribute along the vertical direction while they extend along the other directions.

Force chains can be distinguished into strong and weak force chains according to the force chain thickness and the magnitude of f. They are the strong force chains when $f \geq 1$, while they are the weak force chains when $f < 1$. It is necessary to analyze the proportions of strong force chains at

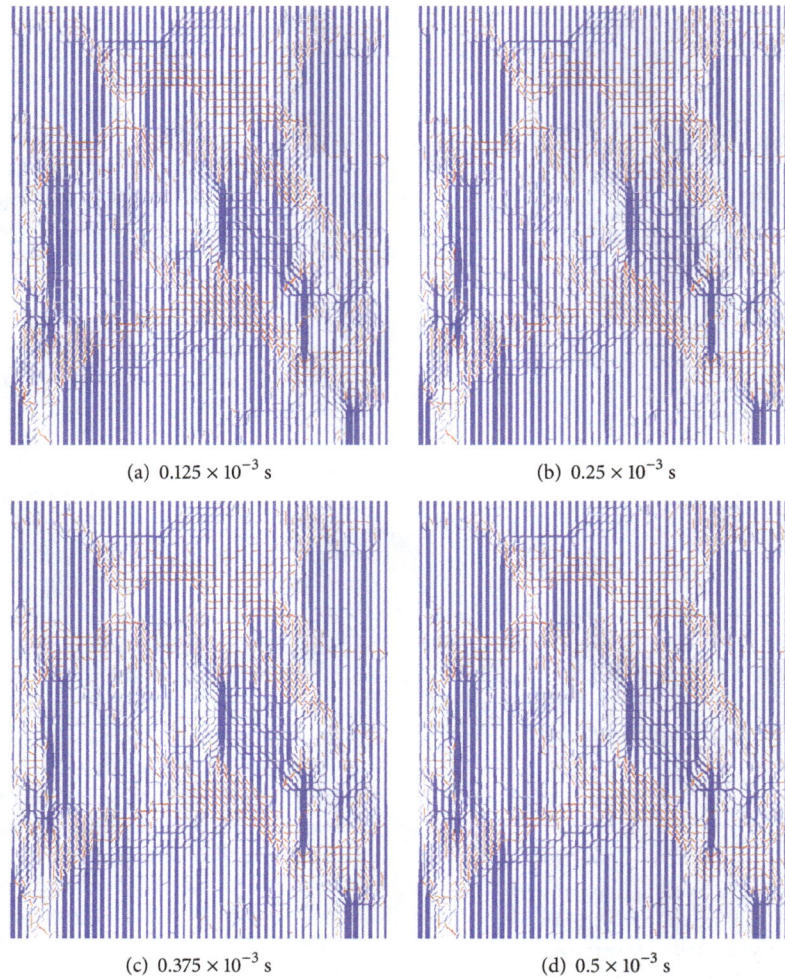

(a) 0.125×10^{-3} s

(b) 0.25×10^{-3} s

(c) 0.375×10^{-3} s

(d) 0.5×10^{-3} s

FIGURE 3: Force chain spatial evolution with loading times.

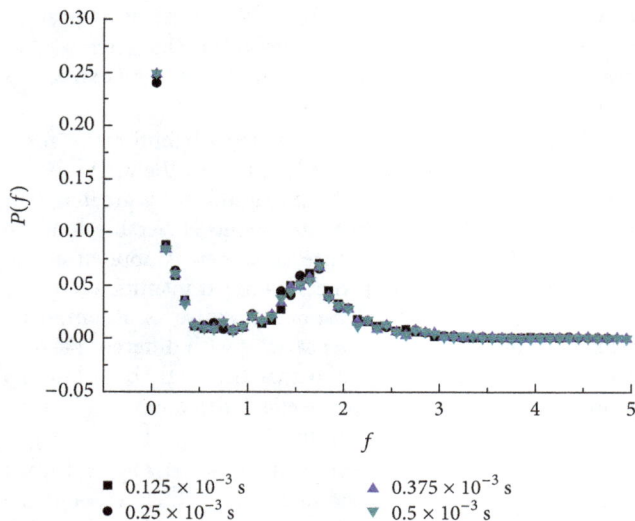

- 0.125×10^{-3} s
- 0.25×10^{-3} s
- 0.375×10^{-3} s
- 0.5×10^{-3} s

FIGURE 4: Probability distribution of normal force chains.

different loading times (shown in Figure 6) due to the strong force chains bearing most of the loading and as the main loading transmission paths.

There are no significant differences in the proportions of strong force chains in Figure 6. The maximum proportion of strong force chains is 51.12% and the minimum proportion of strong force chains is 50.68%, so the maximum difference is 0.44%. The results mentioned in Figure 6 illustrate that the proportions of strong force chains have no changes with the increase of loading time basically after the loading transmits the whole specimen. The strong force chains support most of external loading and become the main parts used to ensure the stability of the whole granular system. Therefore, the interlocked effect from aggregates, especially coarse aggregates, constitutes the skeleton structure, and the aggregates transfer larger loading to form the strong force chain network structure in AC-13 asphalt mixture. However, the fine aggregates and mineral powder particles play a role of filling effect, which are subjected to smaller or zero loading to form the weak force chain network structure.

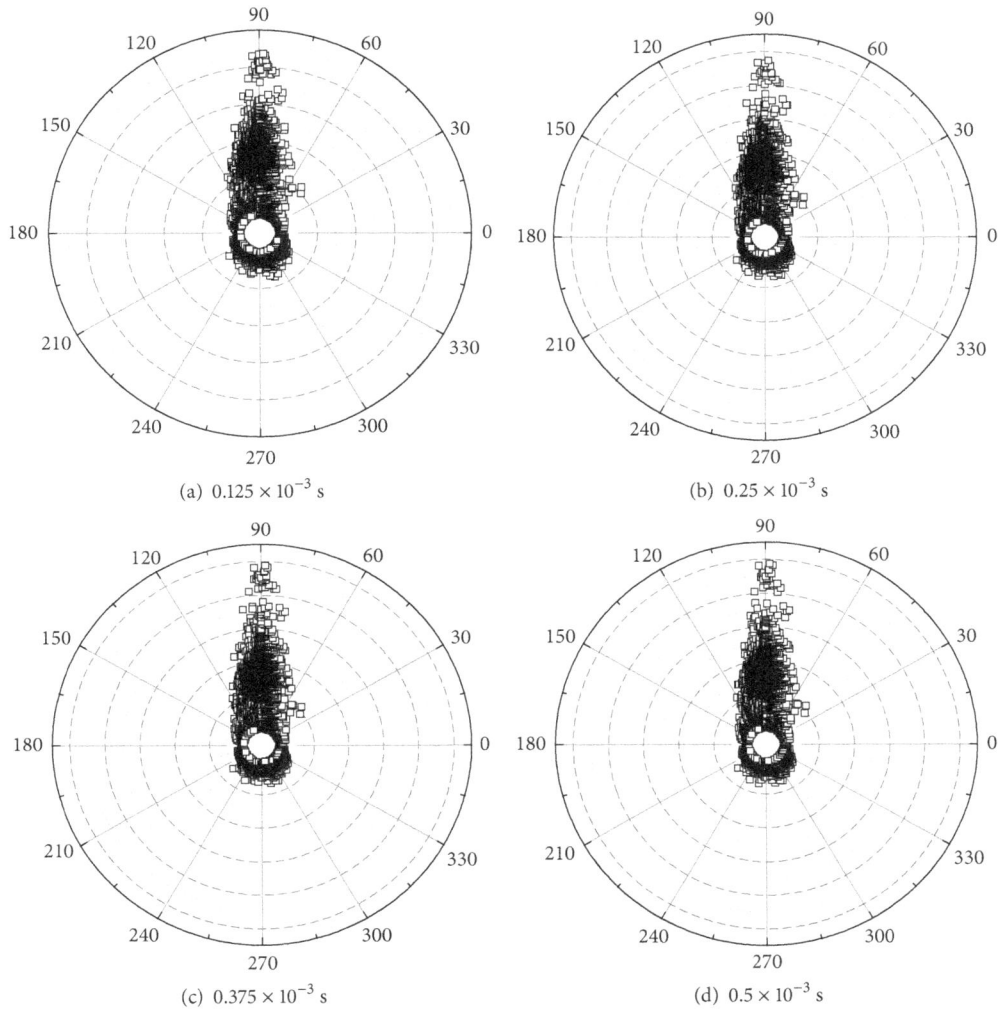

(a) 0.125×10^{-3} s

(b) 0.25×10^{-3} s

(c) 0.375×10^{-3} s

(d) 0.5×10^{-3} s

FIGURE 5: Angle distribution of force chains.

5. Conclusions

The differences between the measured and predicted dynamic modulus with the error of 2.9% and phase angle of 1.8° show that the developed viscoelastic model using the discrete element method is feasible. It is suitable to investigate mesoscopic responses of the AC-13 asphalt mixture.

Force chains have a heterogeneous distribution in asphalt mixture. The acting forces among the contacting particles enhance as the loading time extends, the force chains within aggregates are all the compressive force chains, and the force chains within mastic and at the interfaces of aggregate/mastic are mainly the tensile force chains.

The probability distributions of normal force chains are basically consistent for different f. The probability distributions decrease firstly and then increase. They decrease after reaching the peak at $f = 1.75$ and then stabilize.

The angle distributions of force chains mainly reside nearby 90° and 270°, and the angles distributions in 0°–180° are significantly greater than those in 180°–360°. The proportions of strong force chains are slightly greater than 50%, but the maximum proportion is only 51.12%.

In this paper, we conduct the quantitative analysis of force chains in asphalt mixture under the conditions of 5°C and 20 Hz. In the further study, the evolution and distribution of force chains with different skeleton types can be analyzed under different temperatures and frequencies.

Conflicts of Interest

The authors declare that there are no conflicts of interest regarding the publication of this paper.

Acknowledgments

This work was supported by the National Natural Science Foundation of China (nos. 51408047, 51378073, and 51408043) and the Special Fund for Basic Scientific Research

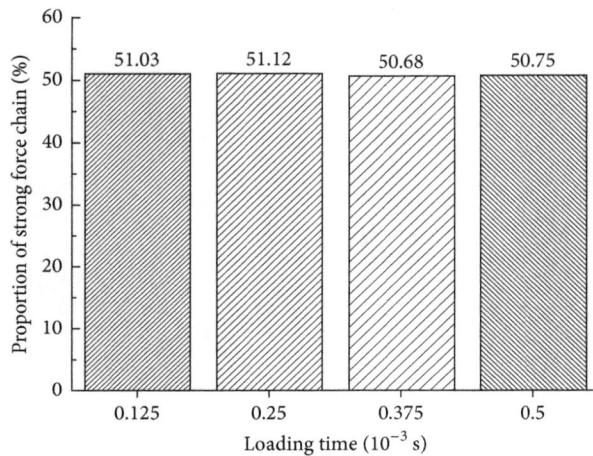

FIGURE 6: Proportion of strong force chain at four loading times.

of Central College of Chang'an University (nos. 310831161002, 310821153502, and 310821152003). The authors gratefully acknowledge their financial support.

References

[1] J.-P. Bouchaud, M. E. Cates, and P. Claudin, "Stress distribution in granular media and nonlinear wave equation," *Journal de Physique I*, vol. 5, no. 6, pp. 639–656, 1995.

[2] J. P. Wittmer, M. E. Cates, and P. Claudin, "Stress propagation and arching in static sandpiles," *Journal de Physique I*, vol. 7, no. 1, pp. 39–80, 1997.

[3] M. E. Cates, J. P. Wittmer, J.-P. Bouchaud, and P. Claudin, "Development of stresses in cohesionless poured sand," *Philosophical Transactions of the Royal Society A: Mathematical, Physical and Engineering Sciences*, vol. 356, no. 1747, pp. 2535–2560, 1998.

[4] M. E. Cates, J. P. Wittmer, J.-P. Bouchaud, and P. Claudin, "Jamming, force chains, and fragile matter," *Physical Review Letters*, vol. 81, no. 9, pp. 1841–1844, 1998.

[5] P. Claudin, J. Bouchaud, M. E. Cates, and J. P. Wittmer, "Models of stress fluctuations in granular media," *Physical Review E*, vol. 57, no. 4, pp. 4441–4457, 1998.

[6] A. Tordesillas, J. Shi, and H. B. Muhlhaus, "Noncoaxiality and force chain evolution," *International Journal of Engineering Science*, vol. 47, no. 11-12, pp. 1386–1404, 2009.

[7] G. W. Hunt, A. Tordesillas, S. C. Green, and J. Shi, "Force-chain buckling in granular media: a structural mechanics perspective," *Philosophical Transactions of the Royal Society A: Mathematical, Physical and Engineering Sciences*, vol. 368, no. 1910, pp. 249–262, 2010.

[8] G. W. Hunt and J. Hammond, "Mechanics of shear banding in a regularized two-dimensional model of a granular medium," *Philosophical Magazine*, vol. 92, no. 28-30, pp. 3483–3500, 2012.

[9] D. Howell, R. P. Behringer, and C. Veje, "Stress fluctuations in a 2D granular Couette experiment: a continuous transition," *Physical Review Letters*, vol. 82, no. 26, pp. 5241–5244, 1999.

[10] T. S. Majmudar and R. P. Behringer, "Contact force measurements and stress-induced anisotropy in granular materials," *Nature*, vol. 435, no. 7045, pp. 1079–1082, 2005.

[11] C.-H. Yi, Y. Liu, T.-D. Miao, Q.-S. Mu, and Y.-L. Qi, "Force transmission in three-dimensional hexagonal-close-packed granular arrays with point defect submitted to a point load," *Granular Matter*, vol. 9, no. 3-4, pp. 195–203, 2007.

[12] T.-D. Miao, C.-H. Yi, Y.-L. Qi, Q.-S. Mu, and Y. Liu, "Force transmission in three-dimensional hexagonal-close-packed granular arrays submitted to a point load," *Acta Physica Sinica*, vol. 56, no. 8, pp. 4713–4721, 2007.

[13] A. Tordesillas, J. Shi, and T. Tshaikiwsky, "Stress-dilatancy and force chain evolution," *International Journal for Numerical and Analytical Methods in Geomechanics*, vol. 35, no. 2, pp. 264–292, 2011.

[14] A. Tordesillas, S. Pucilowski, L. Sibille, F. Nicot, and F. Darve, "Multiscale characterisation of diffuse granular failure," *Philosophical Magazine*, vol. 92, no. 36, pp. 4547–4587, 2012.

[15] Q. C. Sun, F. Jin, G. Q. Wang, and G. H. Zhang, "Force chains in a uniaxially compressed static granular matter in 2D," *Acta Physica Sinica*, vol. 59, no. 1, pp. 30–37, 2010.

[16] Z. W. Bi, Q. C. Sun, J. G. Liu, F. Jin, and C. H. Zhang, "Development of shear band in a granular material in biaxial tests," *Acta Physica Sinica*, vol. 60, no. 3, Article ID 034502, 2011.

[17] Z. You and W. G. Buttlar, "Micromechanical modeling approach to predict compressive dynamic moduli of asphalt mixtures using the distinct element method," *Transportation Research Record*, no. 1970, pp. 73–83, 2006.

[18] Y. Liu and Z. You, "Visualization and simulation of asphalt concrete with randomly generated three-dimensional models," *Journal of Computing in Civil Engineering*, vol. 23, no. 6, pp. 340–347, 2009.

10

Investigation of Cement-Emulsified Asphalt in Plastic Concrete

Xiaohu Yan,[1,2] **Zaiqin Wang** [1,2] **Meijuan Rao** [3] **and Mingxia Li**[2]

[1]*College of Water Conservancy and Hydropower Engineering, Hohai University, Nanjing 210098, China*
[2]*Yangtze River Scientific Research Institute, Wuhan 430010, China*
[3]*State Key Laboratory of Silicate Materials for Architecture, Wuhan University of Technology, Wuhan 430070, China*

Correspondence should be addressed to Meijuan Rao; raomeijuanding@163.com

Academic Editor: Lijing Wang

The mechanical, mesodamage, and the microproperties of cement-emulsified asphalt concrete have been investigated by computed tomography (CT), scanning electron microscopy (SEM), X-ray diffraction (XRD), and thermogravimetric analysis (TG) in this work. Emulsified asphalt delayed the hydration of cement, making the early compressive strength of concrete develop slowly. However, the concrete compressive strength increased rapidly with the demulsification of emulsified asphalt. The damage stages of condense, expansion of volume, rapid crack propagation, and damage by real-time scanning have been observed. The CT mean value of the place near the lower end face suffered a larger decline but a smaller decline to the upper part of the sample. The evolution of concrete suffering damage to failure is a gradual development process, and no sharp expansion of brittle failure. The unhydrated cement, incorporation asphalt, fibrous C–S–H gel, CH, needle-shaped ettringite, and other hydration products were interwoven to constitute emulsified asphalt-cement paste, forming a spatial structure.

1. Introduction

With the development of social productive forces, there is a gradual improvement in the quality of material used for constructing buildings. However, there are still some limitations in the building materials, leading to hindrances in the development of strong structural forms. Superior construction methods cannot be implemented completely due to these limitations of the building material. Many technical problems of construction engineering are currently being solved, thanks to the breakthrough achieved in the synthesis of building materials [1]. According to statistics, seepage plays an important role in the destruction of earth-rock dams, indicating that it is very important for improving the quality of earth-rock dams and preventing leakage. At present, there are two problems that should be solved urgently. These problems are caused by the impervious wall in dyke projects; therefore, the traditional building materials must be updated with sophisticated technology. First problem is to develop antiseepage measures to avoid the reinforcement of dangerous reservoirs. Second problem is to effectively suppress the high pressure of earth-rockfill dam,

which is caused by the deep foundation of the covering layer. To solve all these issues, we should judiciously select the impervious materials.

The plastic concrete is a kind of flexible material with moderate properties, so it can be considered as a building material whose quality is somewhere between the soil and the ordinary concrete. This novel material is used to solve the problems faced while using the common concrete cutoff wall [2]. Presently, most of the impervious core wall materials are made of plastic concrete or asphalt concrete in order to improve the impermeability of the impervious wall and to reduce the elastic modulus of the concrete. In general, plastic concrete always contains certain characteristic toughening components, such as bentonite or clay composite bentonite. By incorporating these toughening components, we substantially improve the crack resistance and impermeability of concrete. With this strategy, we also ameliorate the workability and fluidity of concrete. Thus, we make concerted efforts to reduce the cement consumption and concrete cost [3–6]. The characteristics of plastic concrete are as follows: its elastic modulus is relatively low, and the value of this parameter is very close to the elastic modulus of dam

TABLE 1: Properties of emulsified asphalt.

Items	Results	Technical requirements of hydraulic asphalt SG90 by GB 50092-1996	Performance standards
Needle penetration (25°C, 100 g, 5 s) (0.1 mm)	95.5	80~100	DL/T 5362-2006
Softening point (ring and ball method) (°C)	43.2	42~52	
Ductility (15°C, 5 cm/min) (cm)	160	≥100	
Density (25°C) (g/cm^3)	1.01	Measured data	
Solubility (%)	99.6	≥99.5	
Flash point (°C)	290	≥230	

TABLE 2: Chemical composition of cement, fly ash, and clay (%).

Items	SiO_2	Fe_2O_3	Al_2O_3	CaO	MgO	f·CaO	K_2O	Na_2O	R_2O	SO_3	Loss
Cement	21.7	5.0	4.1	62.1	4.8	0.1	—	—	—	0.7	0.2
Fly ash	59.0	8.8	21.6	5.1	1.5	—	0.8	0.2	0.7	0.6	1.4
Clay	55.6	5.9	15.0	5.6	4.6	—	—	—	—	—	—

body and foundation. Furthermore, the ratio of elastic modulus to strength is generally lower than 500; however, the permeability and durability are limited. The permeability coefficient is generally in the range of 10^{-7} cm/s, and the compression strength generally ranges between 2 MPa and 5 MPa in the ageing period of 28 days. In the traditional impervious wall made from plastic concrete, the compressive strength is not that high. To tackle this issue, we conducted sufficient research on how to improve durability of the material. The traditional material is mostly used for constructing impervious cofferdams that sustain for a temporary period. The traditional material is also used for constructing permanent structures having the foundation of low dam and thin overburden. In this paper, we address the deficiency of traditional plastic concrete in order to synthesize a better material that is a good match of different strength grades of concrete. Our aim is to improve the plasticity of the concrete structure; therefore, the new toughening components of emulsified asphalt have been added to plastic concrete [7]. Based on the results of experimental analysis, we propose that plastic concrete should be used for constructing the permanent impervious wall under deep overburden foundation; the same wall may also be built with plastic concrete of high anticracking site, and so on. Thus, we can effectively solve the major technical problems of hydraulic concrete, such as cracking resistance. Our main purpose is to prevent permeability in order to improve the service life of hydraulic structures.

2. Experimental

2.1. Raw Materials. Cationic emulsified asphalt, Huaxin 42.5 moderate heat Portland cement, clay of a project, Xuanwei grade I fly ash, Jiangsu Bote JM PCA high-efficiency water-reducing agent and GYQ air entraining agent, artificial marble sand, and artificial sandstone rubble were used in this study. The testing result showed that the raw materials meet the relevant technical requirements. Properties of emulsified asphalt and chemical composition of raw materials are shown in Tables 1 and 2.

2.2. Experimental Methods. Experimental methods and data analysis methods in this study were according to SL 237-99 specification of soil test, and SL 352-2006 test code for hydraulic concrete, DL/T5150-2001 specification of concrete cutoff wall used for hydropower and water conservancy project, and DL/T 5330-2015 code for mix design of hydraulic concrete. Different water-binder ratios (0.3, 0.45, and 0.6), asphalt-cement ratios (0, 0.4, and 0.6), and the clay contents (0%, 20%, and 40%) were considered in the test. The content of fly ash was 15%. For explaining its effects on macroscopic mechanical properties, combined with microtesting methods such as CT, XRD, SEM, TG-DSC, and so on [8], microtopography and chemical composition of hydration products of cement-emulsified asphalt slurry were investigated.

CT scanning technology relies on the principle that the degree of attenuation of radiation passing through the medium is proportional to its density. It combined with computer technology and image processing technology; the physical internal density information will be obtained by digital image reflecting. Shape, internal structure, and composition of the material can be recognized based on the density-related image information. Density information of the CT images shows the gray area of information packets.

CT essentially involves a two-dimensional distribution of a physical quantity in the fault area. The physics is that the linear attenuation coefficients $\mu(x, y)$ and $\mu(x, y)$ are directly related to the density of the object. In the early times, CT technology workers were defined as the standard CT number Hp by the attenuation coefficient of water, which was expressed as

$$Hp = \frac{\mu_m - \mu_w}{\mu_w} \times 10^3, \qquad (1)$$

where μ_w and μ_m are the attenuation coefficients of water and the medium, respectively.

The CT number of each pixel can be obtained through (1). Each CT number corresponds to a gray value; thus, a tomographic gray scale digital matrix is formed. The CT testing machine will get a set of CT gray scales by continuously scanning objects in the vertical cross-sectional direction.

TABLE 3: Mix proportion parameters of cement-emulsified asphalt concrete.

Samples	Water-binder ratio	Emulsified asphalt-binder ratio	Clay (%)	Fly ash (%)
Xq1	0.30	0.4	20	15
Xq2	0.30	0.6	20	15
Xq3	0.45	0	0	15
Xq4	0.45	0	20	15
Xq5	0.45	0.4	0	15
Xq6	0.45	0.4	20	15
Xq7	0.45	0.4	40	15
Xq10	0.45	0.6	20	15
Xq12	0.60	0.4	20	15
Xq15	0.60	0.6	20	15

TABLE 4: Mix proportion parameters of cement-emulsified asphalt concrete for the microtest.

Samples	Water-binder ratio	Fly ash (%)	Clay (%)	Emulsified asphalt-binder ratio
X0	0.45	15	0	0
X1	0.45	15	0	0.4
X2	0.45	15	20	0
X3	0.45	15	20	0.4

These gray scales will get the whole composition of the interior of the space in the object through three-dimensional reconstruction. Threshold values for each phase medium and CT number of concrete specimens are hardened cement mortar (2000~2200), aggregate (2200~3071), and interfacial transition zone (1000~1600). Mix proportion parameters of cement-emulsified asphalt concrete and mix proportion parameters of cement-emulsified asphalt concrete for the microtest are shown in Tables 3 and 4.

3. Results and Discussion

3.1. The Influence of Water-Binder Ratio on Properties of Cement-Emulsified Asphalt Concrete. Properties of the cement-emulsified asphalt concrete with the water-cement ratios of 0.3, 0.45, and 0.6, respectively, are shown in Table 5. The results indicate the following effects:

(1) The compressive strength of cement-emulsified asphalt concrete decreased gradually with the steadily increasing water-cement ratio. Compared to the initial phase, the water-binder ratio exerted a slightly greater impact on the compressive strength of concrete at the later stage. This was the result of incorporating emulsified asphalt and clay in order to reduce the dosage of cement in concrete. As a result, there was considerable reduction in the posthydration products. With a large amount of emulsified asphalt, we could significantly alter the strength of concrete. The larger the amount of emulsified asphalt, the lower the compressive strength of concrete would be. The composition and performance of emulsified asphalt was commendable in changing the compressive strength of concrete.

(2) The splitting tensile strength and the tensile strength of emulsified asphalt concrete were reduced by steadily increasing the water-cement ratio. For the concrete, the ratio of compressive strength to the splitting tensile strength was in the range 9.6–11.3. The higher the concrete strength, the greater the ratio would be. In cement-emulsified asphalt concrete, the ratio of compression to tension was slightly lower than the value of ordinary concrete. This indicates that cement-emulsified asphalt concrete had some level of plasticity. The splitting tensile strength was higher than the axial tensile strength. In concrete material, the ratio of splitting tensile strength and axial tensile strength was in the range 1–1.11 on the 28th day; the same ration was in the range 1.08–1.32 on the 90th day. This ratio increased gradually with the increasing age of concrete.

(3) The water-binder ratio had some significant influence on the compressive elastic modulus of cement-emulsified asphalt concrete; the compressive elastic modulus increased when the water-cement ratio decreased steadily; the decline in the water-cement ratio was achieved by increasing the compressive strength of cement-emulsified asphalt concrete.

3.2. Influence of Different Dosages of Emulsified Asphalt on Concrete Properties. The ratio of modulus to strength indicates the internal relationship between the characteristics of tensile strength and elastic modulus of concrete; the modulus is an important index to assess the plasticity of emulsified asphalt concrete [9]. The test results indicate that except for the Xq1 specimen, the elastic moduli of cement-emulsified asphalt concrete specimens were all below 5.2 GPa on the 90th day. The ratio of modulus to strength of concrete was relatively low. In general, this ratio was less than 500. Under some circumstances, it was even below 200. The results indicate that the deformation capacity of emulsified asphalt concrete is greater than the ordinary hydraulic concrete. Therefore, cement-emulsified asphalt concrete has excellent toughness when it is treated with moderate compressive concrete.

Table 6 indicates that the compressive strength, splitting tensile strength, and the tensile strength of cement-emulsified asphalt concrete decreased gradually when we steadily increased the content of emulsified asphalt. In this case, there was sharpest reduction in the compressive strength of concrete; however, the growth rate of long-term strength was higher than that of undoped emulsified asphalt concrete. For emulsified asphalt concrete, the ratio of modulus to strength was significantly lower than the concrete devoid of emulsifying asphalt.

The hydration of cement was delayed by incorporating emulsified asphalt, making concrete own retarding property. This phenomenon can be explained as follows: the emulsified asphalt contained a high amount of water; the water was released after demulsifying the previously emulsified asphalt. The resultant asphalt subsequently participated in the hydration of cement. On the other hand, some bitumen-coated

TABLE 5: Properties of cement-emulsified asphalt concrete under different water-binder ratios.

Samples	Compressive strength (MPa)			Splitting tensile strength (MPa)			Elastic modulus (GPa)			Tensile strength (MPa)	
	14 d	28 d	90 d	14 d	28 d	90 d	14 d	28 d	90 d	28 d	90 d
Xq1	6.7	9.5	15.1	0.64	0.93	1.33	2.96	4.81	10.53	0.88	1.04
Xq6	5.6	8.2	13.3	0.55	0.80	1.19	2.48	3.53	5.14	0.73	0.90
Xq12	4.8	7.2	10.5	0.50	0.70	1.00	1.16	3.04	4.53	0.65	0.85
Xq2	4.1	5.4	7.7	0.38	0.53	0.70	1.98	2.48	3.07	0.50	0.65
Xq10	3.6	5.0	6.9	0.34	0.50	0.68	1.15	2.18	3.16	0.45	0.57
Xq15	3.2	4.0	5.9	0.30	0.38	0.60	0.62	1.01	1.68	0.38	0.50

TABLE 6: Concrete properties under varying dosages of emulsified asphalt (water-cement ratio 0.45).

Samples	Compressive strength (MPa)			Splitting tensile strength (MPa)			Elastic modulus (GPa)			Elastic modulus and compressive strength ratio			Tensile strength (MPa)	
	14 d	28 d	90 d	14 d	28 d	90 d	14 d	28 d	90 d	14 d	28 d	90 d	28 d	90 d
Xq4	13.5	20.8	24.4	0.90	1.60	2.00	12.15	17.79	19.87	900	855	814	1.10	1.78
Xq6	5.6	8.2	13.3	0.55	0.80	1.19	2.48	3.45	5.14	442	430	387	0.75	1.10
Xq10	3.6	5.0	6.9	0.34	0.50	0.68	1.15	2.18	3.16	319	436	485	0.56	0.65

cement particles and hydration products significantly alter the concrete structure, eventually reducing the interfacial bonding strength of the concrete, leading to a decrease in the early strength of the concrete.

3.3. Influence of Varying Dosages of Clay on Concrete Properties.

In plastic concrete, the toughening component is clay as it is able to improve the deformation property of plastic concrete. After mixing emulsified asphalt concrete with clay, we investigated the properties of the mixed material. Thus, we analyzed the influence of varying the dosages of clay on the properties of cement-emulsified asphalt concrete. To obtain the correct mix proportion of emulsified asphalt concrete, we strictly maintained the following parameters: the water-binder ratio of 0.45, the asphalt-cement ratio of 0.4, and the clay content of 0%, 20%, and 40%, respectively. Table 7 presents the test results of this analysis.

The results indicate that clay reduces the elastic modulus of cement-emulsified asphalt concrete. Except for the Xq7 specimen, the ratios of modulus to strength were below 500 for all the specimens. The compressive strength of the Xq6 specimen reached 8.0 MPa, and the ratio of modulus to strength was 430 on the 28th day. When clay is incorporated into plastic concrete, its strength is significantly increased compared to clay-free samples. In both cases, we maintain the same modulus ratio [10].

By increasing the content of clay in cement-emulsified asphalt concrete, it a gradual decrease in the following parameters is observed: the compressive strength, the splitting tensile strength, and the tensile strength. The effect of clay was quite significant on the late compressive strength of concrete. The late growth of splitting tensile strength was faster in cement-emulsified asphalt concrete. In contrast, the late growth of axial tensile strength was slower in cement-emulsified asphalt concrete. Strength was low when cement emulsion asphalt mixed with clay. This is caused by the inert nature of the clay and the retardation of the emulsified asphalt. By adding clay to the concrete, we have significantly reduced the amount of equivalent cement. As a result, the amount of cement hydration produced is also small. These changes weaken the cementing capacity of gravel aggregates and result in a thinner network structure than normal concrete. All these undesirable reactions led to a significant reduction in the strength of concrete.

3.4. Computed Tomography (CT).

The concrete specimen Xq6 was selected to calculate the CT mean value of each section subjected to continuous loading [11–13]. We successfully obtained the CT mean value of each section. Figure 1 illustrates the following developments in the CT scan of the specimen: seven sections were selected along the concrete column; the cross-sectional area was $66.50 \, \text{cm}^2$, and the spacing for each section was 25 mm. Figure 2 illustrates the gray scale CT scan of each concrete section under different stress. Figures 3(a) and 3(b) illustrate the CT mean values curve of each section subjected to continuous load.

As can be seen in Figures 3(a) and 3(b), the three sections of cement-emulsified asphalt concrete were not damaged, but the other four sections were destroyed seriously. This occurred due to the low strength and plasticity of cement-emulsified asphalt concrete. There was relatively small variation in the CT mean values of the first three sections when they were subjected to different loads. This indicates that the state of concrete altered due to compaction and microdilatancy. In the other four sections of the sample, we observed four different concrete damage evolutions: the compaction of each section increased dramatically as we steadily increased the load. This leads to an increase in the density and CT average of cement-emulsified asphalt concrete. When the load was increased to 5.15 MPa, we observed that the CT value of each section started declining slightly. This indicated the initial stage of concrete damage. When we

TABLE 7: Concrete properties under varying dosages of clay.

Samples	Clay (%)	Compressive strength (MPa)			Splitting tensile strength (MPa)			Elastic modulus (GPa)			Elastic modulus and compressive strength ratio			Tensile strength (MPa)	
		14 d	28 d	90 d	14 d	28 d	90 d	14 d	28 d	90 d	14 d	28 d	90 d	28 d	90 d
Xq5	0	9.2	11.1	16.7	1.00	1.10	1.40	1.97	4.02	7.32	213	362	438	0.95	1.28
Xq6	20	5.6	8.2	13.3	0.55	0.80	1.19	2.48	3.53	5.14	442	430	387	0.75	1.10
Xq7	40	3.1	4.2	6.3	0.20	0.40	0.60	0.83	1.57	3.58	268	374	568	0.40	0.55

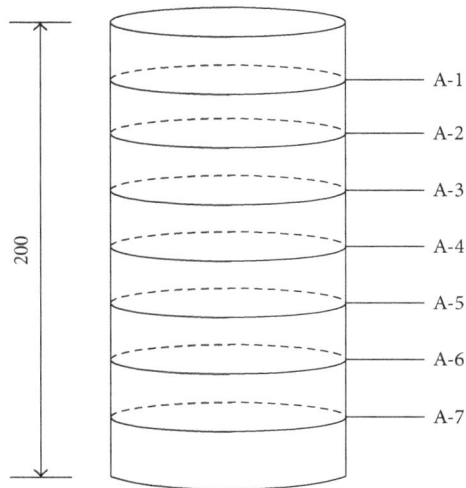

FIGURE 1: Cross sections of cement-emulsified asphalt concrete (size: mm).

increased the limit load to 5.44 MPa, we observed the development of microcrack for the first time. Subsequently, the expansion was substantially enough to accelerate the damage. Then, we observed too many cracks developing in the specimen within a very short period of time. However, there was significant reduction in the volume of specimen expansion, the rapid diminution of the density, and the CT mean value. Finally, we destroyed the lower section of the specimen by increasing the load to 5.30 MPa.

3.5. Analysis of Binary Image. In the mathematical morphology method, we used integral geometry to define geometrical parameters, which were then used to indirectly measure and characterize irregular geometric shapes. Moreover, the random nature of the image was comprehended by random set theory of the method [14–17]. Therefore, it can be used to quantitatively describe CT images of concrete structure. To determine the variation in each pixel gray value and to estimate the crack criterion for each pixel gray value, we used the following equation:

$$H_{i,j} = \max\left(H_{i,j}, H_{i+1,j}, H_{i,j+1}, H_{i+1,j+1}\right), \qquad (2)$$

where $H_{i,j}$ is the CT gray value of the i line and j column pixel, $i, j = 1, 2, 3, \ldots, 1024$. All the image data are represented by 0 and 1; therefore, you can extract the crack by setting a gray threshold as follows: when the intensity is greater than the threshold indicated by 1 or is less than the

threshold indicated by 0. The data matrix of a gray scale image representing 1-pixel size was considered as a unit for statistics; it was substituted into the equation to extract threshold ξ at a certain crack, so it can be realized through binarization of the cracked image, namely, crack extraction. When $H_{i,j} < \xi$, the points are included in the crack area; when $H_{i,j} < \xi$, the points are included in the noncracked area. Figure 4 and Table 8 present the test results of this analysis.

Figure 5 illustrates that there are many cavities in the cement-emulsified asphalt concrete specimens; these cavities have an uneven size in the specimens. In the cement-emulsified asphalt concrete specimens, stress concentration occurs easily and a crack initiation point develops. These undesirable events ultimately lead to widespread cracks that damage the specimen completely.

By examining the CT scan binary image, we can summarize the development trend of the crack in each section of the specimen. In the concrete section A-7, we observed continuous cracks at the microscopic level. However, other sections did not form continuous crack until the loading was increased to 4.9 MPa. In this period, the concrete exhibited alternating stages of compaction and microdilatancy. These observations complied with the results of CT number analysis. When the load was increased continuously, we observed the gradual development of crack in section A-7. First, only two continuity cracks were observed at the edge of the section. The crack width subsequently increased. At the same time, scattered fine cracks were observed in the middle of the cross section, which correlates with the formation at the crack edge. All these cracks substantially destroyed the section of concrete. Based on these observations, we infer that the concrete crack was caused by the hole present in sections A-4 and A-6. When the load was increased to 5.44 MPa in section A-4, we observed intermittent microcracks. Furthermore, we observed continuous cracks when the load was increased to 5.30 MPa. Near the hole, stress concentration appeared when the section A-6 was subjected to stress. Thus, the development of cracks was rapid, extending into the weak zone of the concrete structure. All these developments resulted in the separation of cementitious materials and aggregates at the interface. Such closely connected, adjacent cracks were observed in an interlaced and interconnected fashion, leading to the development of crack propagation. Crack stress unleashed rapidly with the propagation of cracks. When the load was increased in definite increments, we could not observe continuous cracks in the section A-2; the section was minutely examined by the CT image intuitive method.

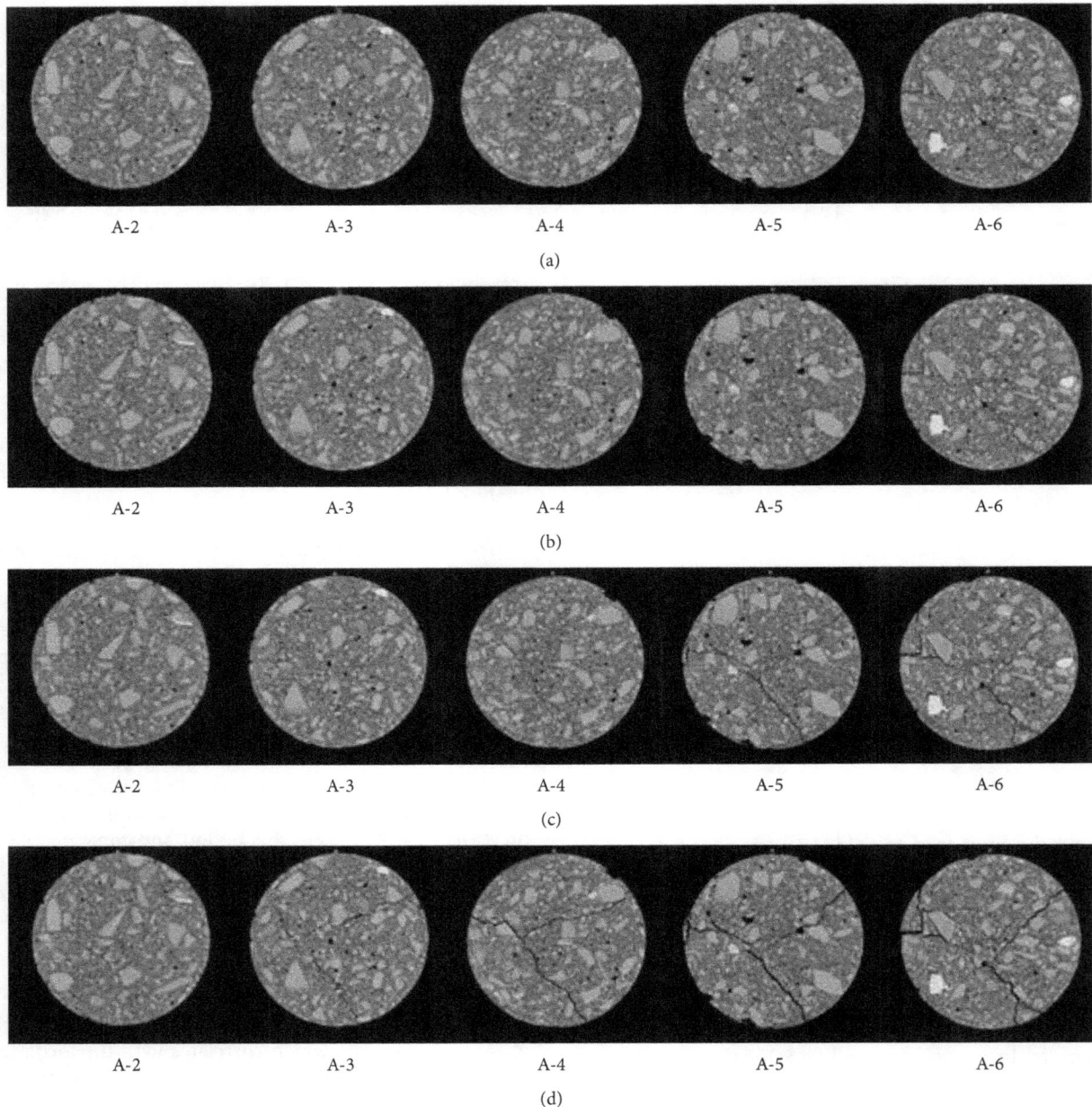

Figure 2: Gray CT picture of concrete cross section under various stresses (28d). (a) $\sigma = 3.18\,\text{MPa}$, (b) $\sigma = 5.15\,\text{MPa}$, (c) $\sigma = 5.44\,\text{MPa}$, and (d) $\sigma = 5.30\,\text{MPa}$.

However, a small piece of continuous crack was detected when we implemented the binary image analysis method on the section exposed to the ultimate load, and the continuous crack developed due to low compressive strength and a certain amount of toughness in cement-emulsified asphalt concrete.

When the cement-emulsified asphalt concrete was subjected to the ultimate load, we observed fine cracks in the upper section; however, these cracks were observed because the upper part was experiencing alternating states of compaction and expansion. Based on these observations, we conclude that a lot of voids had developed in the cement-emulsified asphalt concrete. These cracks were immensely useful in dissipating the external pressure. Furthermore, the toughness of the material increased tremendously when the concrete was mixed with the emulsified asphalt and the clay-toughening component; therefore, the novel hybrid material was experiencing alternating states of compaction, expansion, reconsolidation, and reexpansion when subjected to increasing amounts of load. In the specimen, we mainly observed small cracks and discontinuous cracks. In other words, the specimen does not crack even when subjected to an increased load. In the middle and lower parts, we observed a large connectivity crack under the action of failure load; however, penetrating crack was not observed in the upper and lower layers.

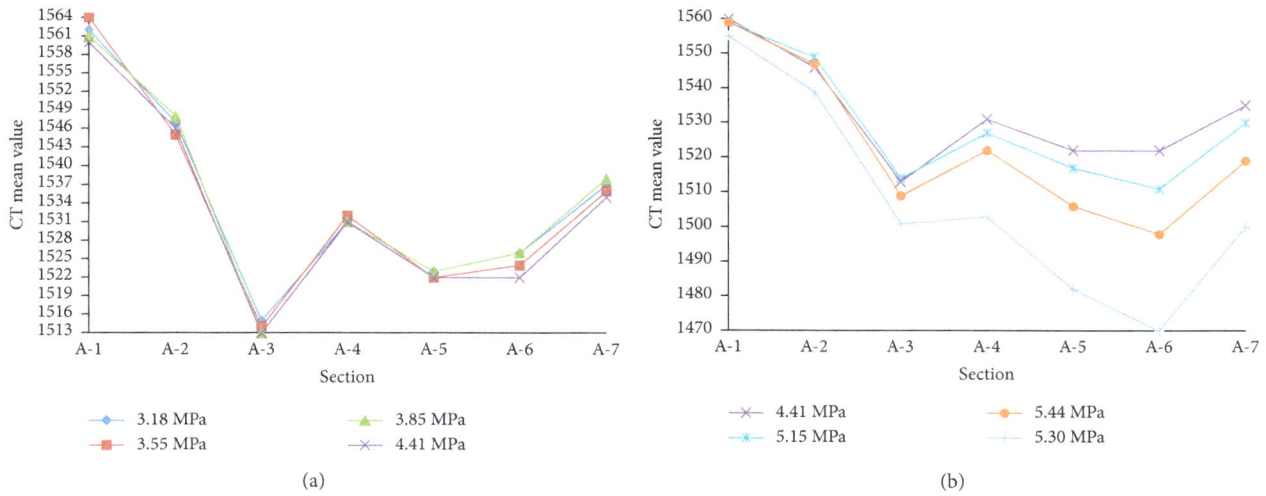

(a)

(b)

FIGURE 3: (a) CT mean value comparison of each cross section under continuous loading (3.18 MPa, 3.55 MPa, 3.85 MPa, and 4.41 MPa). (b) CT mean value comparison of each cross section under continuous loading (4.41 MPa, 5.15 MPa, 5.44 MPa, and 5.30 MPa).

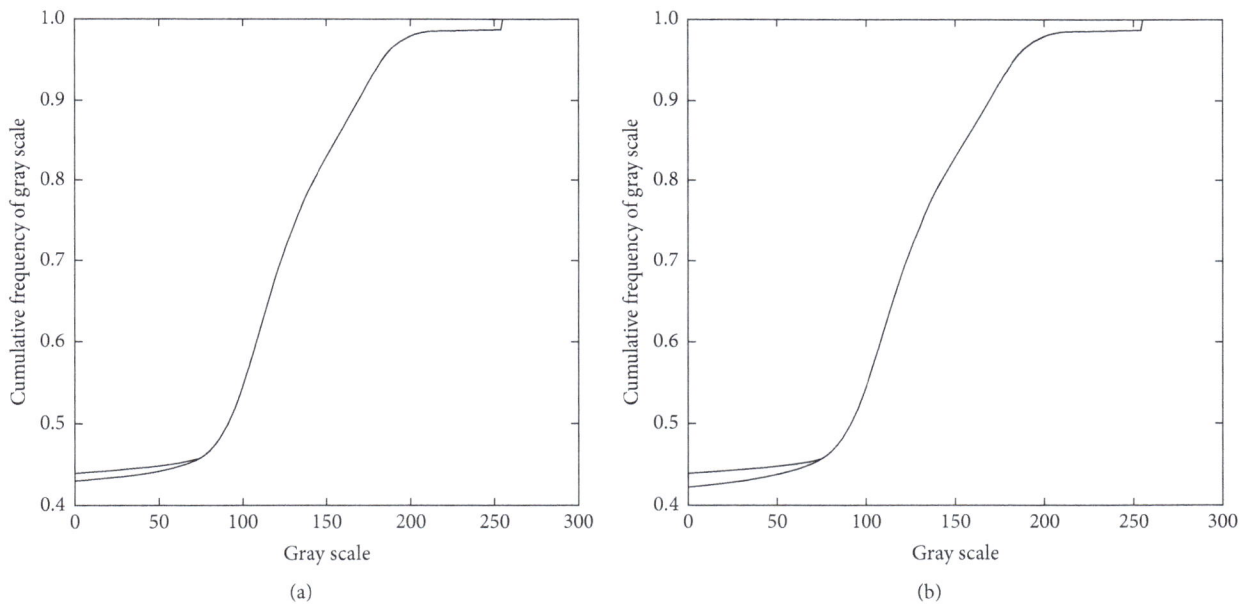

(a)

(b)

FIGURE 4: (a) Cumulative frequency of gray scale with the load of 5.44 MPa and 3.18 MPa under section A-7 of Xq6. (b) Cumulative frequency of gray scale with the load of 5.30 MPa and 3.18 MPa under section A-7 of Xq6.

TABLE 8: Binary analysis threshold of each section of Xq6.

Sections	5.15 MPa to 3.18 MPa	5.44 MPa to 3.18 MPa	5.30 MPa to 3.18 MPa	Mean
A-2	—	—	75	75
A-4	—	73	69	71
A-6	71	65	62	66
A-7	84	79	81	81

3.6. Scanning Electron Microscope (SEM) Analysis. The SEM images of emulsified asphalt-cement paste specimens' hydration products at 7th and 28th day of different dosage combinations as in Table 4 are shown in Figure 6.

As shown in Figure 6(a), C–S–H gel, Ca(OH)$_2$, AFt, AFm, and other hydration products were observed along with the unhydrated cement clinker particles in the undoped emulsified asphalt and clay cement at the 7th day. Moreover, the slurry was highly porous in nature. In addition, some products had not yet developed into a crystalline form completely at the 7th day: C–S–H gel was mainly type II, but it was difficult to find C–S–H gel particles of type III. There were some needle-shaped ettringite (AFt) in the holes because gypsum and lime stimulated the samples to form ettringite [18, 19] (AFt). When we increased the ageing period in definite increments, we observed that there was a decrease in unhydrated cement particles; however, the

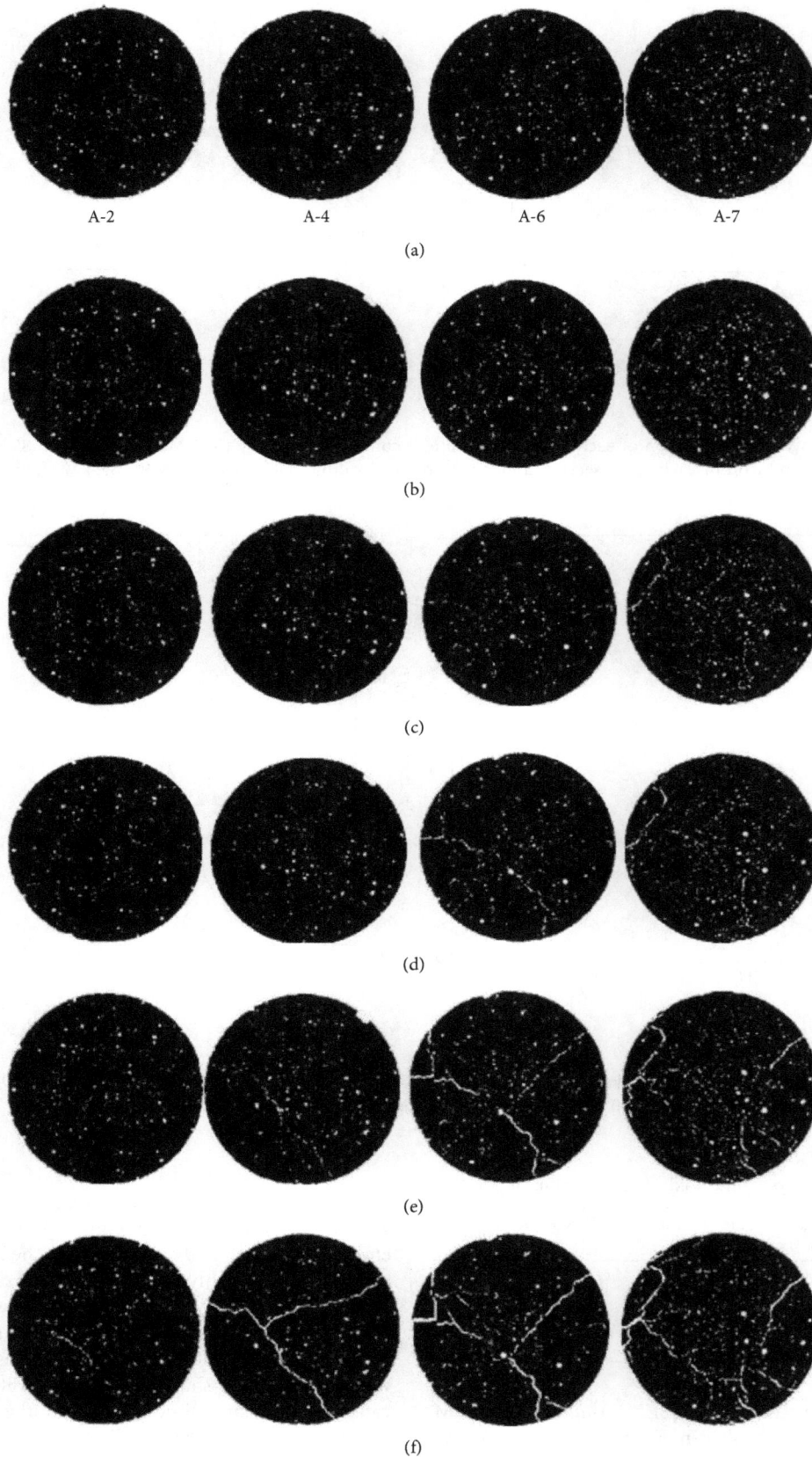

FIGURE 5: Binary image of the Xq6 specimen under different stress sections (28 d). (a) $\sigma = 3.18$ MPa, (b) $\sigma = 3.55$ MPa, (c) $\sigma = 4.9$ MPa, (d) $\sigma = 5.15$ MPa, (e) $\sigma = 5.44$ MPa, and (f) $\sigma = 5.30$ MPa.

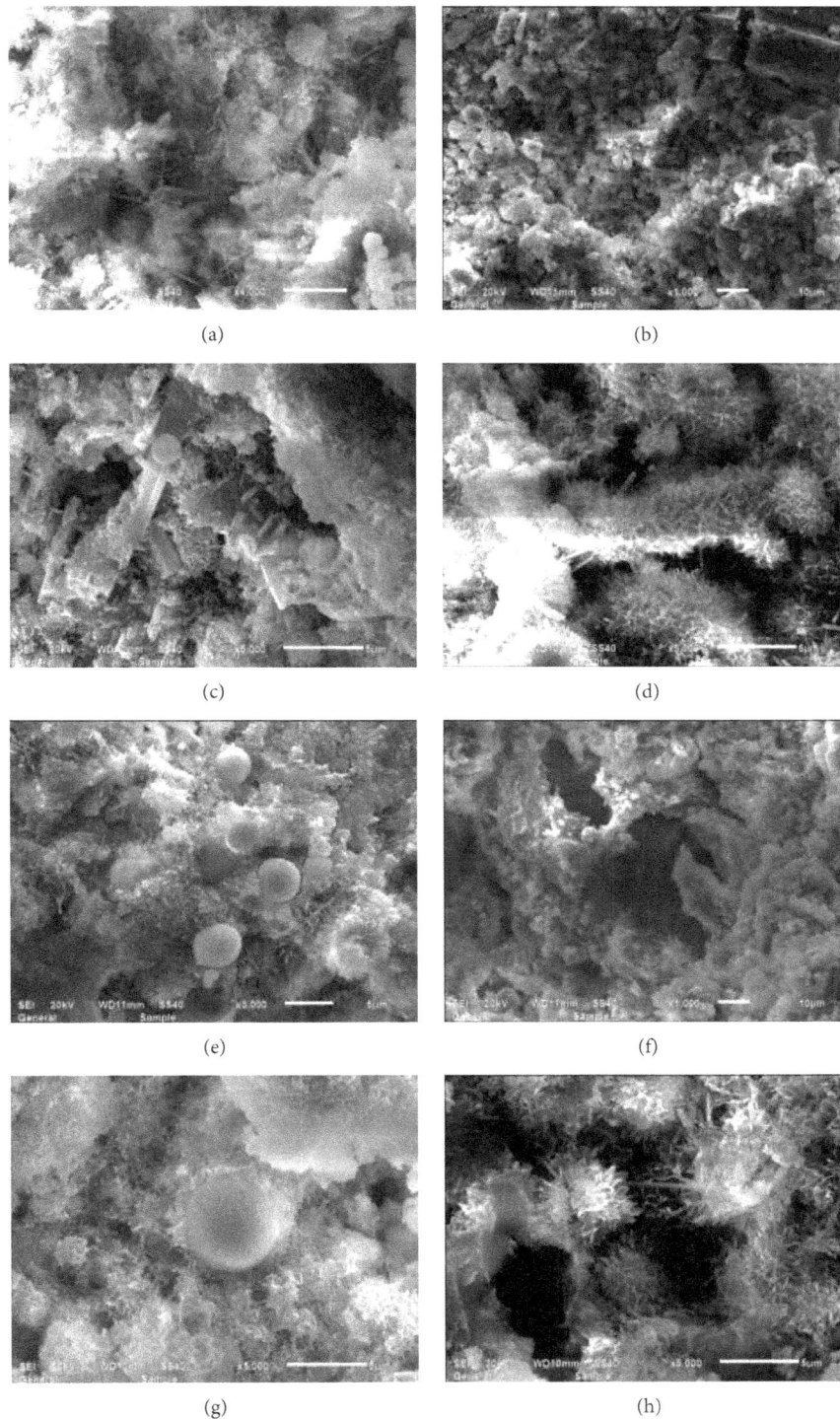

(a)

(b)

(c)

(d)

(e)

(f)

(g)

(h)

FIGURE 6: SEM images of cement-emulsified asphalt paste samples: on the 7th day, (a) without emulsified asphalt and clay, (b) with emulsified asphalt, (c) with clay, and (d) with emulsified asphalt and clay; on the 28th day, (e) without emulsified asphalt and clay, (f) with emulsified asphalt, (g) with clay, and (h) with emulsified asphalt and clay.

amount of hydration products increased simultaneously, and the structure gradually became dense. The shape of hydration products C–S–H of cement paste was approximately the same when the cement paste was mixed with single-doped emulsified asphalt [20], single-doped clay, and pure cement; the product was mainly appearing in the form of spherical C–S–H gel. Figure 6(c) illustrates the part where the clay particles fill the void, leading to the formation of a structure that is more compact than the single-doped emulsified asphalt-cement paste. Furthermore, we also observed a small amount of needle-shaped ettringite (AFT), small pieces of plate-type Ca(OH)$_2$ crystals, and fly ash particles; the hydration products had completely wrapped these minor products. Figures 6(b) and 6(d) illustrate that the emulsified asphalt was not involved in the hydration reaction of cement; however, many pores were observed in the cement-emulsified asphalt paste. These pores were caused by the "vacancy" of bubble burst and water evaporation during the forming process. The surface of the slurry is not smooth; it is very uneven and there is a lot of bump [19, 21–23]. Asphalt particles were round in shape, with a significant portion of particles being encased with cement particles or clay particles. These factors decelerated the cement hydration rate, leading to the slow development of concrete strength.

Figure 6(e) shows the nature of products on the 28th day of ageing: hydration products were fibrous (type I), meshy (type II), and granular (type III) C–S–H gels (some type I of C–S–H gel shape were flowers). Furthermore, we also observed Ca(OH)$_2$ crystal and monosulfate calcium sulfoaluminate hydrates in the form of hexagonal flakes. Figure 6(g) shows that the clay particles were gradually surrounded by fibrous, flocculent, and layered hydration products. In addition, C–S–H gel and hydrated sulfoaluminate had intertwined with each other to form a relatively dense structure, leading to a reduction in the volume of pores. However, crystals of hydration products appeared in a haphazard arrangement. As shown in Figures 6(f) and 6(h), emulsified asphalt contains water that initiates and promotes hydration. As a result, the hydrate content gradually increases. The shape of asphalt particles had changed from a regular sphere to an irregular ball pie. These particles had got attached to the cement hydration products.

3.7. X-Ray Diffraction Analysis (XRD).

The XRD patterns of emulsified asphalt-cement paste specimens' hydration products at 7th day and 28th day of different dosage combinations as in Table 4 are shown in Figures 7 and 8.

As shown in Figures 7 and 8, the hydration products of cement-emulsified asphalt paste were substantially similar even in different proportions; the main hydration products were C$_2$S, C$_3$S, Ca(OH)$_2$, and ettringite. The amount of hydration products were increasing gradually with the progress of ageing process, but a different number of hydration products were generated in different cementitious systems. The ettringite diffraction peaks of X2 and X3 specimens are higher than the peaks of X0 and X1 specimens on the 7th and 28th days of ageing.

The patterns clearly illustrate that kaolinite, quartz, and calcite were the main components of clay. By comparing the specimens X0, X2, and X3, we found that this clay was a kind of inert admixture, and it was not involved in the hydration reaction. The reaction played a significant role in generating products, which acted as fillers in the concrete. Thus, the concrete pore structure improved tremendously, and there was a sharp decline in the rate at which the load damaged the concrete. By comparing the X-ray diffraction patterns of both types (doped and undoped) of emulsified asphalt

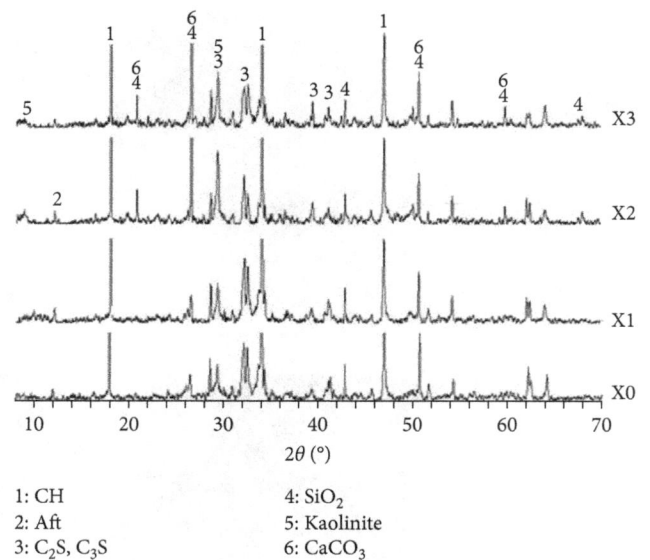

FIGURE 7: X-ray diffraction pattern of cement-emulsified asphalt paste under different proportions for 7 d.

1: CH   4: SiO$_2$
2: Aft   5: Kaolinite
3: C$_2$S, C$_3$S   6: CaCO$_3$

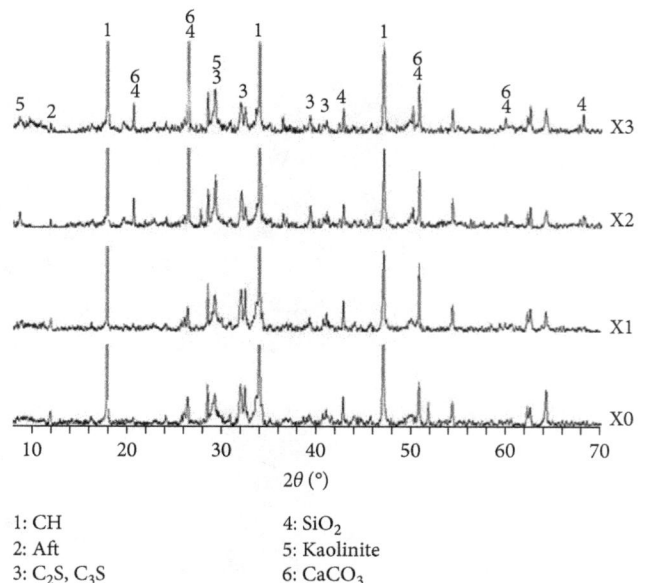

FIGURE 8: X-ray diffraction pattern of cement-emulsified asphalt paste under different proportions for 28 d.

1: CH   4: SiO$_2$
2: Aft   5: Kaolinite
3: C$_2$S, C$_3$S   6: CaCO$_3$

specimens, we found that there was no chemical reaction between asphalt and cement. Moreover, no new mineral phases were formed in the cement hydration products. Thus, only Ca(OH)$_2$ and other characteristic peaks were observed in the patterns of cement-emulsified asphalt; the patterns included all the characteristic peaks of hydration products obtained from cement. Furthermore, characteristic peaks representing new material were not observed in the patterns. When we used different proportions of specimens, the intensity of diffraction peaks representing the specimens was different at different ages; the diffraction peak of X0

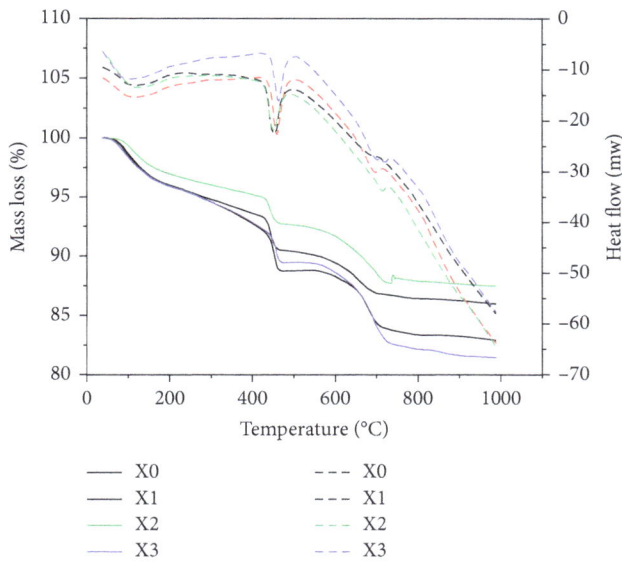

FIGURE 9: TG-DTA curves of paste specimens with different mix ratios for 7 d.

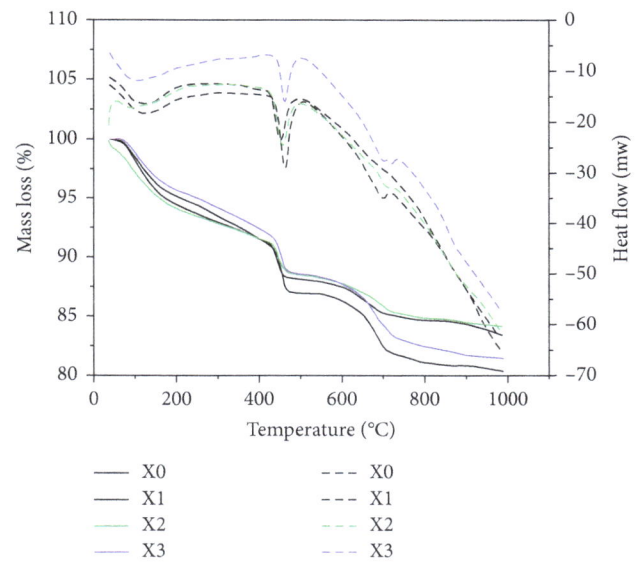

FIGURE 10: TG-DTA curves of paste specimens with different mix ratios for 28 d.

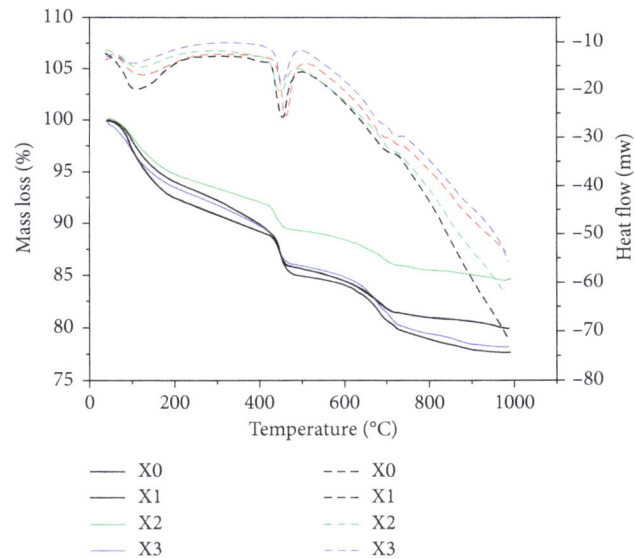

FIGURE 11: TG-DTA curves of paste specimens with different mix ratios for 90 d.

was higher than that of X1 on the 7th day of ageing; however, the diffraction peak of X0 was slightly lower than that of X1 on the 28th day of ageing. Both these effects were related to the retarding property of the dopant in the emulsified asphalt. Furthermore, there were obvious amorphous peaks of components in the diffraction pattern, and these peaks were observed due to the scattering of asphalt.

3.8. Differential Thermal Analysis (TG-DTA). Table 4 shows the results of differential thermal analysis that was performed on mix proportions of cement-emulsified asphalt concrete. Figures 9–11 illustrate the TG-DTA curves of the cement-emulsified asphalt with different mix ratios on the 7th, 28th, and 90th day of ageing.

The TG curves clearly manifest that the changes in the intensity of peak with respect to the three ages; these changes in peak intensity completely agreed with the changes observed with the variation of temperature. A relatively small loss of mass is observed when the X2 specimen is mixed with single clay. In contrast, a massive loss of mass is observed when the X1 specimen is mixed with single-emulsified asphalt. Moreover, the mass loss gradually increased with the increasing ages, indicating that the hydration reaction of paste samples was becoming more adequate when they were allowed to age for a longer period of time. In the graph, a weak peak appeared from 80°C to 100°C; the peak characterized the dehydration of calcium silicate hydrate and ettringite. A sharp peak was observed between 400°C and 500°C, and this peak represented the endothermic valley of Ca(OH)$_2$ dehydration. Finally, a sharp peak was observed between 650°C and 700°C, and it represented the endothermic valley of CaCO$_3$ decomposition. Despite using different proportions, the hydration products of cement paste were

basically the same as that of X0, but the number of products produced was slightly different.

The DTA curve illustrates that there was a steady decrease in the content of Ca(OH)$_2$ crystals in the X0 specimen at the late stages of analysis. This indicates that Ca(OH)$_2$ is involved in the secondary hydration reaction. In the X1 specimen containing single emulsion asphalt and the X3 composite specimen containing the dopant, the content of Ca(OH)$_2$ increased at the late period. The heat flow of X1 was 23 mw, 30 mw, and 25 mw on the 7th, 28th, and 90th day of ageing, respectively. This indicates that the hydration rate of Ca(OH)$_2$ was relatively slow at the early stage; however, it increased rapidly at the late stage of analysis. This effect has been attributed to the retarding effect of emulsified asphalt, which is in complete agreement with the above analysis results; the results represent the mechanical properties of concrete.

4. Conclusions

(1) Emulsified asphalt incorporation in concrete delayed the hydration of cement, resulting in relatively slow development of early compressive strength of concrete; the late strength of concrete increased rapidly with the demulsification of emulsified asphalt and hydration of cement. The elastic modulus of plastic concrete is small, and the compressive strength is low. Mixed with emulsified asphalt and clay, the concrete whose compressive strength of 28 d could reach 9 MPa was with lower elastic modulus; the ratio of modulus to strength of the concrete was generally less than 500 and even some below 200, and the deformation capacity was greater than that of the ordinary hydraulic concrete.

(2) The CT mean value analysis of cement-emulsified asphalt concrete specimens showed that the failure process of cement-emulsified asphalt concrete could be divided into 4 stages: compaction, dilatancy, crack propagation, and failure. The volume expansion of each section was not consistent, and the variation of the CT mean value of each section was different; the CT mean value of the place near the lower end face suffered a larger decline but a smaller decline to the upper part of the sample. The evolution of concrete suffering damage to failure is a gradual development process, and no sharp expansion of brittle failure, which further showed that the cement-emulsified asphalt concrete had certain toughness.

(3) There were many pores in the cement-emulsified asphalt paste and a lot of uneven bump on slurry surface. A large number of cement hydration products and some asphalt films which cladded on the surface of the hydration product formed the hardened paste skeleton. The unhydrated cement, asphalt, fibrous C–S–H gel, CH, needle-shaped ettringite, and other hydration products were interwoven to constitute emulsified asphalt-cement paste and to form a spatial structure. There was no chemical reaction between cement and asphalt which produced by demulsification of emulsified asphalt, but there was a retarding effect of emulsified asphalt. The cement-emulsified asphalt concrete has certain plasticity, so we can explore it using in the permanent seepage control engineering under deep overburden.

Conflicts of Interest

The authors confirm that the mentioned funding did not lead to any conflicts of interest regarding the publication of this manuscript. Also there are no other possible conflicts of interest in the manuscript.

Acknowledgments

This work was supported by the National Natural Science Foundation of China (No. 51539002).

References

[1] Y. Li, *Building Materials*, China Water Power Press, Beijing, China, 2001.

[2] Q. Wang, W. Sun, and H. Xiong, *The Plastic Concrete Cut-Off Wall*, China Water Power Press, Beijing, China, 2008.

[3] J. Ouyang, Y. Tan, Y. Li, and J. Zhao, "Demulsification process of asphalt emulsion in fresh cement-asphalt emulsion paste," *Materials and Structures*, vol. 48, no. 12, pp. 3875–3883, 2015.

[4] A. Mahboubi and A. Ajorloo, "Experimental study of the mechanical behavior of plastic concrete in triaxial compression," *Cement and Concrete Research*, vol. 35, no. 2, pp. 412–419, 2005.

[5] C. Zhang, Y. Chen, and Y. Guo, "Clay concrete for cutoff wall in earth dam," *Journal of Hydraulic Engineering*, vol. 36, no. 12, pp. 1464–1469, 2005.

[6] S. Hinchberger, J. Weck, and T. Newson, "Mechanical and hydraulic characterization of plastic concrete for seepage cut-off walls," *Canadian Geotechnical Journal*, vol. 47, no. 4, pp. 461–471, 2010.

[7] A. M. Sha and Z. J. Wang, "Microstructure of mastics-aggregate interface in cement emulsified asphalt concrete," *Journal of Changan University*, vol. 4, p. 28, 2008.

[8] S. F. Brown and D. Needham, "A study of cement modified bitumen emulsion mixtures," in *Association of Asphalt Paving Technologists Proceedings of the Technical Sessions 2000*, White Bear Lake, MN, USA, 2000.

[9] S. Oruc, F. Celik, and M. V. Akpinar, "Effect of cement on emulsified asphalt mixtures," *Journal of Materials Engineering and Performance*, vol. 16, no. 5, pp. 578–583, 2007.

[10] J. Liu, X. Zheng, S. Li et al., "Effect of the stabilizer on bubble stability and homogeneity of cement emulsified asphalt mortar in slab ballastless track," *Construction and Building Materials*, vol. 96, pp. 135–146, 2015.

[11] J. Zhou, F. Dang, H. Chen et al., "Breakage meso mechanism of concrete based on CT test under uniaxial compression," *Journal of Xi'an University of Technology*, vol. 22, no. 4, pp. 335–360, 2006.

[12] Z. X. Wen and F. W. Liu, *Concrete Quality Inspected by Acoustic Nondestructive CT*, Vol. 9, China Three Gorges Construction, Beijing, China, 2002, in Chinese.

[13] W. Tian and F. Dang, "CT image analysis of meso fracture process of concrete," *Engineering Journal of Wuhan University*, vol. 41, no. 2, pp. 69–74, 2008.

[14] Y. Wu, W. Ding, and G. Cao, "Observation and detection of evolution process of rock crack on CT scale under uniaxial and triaxial compression condition," *Journal of Xi'an University of Technology*, vol. 19, no. 2, pp. 115–119, 2003.

[15] J. Ren, X. Ge, Y. Pu, W. Ma, and Y. Zhu, "Primary study of real-time CT testing of unloading damage evolution law of rock," *Chinese Journal of Rock Mechanics and Engineering*, vol. 19, no. 6, pp. 697–701, 2000.

[16] H. Peng and J. Zhang, "The boundary element method for bond—anchorage length," *International Journal Hydroelectric Energy*, vol. 19, no. 2, pp. 85–88, 2001.

[17] A. M. Nevelle, *Properties of Concrete*, Person Education Press, London, UK, 4th edition, 2008.

[18] F. M. Nejad, M. Habibi, P. Hosseini, and H. Jahanbakhsh, "Investigating the mechanical and fatigue properties of sustainable cement emulsified asphalt mortar," *Journal of Cleaner Production*, vol. 156, pp. 717–728, 2017.

[19] T. Rutherford, Z. Wang, X. Shu, B. Huang, and D. Clarke, "Laboratory investigation into mechanical properties of cement emulsified asphalt mortar," *Construction and Building Materials*, vol. 65, no. 13, pp. 76–83, 2014.

[20] Z. Zhao, "Lime active decay law is studied in the lime soil," *Transportation Science and Technology*, vol. 4, pp. 98–100, 2005.

[21] K. J. Mun, W. K. Hyoung, C. W. Lee, S. Y. So, and Y. S. Soh, "Basic properties of non-sintering cement using phospho-gypsum and waste lime as activator," *Construction and Building Materials*, vol. 21, no. 6, pp. 1342–1350, 2007.

[22] Z. Wang, *Structure and Performance of the Interface Zone between Mortar and Aggregate in Cement Emulsified Asphalt Concrete*, Chang'an University, Xi'an, China, 2007.

[23] Z. Wang, Q. Wang, and T. Ai, "Comparative study on effects of binders and curing ages on properties of cement emulsified asphalt mixture using gray correlation entropy analysis," *Construction and Building Materials*, vol. 54, pp. 615–622, 2014.

11

Influence of Buton Rock Asphalt on the Physical and Mechanical Properties of Asphalt Binder and Asphalt Mixture

Yafei Li,[1] Jing Chen,[1] Jin Yan,[1] and Meng Guo (ID)[2]

[1]*Research and Consulting Department of Engineering Structure and Materials Research Center, China Academy of Transportation Sciences, 240 Huixinli, Chaoyang District, Beijing 100029, China*
[2]*College of Architecture and Civil Engineering, Beijing University of Technology, Beijing 100124, China*

Correspondence should be addressed to Meng Guo; mguo@ustb.edu.cn

Academic Editor: Ana María Díez-Pascual

In order to study the effect of different rock asphalt contents on the physical and mechanical properties of an asphalt binder and asphalt mixture, the physical and mechanical tests and analysis were conducted. An on-site case was investigated to verify the effectiveness of rock asphalt-modified pavement. The results show that the activation treatment can effectively enhance the molecular polarity of Buton rock asphalt. The "wet process" was used to prepare the Buton rock asphalt-modified asphalt binder, and the high-temperature performance and aging resistance were significantly improved. The modified asphalt prepared by mixing 30% rock asphalt shows the optimum balance between service performance and segregation. The on-site full-scale application of the Buton rock asphalt-modified asphalt pavement showed the good workability and service performance. This research demonstrated the ability of rock asphalt improving asphalt pavement on multiscales. It is helpful for the broader application of rock asphalt in asphalt pavement.

1. Introduction

Rock asphalt (RA) is a kind of asphalt coming from the petroleum flowing into the split cracks of rock. It is formed after billions of years of accumulation and changes under the combined action of heat, pressure, oxidation, catalyst, and bacteria. As a kind of natural asphalt, it is a kind of green, energy saving, environmental new pavement material because it has high degree of fusion with asphalt and does not require chemical processing. When it is used in the modified asphalt, it can improve the road performance of the modified asphalt, especially high-temperature stability, water resistance, and durability, with remarkable social and economical benefits. The Buton rock asphalt (BRA), as a representative product of rock asphalt, is produced in the Buton island of Indonesia. It comes from the sedimentation of the Jurassic marine animal fossils and is characterized by high asphalt content and high nitrogen content, being resiniferous and nonwaxy [1–5]. It is added to ordinary asphalt mixture as external admixture at home and abroad so as to improve the high-temperature performance and water stability performance of the asphalt mixture. This method is commonly known as "dry process." However, as seen from the application effect, the Buton rock asphalt cannot play its maximum role due to construction variability such as the mixing uniformity of construction [6–10].

Zhong et al. found that the addition of rock asphalt improved the high-temperature performance of petroleum bitumen binders and mixtures. The moisture damage resistance, tensile strength, and fatigue performance of petroleum mixture were enhanced as well. The low-temperature performance was slightly weakened [11]. Li et al. evaluated the potential impact of different types of rock asphalts on the performance of asphalt composites. They avoided the extraction of the asphalt binder from rock asphalt and simplified the process of evaluating the potential impact of rock asphalts on mixture performance. They found that addition of rock asphalts increases material stiffness and slightly reduces relaxation potential of asphalt composites at low-temperatures [12]. Zou and Wu studied the rheological

TABLE 1: Properties of the BRA used in this research.

Items	Unit	Test results	Technical requirements
Colour character	—	Brown powder	Black or brown powder
Ash content	%	61.7	≤80
Moisture content	%	2	≤2
Asphalt content	%	27	—
Particle size range	—	—	—
4.75 mm	%	100	100
2.36 mm	%	95.8	95~100
1.18 mm	%	82.8	>80

properties and field applications of the Buton rock asphalt. They found that with increasing BRA content, the binder's penetration decreased, softening point increased, dynamic viscosity at 60°C increased, and complex modulus increased. The BRA-modified asphalt concrete mixtures had better rutting performance as compared to the control asphalt concrete mixture sample [13]. Li et al. investigated the relationship between the microstructure and the performance of the Buton rock asphalt by using surface free energy and an infrared spectrum analysis. They found that mixing the BRA was a physical modification. An increase of rocking asphalt content can result in the increase of the hydrophobicity of the asphalt mixtures [14]. Rock asphalt can be seen as a composite consisting of asphalt binder and fillers. Therefore, for the interfacial behavior between rock asphalt and asphalt binder (or aggregate), refer Guo et al.'s research [15–18].

In view of the construction variability which is inevasible for the production of the Buton rock asphalt-modified asphalt by "dry process," this research innovatively put forward the "wet process." The rock asphalt is firstly pretreated by the activation process for preliminary grinding and activation. Then, the activated rock asphalt is mixed with the matrix asphalt. Finally, the mix asphalt is grinded by colloid mill to produce the modified rock asphalt. This process can fully integrate the rock asphalt with the matrix asphalt, thus effectively promoting the cross-linking polymerization of the polar functional groups in the rock asphalt and the active groups (carboxyl, aldehyde, carbonyl, and naphthalene) in the matrix asphalt, improving the arrangement mode and net structure (node and strength) of the matrix asphalt molecules and enhancing the asphalt cohesion. In this way, it significantly improves the antifluidity, antioxidation, adhesion, and temperature susceptibility of the modified asphalt, thus improves the high-temperature resistance, water damage resistance, and fatigue performance of the mixture of rock asphalt and modified asphalt, and makes the mixture more suitable for large-scale production.

2. Materials and Methods

2.1. Raw Materials. The raw material of the Buton rock asphalt (BRA) is the rock asphalt powder produced by Hubei Zhengkang Asphalt Technology Co., Ltd. The specific performance indicators are shown in Table 1, and all technical indicators meet the specification requirements [19].

2.2. Activation Pretreatment. The activation pretreatment aims at improving the degree of fusion between the rock asphalt and the matrix asphalt and thus fully exerts the modification ability of the rock asphalt [13]. The concrete process is as follows. The rock asphalt is firstly broken up and dehydrated and then is grinded at high temperature. According to microscopic image characterization, the rock asphalt molecules have extremely strong polarity after high-temperature "activation." They connect the asphalt molecules in the matrix asphalt and the resin and ash content in the rock asphalt together to form a stable multidimensional net structure which effectively improves the performance of the matrix asphalt, as shown in Figures 1 and 2.

2.3. "Wet Process" of the Buton Rock Asphalt. Specifically, the "wet process" of the Buton rock asphalt developed in this paper includes the following: the matrix asphalt is firstly preheated from 150°C to 160°C by a heating system and is then pumped to the asphalt tank. At the same time, the Buton asphalt mixture, which has been broken and dehydrated, is added to the feed inlet and is slowly added to the matrix asphalt. Then, the mixing device is started. After stirring for 0.5 h to 1 h, the premix is pumped into the self-developed colloid mill for full grinding. The rock asphalt particles are grinded to less than 100 mesh. When they uniformly suspend in the matrix asphalt, the modified rock asphalt can be obtained. Finally, the prepared modified Buton rock asphalt is stored in the storage tank for proper storage. The specific equipment design drawings and the entity diagram of the equipment are shown in Figures 3 and 4.

2.4. Sample Preparation. In order to compare the influence of different mixing amounts of the Buton rock asphalt on the performance of modified asphalt and asphalt mixture and determine the optimum mixing amount of rock asphalt, 10%, 20%, 30%, 40%, and 50% of the activated rock asphalt are, respectively, added to the No. 70 road petroleum asphalt to produce the modified Buton rock asphalt according to the "wet process." The performance of modified asphalt and modified asphalt mixture is tested.

2.5. Testing Protocols

2.5.1. Asphalt Binder. Penetration test, softening point test, and ductility test were conducted according to the Test Specification of Asphalt and Asphalt Mixture (JTG E20–2011) [20].

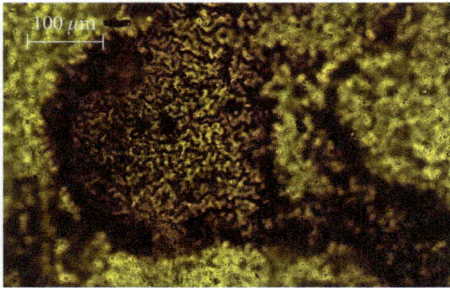

FIGURE 1: Rock asphalt after activation pretreatment.

FIGURE 2: Multidimensional network of rock asphalt.

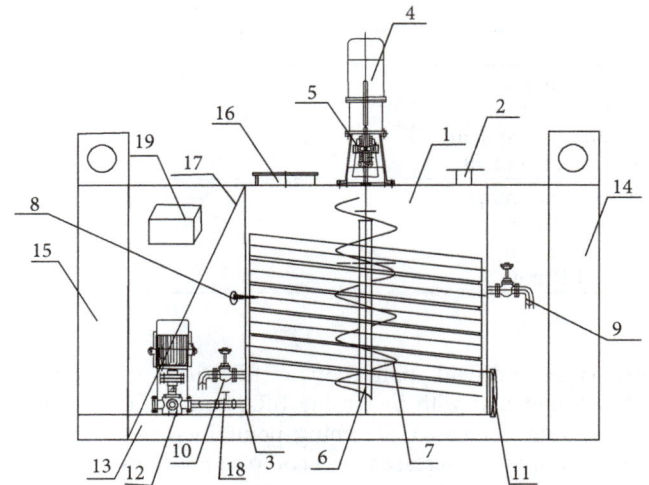

FIGURE 3: Schematic of the device used for preparation of rock asphalt-modified asphalt binder. 1: agitator tank; 2: feed port; 3: discharge port; 4: mixing motor; 5: reducer; 6: stirring shaft; 7: helical blade; 8: thermometer; 9: first high-temperature gate valve; 10: second high-temperature gate valve; 11: service port; 12: production delivery pump; 13: baseboard; 14: first side plate; 15: second side plate; 16: man way; 17: ladder stand; 18: insulation two-way; 19: distribution box.

The gravimetric capillary method was used to measure the kinematic viscosity of the asphalt binder at 135°C. Measurements using capillary viscometers were based on the relation between viscosity and time. The more viscous the asphalt, the longer it will take to flow through a capillary under the influence of gravity alone. There are several standardized capillaries in use today. Most laboratory instruments employ glass capillaries or "tubes." A more recent advancement for field measure of kinematic viscosity employs a split aluminum cell capillary. In this research, the manual constant temperature bath system consisting of a very precise temperature-controlled bath was used. A sample of the asphalt binder was suctioned into the tube until it reaches the start point. The suction was then released, and the asphalt binder flowed by gravity through the controlled capillary section of the tube. Two or three marks were visible on the tube. We watched the meniscus of the asphalt binder as it passes the start point. At this point, we recorded the time it took the asphalt binder to pass the final mark. The tubes were selected such that the test would take a minimum of 200 secs to complete. This made it easier for manual timekeeping. More details can be found in ASTM D445.

2.5.2. *Asphalt Mixture.* Dynamic stability and residual stability can be obtained from the basic Marshall tests of asphalt mixtures [20]. Regarding freeze-thaw splitting tests, the compaction method of the specimen of the freeze-thaw cycle test was gyratory compacting. The diameter of the specimen was 100 mm, and the height was 63.5 mm ± 1.3 mm. The procedure of the free-thaw cycle is as follows:

(1) The specimens were randomly divided into two groups. The first group of the specimen was kept on the platform at room temperature; the second group of the specimen was immersed in water with 97.3~98.7 kPa

FIGURE 4: Picture of the device used for preparation of the rock asphalt-modified asphalt binder.

for 15 minutes, then the valve was opened, and the specimen was kept in the water without pressure for 0.5 hours.

(2) The second group of specimens was removed into a plastic bag with 10 ml of water, and then the condition temperature was maintained at −18°C ± 2°C for 16 hours.

(3) The specimens removed from the low-temperature case were immediately put in the water tank at the temperature of 60°C ± 0.5°C for 24 hours.

The specimens were kept in 25°C ± 0.5°C constant temperature water tank for 2 hours and then were removed, and immediately splitting loading by the MTS machine was applied. The loading rate is 50 mm/min. The indirect tensile strength is calculated according to the following equation, and the TSR is calculated according to (2):

TABLE 2: Influence of different rock asphalt contents on the properties of the asphalt binder.

Test items	Sample types					
	Pure asphalt	Pure asphalt + 10% BRA	Pure asphalt + 20% BRA	Pure asphalt + 30% BRA	Pure asphalt + 40% BRA	Pure asphalt + 50% BRA
Penetration (25°C, 100 g, 5 s) (0.1 mm)	73	66	58	55	46	32
Softening point (°C)	46	48	50	52	58	60
Ductility (15°C, 5 cm/min) (cm)	100	68	42	34	21	12
Kinematic viscosity at 135°C (Pa·s)	0.452	0.476	0.512	0.625	0.856	1.343
Difference of softening points between 0 and 48 hours (°C)	—	0.5	1	1.5	2.5	4
Ratio of penetrations before and after RTFOT test	71	75	81	86	83	82

$$R_T = \frac{0.006287 P_T}{h}, \qquad (1)$$

where R_T is the indirect tensile strength (MPa), P_T is the maximum value of the test load (N), and h is the specimen height (mm).

$$\text{TSR} = \left(\frac{R_{T2}}{R_{T1}}\right) \times 100, \qquad (2)$$

where TSR is the tensile strength ratio (%), R_{T1} is the indirect tensile strength before freeze-thaw cycle (MPa), and R_{T2} is the indirect tensile strength after freeze-thaw cycle (MPa).

This study used the low-temperature blending test to evaluate the low-temperature anticracking performance of asphalt mixtures. The blending failure strain was selected as the evaluation index of low-temperature performance of asphalt mixtures. The bigger the blending failure strain, the better the low-temperature performance of asphalt mixtures. The diameter of specimens was 250 mm × 30 mm × 35 mm. The test temperature was −10°C, and the loading rate was 1 mm/min.

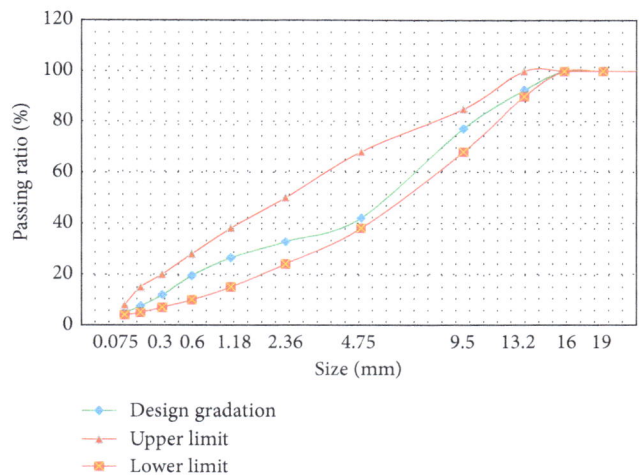

FIGURE 5: Gradation of selected AC-13 asphalt mixture.

3. Results and Discussion

3.1. Influence of Different Rock Asphalt Contents on the Physical and Mechanical Properties of Asphalt Binder. Related tests are carried out on the No. 70 road petroleum asphalt and modified rock asphalt with different mixing amounts according to the Test Specification of Asphalt and Asphalt Mixture (JTG E20–2011) [20]. The test results are shown in Table 2.

It can be seen from Table 2 that the penetration value decreased and the softening point increased with the increase of BRA content. It indicates that BRA can improve the permanent deformation resistance of the asphalt binder, further to improve the high-temperature service performance of BRA-modified asphalt pavement. Table 2 also shows that the ductility decreased dramatically after BRA modification. It means the elongation ability of asphalt approached a limit, and most probably it was the same for elasticity of asphalt that was a fundamental property for lasting serving life of flexible pavement. Regarding viscosity, the more viscous asphalt requires more energy consumption in every single process of asphalt industry including storage, transporting, placing, and compacting. Therefore, a greater viscosity could be disadvantageous for construction of

asphalt pavement. However, a greater viscosity also can help the asphalt pavement resisting the high-temperature rutting. Selecting a proper BRA/asphalt binder ratio to obtain a suitable viscosity to balance the workability and service performance of asphalt pavement is a key of BRA application. At the same time, after the aging of the rotating film, the residual penetrations of the modified Buton rock asphalt with five different mixing amounts are all higher than that of the matrix asphalt, indicating that the antiaging performance of the modified Buton rock asphalt has obvious improvement. In addition, the storage stability of the modified Buton rock asphalt decreases with the increase of rock asphalt content. After mixing 40% and 50% of rock asphalt, the differences of softening points of BRA-modified asphalt between 0 and 48 hours were 2.5°C and 4°C, respectively. It indicates that the modified Buton rock asphalt has relatively serious segregation. According to the requirement of Technical Specifications for Construction of Highway Asphalt Pavements (JTG F40-2004), the difference of softening points of modified asphalt between 0 and 48 hours should be less than 2.5°C [21]. Thus, 40% and 50% are not recommended as the best mixing amount of rock asphalt.

3.2. Influence of Different Rock Asphalt Contents on the Physical and Mechanical Properties of Asphalt Mixture.

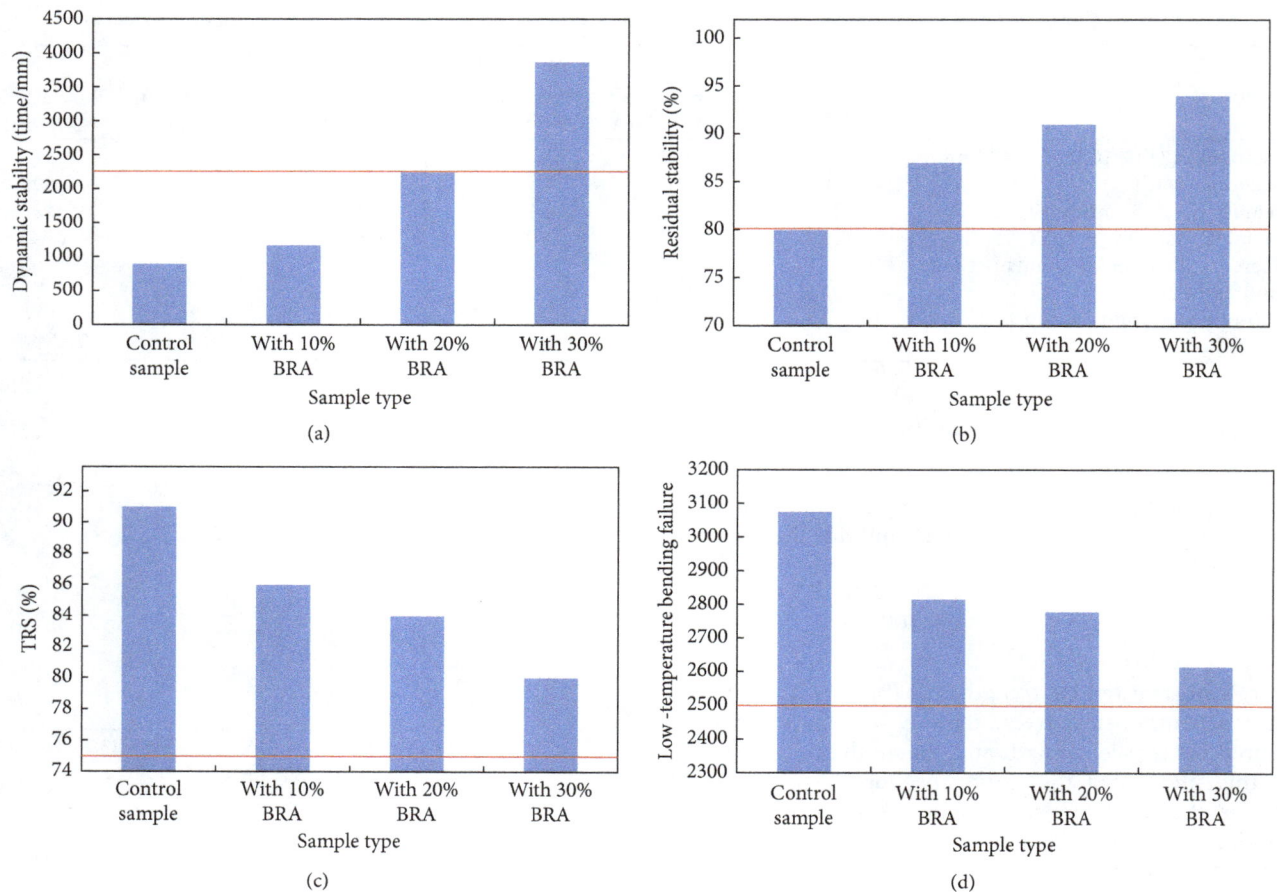

FIGURE 6: Effect of different rock asphalt contents on the performance of asphalt mixture. (a) Dynamic stability. (b) Residual stability. (c) Freeze-thaw split strength ratio TSR. (d) Low-temperature bending failure strain.

According to the above analysis, the modified Buton rock asphalt with a mixing amount of 40% or 50% has more serious segregation. Therefore, in the performance analysis of the modified Buton rock asphalt, this paper only compares the performances of the modified Buton rock asphalt mixture with a mixing amount of 10%, 20%, and 30%, respectively. And based on this, the best mixing amount of the modified Buton rock asphalt is determined. The AC-13-type asphalt mixture is selected as the test object with gradation shown in Figure 5.

The Marshall test is carried out to determine the best asphalt-aggregate ratio of the modified Buton rock asphalt with different mixing amounts. In addition, the high-temperature rutting test, low-temperature beam bending test, immersion Marshall test, and freeze-thaw splitting test are also carried out according to the specification [20]. The detailed test procedure can be found at Section 2.5. According to the technical specifications for modified asphalt pavements, the dynamic stability should be larger than 2400 time/mm, the residual stability should be larger than 80%, and the TSR value should be larger than 75% [21]. The detailed tests results are shown in Figure 6.

It can be seen from Figure 6 that the dynamic stability and residual stability of the Buton rock asphalt mixture increase with the increase of the mixing amount of rock asphalt, and the beam bending failure and freeze-thaw splitting strength ratio

FIGURE 7: The connection between the rock asphalt modification equipment and the mixing plant.

increase with the increase of the mixing amount of rock asphalt, indicating that the high-temperature resistance and antistrip performance of modified asphalt mixture increase significantly with the increase of rock asphalt content, and its low-temperature crack resistance and low-temperature freezing resistance decrease with the increase of the mixing amount of rock asphalt. When the mixing amount of rock asphalt reaches 30%, the modified Buton asphalt mixture has the best high-temperature antirutting performance and antistrip

TABLE 3: Properties of the rock asphalt-modified asphalt binder in the field.

Test items	Penetration (25°C, 100 g, 5 s) (0.1 mm)	Softening point (°C)	Ductility (15°C, 5 cm/min) (cm)	Kinematic viscosity at 135°C (Pa·s)	Difference of softening point between 0 and 48 hours (°C)	Ratio of penetrations before and after RTFOT test (%)
Value	53	51	33	0.575	1.5	83

TABLE 4: Physical properties of rock asphalt-modified asphalt mixture in the field.

Test items	Asphalt-aggregate ratio (%)	Theoretical maximum relative density	Bulk relative density	Void content (VV) (%)	Voids in mineral aggregate (VMA) (%)	Voids of mineral aggregate that are filled with asphalt (VFA) (%)
Value	4.8	2.544	2.451	3.7	14.1	74.2

TABLE 5: Mechanical properties of rock asphalt-modified asphalt mixture in the field.

Test items	Stability (MS) (kN)	Flow number (FL) (mm)	Dynamic stability (time/mm)	Residual stability (%)	Freeze-thaw split strength ratio (%)	Low-temperature bending failure strain ($\mu\varepsilon$)
Value	12.64	3.5	3376	91	81	2683

performance, and its low-temperature crack resistance and low-temperature freezing resistance also meet the regulatory requirements. Thus, it is recommended to set 30% as the best mixing amount of the rock asphalt.

3.3. Case Application.

The rock asphalt-modified asphalt pavement was implemented in the major maintenance project of asphalt pavement of the Huoqiu section of S310 Linye Road within Lian City of Anhui Province. This section was built according to the secondary highway standard, with a design speed of 60 kilometers per hour. After milling, the AC-13 asphalt mixture with a thickness of 4 cm was paved. The asphalt is the Ssangyong virgin asphalt produced in Jiangyin, South Korea The coarse and fine aggregates come from the Chaohu Zhongcai Limestone Gravel Plant, and it is composed of limestone; the mineral powder is produced by Lvan Traffic Industry Co., Ltd. The connection between the rock asphalt modification equipment and the mixing plant is shown in Figure 7.

According to the local climate and traffic characteristics, 30% rock asphalt was mixed in the matrix asphalt for processing modified rock asphalt. More details can be found in Technical Specifications for Construction of Highway Asphalt Pavements. JTG F40-2004 [21]. The specific properties are shown in Table 3.

According to the Marshall test, the best asphalt-aggregate ratio was determined. And the related road performance is tested. The specific data are shown in Tables 4 and 5. The test methods were all standard procedure. They can be found in Technical Specifications for Construction of Highway Asphalt Pavements (JTG F40-2004) [21]. As can be seen from the tables, the design requirements are met.

According to the test results, the heating temperature of the modified Buton rock asphalt is 140°C to 150°C. The aggregate heating temperature is 170°C to 180°C. The dry mixing time of the aggregate is 7 s, and the wet mixing time is 42 s. The out-feeding temperature is determined as 150°C to 160°C. The on-site construction technology includes two times of rolling by the single vibratory road roller with a weight of 13 tons, 6 times of rolling by the rubber-tired roller with a weight of 26 tons and one time of rolling for leveling carried out by the single vibratory road roller with a weight of 11 tons.

The mixture paved on the site is uniform without segregation, and the cored sample demonstrates that the compaction degree can reach 98%.

4. Conclusions

The objective of this study was to investigate the effect of different rock asphalt contents on the physical and mechanical properties of the asphalt binder and asphalt mixture by conducting mechanical tests and microstructure analyzation. The following is a summary of conclusions that can be drawn based on the aforementioned results and discussion:

(1) The activation treatment is carried out on the Buton rock asphalt powder, which effectively enhances the molecular polarity of rock asphalt, promotes the coupling of resin and ash content in the rock asphalt and the matrix asphalt, and thus forms a stable multidimensional net structure to improve the performance of the matrix asphalt.

(2) In this study, the self-developed on-site modification equipment is innovatively adopted. The activated Buton rock asphalt is added to the matrix asphalt by "wet process" to prepare the modified Buton rock asphalt which has significantly improved high-temperature resistance and aging resistance.

(3) The modified asphalt prepared by mixing 40% rock asphalt has more serious segregation. According to the comparison test of road performance, it is concluded that the modified asphalt prepared by mixing 30% rock asphalt has more balanced properties. Thus, 30% is recommend as the best mixing amount for the production of modified rock asphalt by "wet process."

(4) According to the verification of entity engineering, the modified Buton rock asphalt mixture prepared by "wet process" has better application property and workability and is suitable for large-scale production. However, due to the short completion time of the test section, the long-term performance of pavement cannot be reflected completely. The long-term performance observation of the test section will be carried out in the further work.

Conflicts of Interest

The authors declare no conflicts of interest.

Authors' Contributions

Yafei Li and Jing Chen conceived and designed the experiments; Yafei Li performed the experiments; Yafei Li and Meng Guo analyzed the data; Jin Yan contributed reagents/materials/analysis tools; Meng Guo wrote the paper.

Acknowledgments

This study was supported by Beijing Natural Science Foundation (8174071) and Quota Funds for Promoting the Connotation Development of Universities, Beijing University of Technology (004000514118017).

References

[1] N. M. Jackson and E. L. Dukatz, "An evaluation of the suitability of the superpave system to the design of limestone rock asphalt (LRA) mixes," in *Symposium on Progress of Superpave (Superior Performing Asphalt Pavement)-Evaluation and Implementation*, vol. 1322, pp. 118–125, ASTM, West Conshohocken, PA, USA, 1997.

[2] B. S. Subagio, B. Siswosoebrotho, and R. Karsaman, "Development of laboratory performance of Indonesian rock asphalt (ASBUTON) in hot rolled asphalt mix," *Proceedings of the Eastern Asia Society for Transportation Studies*, vol. 4, no. 1-2, pp. 436–449, 2003.

[3] H. L. Li, T. Liu, and H. W. Xie, "Experimental investigation on asphalt modified by Xinjiang Wuerhe rock asphalt," in *Proceedings of the 5th China/Japan Workshop on Pavement Technologies*, Xi'an, China, p. 179, 2009.

[4] D. W. Cao and Y. F. Yi, "Study of colloidal properties of natural rock asphalt," in *Proceedings of the International Workshop on Energy and Environment in the Development of Sustainable Asphalt Pavements*, Xi'an, China, p. 130, 2010.

[5] S. W. Du and C. F. Liu, "Performance evaluation of high modulus asphalt mixture with Button rock asphalt," *Advanced Materials Research*, vol. 549, pp. 558–562, 2012.

[6] F. Ma and C. Zhang, "Road performance of asphalt binder modified with natural rock asphalt," *Advanced Materials Research*, vol. 634–638, pp. 2729–2732, 2013.

[7] W. T. Huang and G. Y. Xu, "Experimental study of high temperature properties and rheological behavior of Iranian rock asphalt," *Advanced Materials Research*, vol. 671–674, pp. 1277–1281, 2013.

[8] G. Rusbintardjo, "Utilization of Buton natural rock asphalt as additive of bitumen binder in hot mix asphalt mixtures," *Advanced Materials Research*, vol. 723, pp. 543–550, 2013.

[9] J. Li, X. N. Zhang, Y. Liu, and C. H. Wu, "Study on the elemental analysis and relative molecular weight of 5U rock asphalt," *Advanced Materials Research*, vol. 734–737, pp. 2132–2135, 2013.

[10] A. Kawakami, I. Sasaki, K. Kubo, S. Ueno, M. Hermadi, and W. Pravianto, "Possibility to utilize new natural rock asphalt for guss asphalt," in *Asphalt Pavements*, pp. 1513–1520, CRC Press, Boca Raton, FL, USA, 2014.

[11] K. Zhong, X. Yang, and S. Luo, "Performance evaluation of petroleum bitumen binders and mixtures modified by natural rock asphalt from Xinjiang China," *Construction and Building Materials*, vol. 154, pp. 623–631, 2017.

[12] R. X. Li, P. Karki, P. W. Hao, and A. Bhasin, "Rheological and low temperature properties of asphalt composites containing rock asphalts," *Construction and Building Materials*, vol. 96, pp. 47–54, 2015.

[13] G. L. Zou and C. Wu, "Evaluation of rheological properties and field applications of Buton rock asphalt," *Journal of Testing and Evaluation*, vol. 43, no. 5, article 20130205, 2015.

[14] J. Li, X. X. Guo, Y. Z. Jiang, N. N. Bai, Y. Liu, and C. H. Wu, "A micro-analysis of Buton rock asphalt," *Journal of Computational and Theoretical Nanoscience*, vol. 12, no. 9, pp. 2751–2756, 2015.

[15] M. Guo, A. Bhasin, and T. Q. Tan, "Effect of mineral fillers adsorption on rheological and chemical properties of asphalt binder," *Construction and Building Materials*, vol. 141, pp. 152–159, 2017.

[16] M. Guo, Y. Q. Tan, J. X. Yu, Y. Hou, and L. B. Wang, "A direct characterization of interfacial interaction between asphalt binder and mineral fillers by atomic force microscopy," *Materials and Structures*, vol. 50, p. 141, 2017.

[17] Y. Q. Tan and M. Guo, "Using surface free energy method to study the cohesion and adhesion of asphalt mastic," *Construction and Building Materials*, vol. 47, pp. 254–260, 2013.

[18] M. Guo, Y. Q. Tan, and S. W. Zhou, "Multiscale test research on interfacial adhesion property of cold mix asphalt," *Construction and Building Materials*, vol. 68, pp. 769–776, 2014.

[19] Ministry of Transport of the People's Republic of China, *Modifiers for Asphalt Mixture–Part 5: Natural Asphalt*, JT/T860.5-2014, Ministry of Transport of the People's Republic of China, Beijing, China, 2014.

[20] Ministry of Transport of the People's Republic of China, *Standard Test Methods of Bitumen and Bituminous Mixture for Highway Engineering*, JTG E20-2011, Ministry of Transport of the People's Republic of China, Beijing, China, 2011.

[21] Ministry of Transport of the People's Republic of China, *Technical Specifications for Construction of Highway Asphalt Pavements*, JTG F40-2004, Ministry of Transport of the People's Republic of China, Beijing, China, 2004.

12

Comparison of Asphalt Mixtures Designed Using the Marshall and Improved GTM Methods

Jianbing Lv⑩,[1] Xu Zhancheng,[1] Yin Yingmei,[1] Zhang Jiantong,[1] Sun Xiaolong,[1] and Wu Chuanhai[2]

[1]*Guangdong University of Technology, Guangzhou 510006, Guangdong, China*
[2]*Guangdong Hualu Traffic Technology Co. Ltd, Guangzhou 510006, Guangdong, China*

Correspondence should be addressed to Jianbing Lv; ljbzh@126.com

Guest Editor: Ghazi G. Al-Khateeb

The Marshall method is today considered the standard method of asphalt mixture design for practical engineering applications. By using this method, engineering designers reap the benefits of its easy implementation and inexpensive equipment requirements. However, the Marshall method also has shortcomings and limitations, such as the difficulty in simulating the actual working conditions of a road under heavy load. Therefore, it is desirable to develop alternative methods for designing asphalt mixtures that can simulate the actual conditions under which the road will be used and so enable technically superior road construction. The emergence of the gyratory testing machine (GTM) method represents a new direction in asphalt mixture design that could plan more effectively for heavy loads in a hot and humid environment. In this paper, the two design methods are compared on the basis of the oil-stone ratio, high-temperature stability, water stability, and rutting resistance of the mixes they recommend. We put forward an improved GTM method suitable for the high temperatures and heavy traffic in Guangdong Province. This work provides a foundation for the large-scale popularization and application of the GTM method.

1. Introduction

The premature destruction of asphalt pavement in high-grade Chinese highways mainly occurs through the formation of grooves, oil pan, and water damage. Studies have shown that these early failure phenomena are attributable to a high asphalt content, the low density of the mixture, the degree of compaction, high porosity, or poor gradation [1–4]. At present, the most commonly used asphalt mixture design methods are the Marshall method, the Wim method, the superpave volume method, and the gyratory testing machine (GTM) method. The formation process with the GTM method simulates the actual conditions experienced by the road, enabling the design of an asphalt mixture with good antirutting performance. Due to this major advantage, GTM is gaining increasing attention in national road engineering circles [5–10].

At present, asphalt mixtures designed using the Marshall method cannot control the density of the final specimen formed, which means that the porosity cannot be adequately controlled. In theory, GTM design takes the final density of the pavement mixture as a design constraint. This significantly remedies some of the flaws inherent in the Marshall design method. The early GTM design method was mainly aimed at preventing the deformation of the rut and did not pay special attention to the durability, aging resistance, and fatigue resistance of the pavement structure. And the GTM method has not proposed a special method for the selection of aggregate gradation; hence, only the traditional grading specifications and determining methods were used (used in the Marshall design method). In addition, it is still controversial for how to use GSI and GSF indicators to determine the best asphalt ratio of asphalt mixtures. Therefore, the early GTM design method is necessary to be improved.

The density of a GTM-designed asphalt mixture at equilibrium is determined by instrumental parameters such as the machine angle, vertical pressure, and test temperature. However, it can be challenging to determine the optimal oil-stone ratio for the GTM method due to a lack of consensus as to the appropriate gyratory stability index (GSI), with some scholars advocating for using a GSI of 1, and others, a GSI of 1.03. Additionally, the performance of mixtures designed using the GTM method is not demonstrably superior, indicating that the method still needs improvement.

2. A Comparative Analysis of the Improved GTM and Marshall Methods

2.1. Selection of Raw Materials

2.1.1. Selection of Asphalt and Minerals. The asphalt used in this study is Grade A No. 70 asphalt produced by the China National Petroleum Company (CNPC). Its technical indicators are in accordance with the requirements of current Chinese regulations [11] (Table 1). The coarse aggregate is granite gravel produced by Qingyuan (stone specifications: 11–22 mm, 11–16 mm, 6–11 mm, and 3–6 mm), and fine aggregates are granite produced by Qingyuan. The filler is ground limestone produced by Conghua, and the anti-stripping agent is a cement produced in Pingtang Town. Its technical indicators are in accordance with the requirements of current Chinese regulations [11] (Table 2). To minimize variability in the test data, the aggregates were washed and sieved and then backmatched.

2.1.2. Selection of Mineral Aggregate Gradations. GTM rotary compaction and Marshall compaction tests were carried out on four kinds of AC-16-type asphalt mixtures commonly used in Guangdong Province. The high-temperature stability and water stability of the mixes produced using the two methods were compared and analyzed. The gradations of the mineral aggregate tested are listed in Table 3.

2.2. Determination of the Best Oil-Stone Ratio Using the Marshall Method.
Marshall asphalt mixture tests were carried out according to the current standard practice in China, being compacted 75 times on both sides at a compaction temperature of 140°C–150°C. The best oil-stone ratios for each gradation were determined by plotting the data and are shown in Table 4. Both light and heavy traffic were considered in determining the optimum oil-stone ratio, for which design porosities of 4.0% and 5.0%, respectively, were adopted.

Table 4 indicates the following:

(1) The characteristics of gradations 1 and 2 are very similar to each other and are consistent with past experience; gradations 3 and 4 also show very similar characteristics to each other but differ significantly from 1 and 2. It is necessary to use a higher proportion of asphalt to achieve the same porosity for gradations 3 and 4.

(2) The VMA (voids in mineral aggregate, calculated by theoretical maximum relative) with the best oil-stone ratio for gradations 1 and 2 does not meet the requirements of the specification [11]. To meet the requirements, the porosity would need to be reduced to 3.5%, significantly increasing the amount of asphalt required.

(3) New technical specifications for asphalt pavement construction have recently been published [11]. Because the absorbency of asphalt is taken into account in the calculation of mineral aggregate porosity, the calculated VMA is ~1–2% lower than that with the previous method. This has rendered it the most difficult requirement to meet in gradation design. For the materials used in this case, the composition of gradations 1 and 2 should be adjusted according to the actual materials that would be used. Adjusting the size of the gradation is the simplest way to make the indicators of the Marshall test meet the requirements.

(4) When designing for heavy traffic, the best oil-stone ratio of AC-16 asphalt was reduced by 0.3–0.5%. However, only the porosity and not the saturation and mineral aggregate gap met the requirements [12].

2.3. Improved Optimal Design of the Oil-Stone Ratio Using the GTM Method.
In the GTM test, each asphalt mixture was molded according to ASTM D3387. The rotation parameters were set to a vertical pressure of 0.7 MPa and a machine angle of 0.8° (oil pressure gauge); the specimen model was controlled as a limit equilibrium. The sample diameter was set at 101.6 mm and the mold temperature at 60°C, and the initial temperature of compaction was 140–150°C. The test results are shown in Figures 1–6.

Figures 1–6 indicate that the gyratory shear factor (GSF) of the asphalt mixtures tested is greater than 1.3 and the rotation stability coefficient, GSI, is less than 1.05. It is not, however, possible to meet the requirements of the design [11] (Table 5) using either the initial GTM design method (when the gyratory stability index, GSI, is close to 1.0, the corresponding amount of asphalt is the maximum amount of asphalt in the mixture, and when the GSF of the mixture is greater than 1.0, the mixture density reaches the maximum value) or the results of relevant research (GSI = 1.05, oil-stone ratio for GSF >1.3) as the best asphalt mix dosage standards.

It is therefore necessary to find an alternative method for determining the optimum amount of asphalt for the GTM mixture. We have devised an improved GTM asphalt mixture design methodology for selecting the best oil-stone ratio, as follows:

(1) As in Figures 1–6, data are plotted using the oil-stone ratio or weight of asphalt as the abscissa and the volume index and mechanical indicators of the GTM specimens on the vertical axis. A smooth curve is plotted to fit the results.

(2) Firstly, the asphalt dosage range OAC_{min}–OAC_{max} (OAC = optimum asphalt content) that would meet the technical standards for GTM design of an asphalt

TABLE 1: No. 70 asphalt test results.

Test	Method	Specified value	Measured value
Penetration (0.1 mm)	25°C, 100 g, 5 s	60–80	61
	15°C, 100 g, 5 s	—	20
	30°C, 100 g, 5 s	—	113
Penetration index (PI)	—	−1.5 to +1.0	−1.42
Ductility (cm)	5 cm/min, 15°C	≮100	>100
	5 cm/min, 10°C	≮15	>100
Softening point (°C)	Ring and ball method	≮46	47.0
Dynamic viscosity (Pa·s)	60°C	≮180	191.9
Kinematic viscosity (Pa·s)	135°C	—	0.365

TABLE 2: Aggregate physical and mechanical indicators.

Material name	Apparent relative density (g/cm^3)	Gross volume relative density (g/cm^3)	Needle particle content (%)	Crushing value (%)	<0.075 particle content (%)	Water absorption (%)
11–22 mm gravel	2.740	2.699	5.5	—	0.3	0.56
11–16 mm gravel	2.744	2.701	4.8	11.7	0.2	0.63
6–11 mm gravel	2.752	2.692	—	—	0.4	0.82
3–6 mm gravel (≥2.36 mm)	2.726	2.652	—	—	2.1	1.02
3–6 mm gravel (<2.36 mm)	2.705	2.644	—	—		0.96
0–3 mm gravel (≥2.36 mm)	2.725	2.639	—	—	4.8	1.21
0–3 mm gravel (<2.36 mm)	2.714	2.639	—	—		1.06
Filler	2.784	—	—	—	—	—
Cement	3.099	—	—	—	—	—

TABLE 3: Selected AC-16 asphalt mixture design gradations.

Gradation number	Sieve hole (mm) pass rate (%)											
	26.5	19	16	13.2	9.5	4.75	2.36	1.18	0.6	0.3	0.15	0.075
1	100	100	95.0	80.0	60.0	40.0	29.5	22.0	17.5	13.0	9.5	6.5
2	100	100	97.5	80.0	60.0	35.0	26.5	22.0	17.5	13.0	9.5	6.5
3	100	100	97.5	82.5	68.0	52.5	41.0	29.5	22.0	16.0	11.0	6.0
4	100	100	97.3	78.5	56.9	30.0	23.0	19.3	15.6	11.9	9.0	6.5

TABLE 4: Summary of the oil-stone ratios for asphalt mixtures selected using the Marshall method.

Gradation number		Selected oil-stone ratio (%)	Theoretical maximum relative density (g/cm^3)	Measured density (g/cm^3)	Porosity (%)	Mineral material clearance rate (%)	Saturation (%)	Marshall stability (kN)	Flow value (0.1 mm)
1	Light load	4.47	2.502	2.401	4.0	13.0	68.9	13.05	35.0
	Overload	4.17	2.512	2.387	5.0	13.2	62.3	13.20	33.5
2	Light load	4.56	2.504	2.405	4.0	13.0	69.6	11.70	33.0
	Overload	4.15	2.518	2.391	5.0	13.1	61.5	12.50	28.0
3	Light load	4.84	2.488	2.388	4.0	13.5	70.2	13.30	23.9
	Overload	4.50	2.499	2.374	5.0	13.7	63.5	14.80	24.4
4	Light load	4.98	2.488	2.388	4.0	14.0	71.3	9.60	34.9
	Overload	4.60	2.501	2.375	5.0	14.2	64.5	9.70	34.9

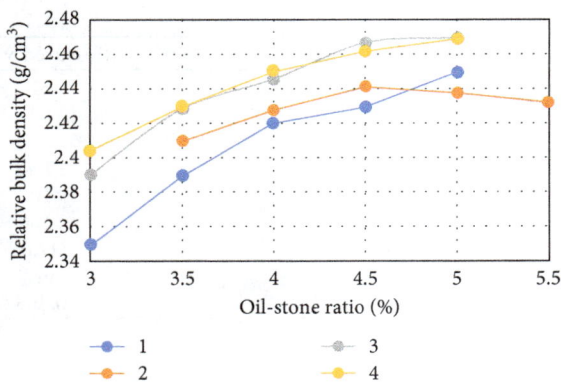

FIGURE 1: Relation between AC-16 relative bulk density and oil-stone ratio.

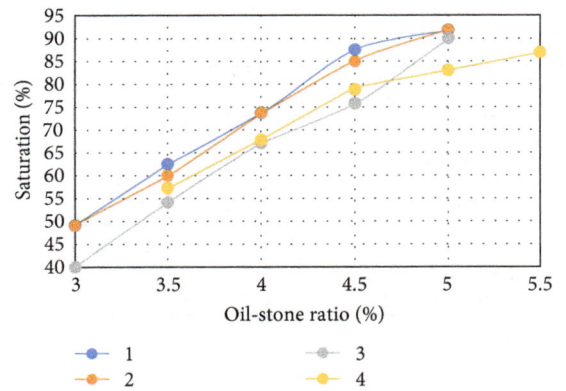

FIGURE 2: Relation between AC-16 porosity and oil-stone ratio.

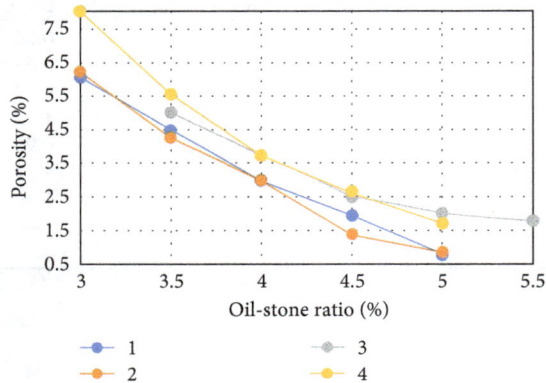

FIGURE 3: Relation between AC-16 mineral material clearance rate and oil-stone ratio.

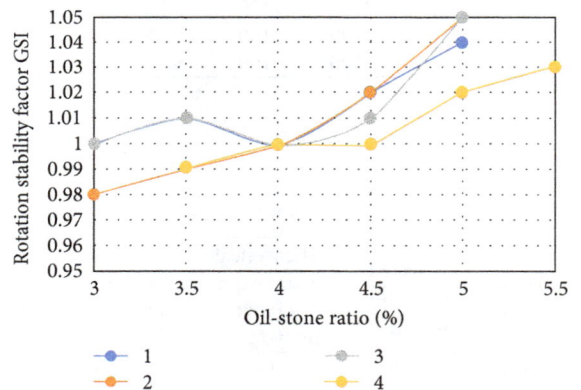

FIGURE 4: Relation between AC-16 saturation and oil-stone ratio.

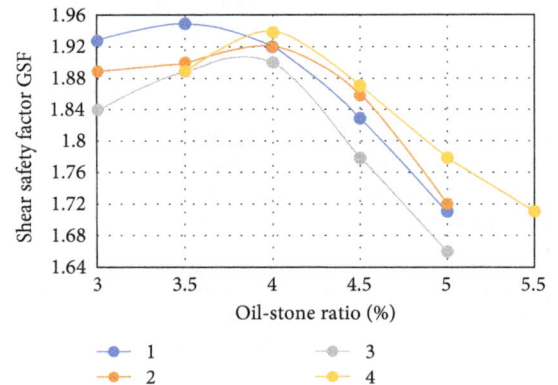

FIGURE 5: Relation between AC-16 shear safety factor and oil-stone ratio.

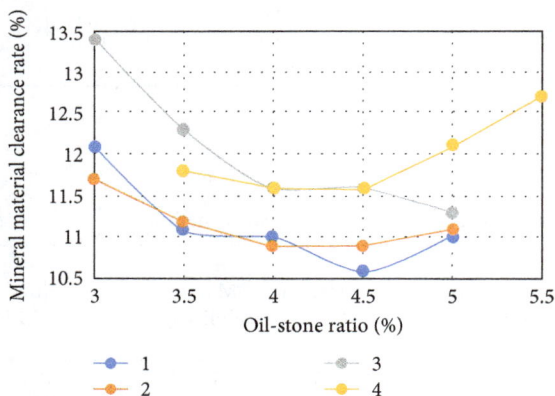

FIGURE 6: Relation between AC-16 rotation stability factor and oil-stone.

mixture (Table 5) is determined. The selected range of bitumen usage must cover the full range of porosity. Furthermore, it should cover as much of the range in the asphalt saturation requirements as possible and produce a peaked GSF curve. If the full range of design porosity is not covered, the test must be repeated.

(3) The maximum density, a_1, the maximum shear safety factor GSF, a_2, the target void fraction (or median), a_3, and the asphalt dosage, a_4, in the asphalt saturation range are taken from the curves. If the range of asphalt used in the test fails to cover the required range of asphalt saturation, the average value of a_1, a_2, and a_3 is taken as OAC_1. If the GSF or the density

TABLE 5: Technical standards for an asphalt mixture designed by the GTM method (0.7 Mpa).

Pilot projects	Technical indicators
Dimensions of the specimen (mm)	101.6 × 100
Standard density	GTM final density about rotational compaction
Porosity (%)	2.0~4.0
Saturation VFA (%)	60~80
Rotation stability GSI	≯1.05
Shear stability GSF	≮1.30
The minimum VMA requirement corresponding to the nominal particle size (mm)	
Nominal particle size (mm)	26.5 19.0 16.0 13.2
Mineral material clearance rate VMA (%)	10.0 10.0 10.5 11.0

does not reach a peak value, we take the goal porosity corresponding to bitumen quantity a_3 as OAC_1. OAC_1 must be in the range OAC_{min}–OAC_{max}, or else the design should be remixed.

(4) The median value of OAC_{min}–OAC_{max} with indicators in line with technical standards (excluding VMA) is used for OAC_2.

(5) The median of OAC_1 and OAC_2 is used as the best asphalt OAC.

(6) On the basis of the optimum amount of asphalt, we determine the voidage and check whether the VMA meets the technical requirements.

Applying this improved GTM design methodology to the experimental results shown in Figures 1–6, we determined the best oil-stone ratio for each gradation (Table 6).

2.4. Contrastive Analysis of the Two Design Methods

(1) For an asphalt mixture with the same proportions, the GTM specimen density was 1.52–3.36% higher than that from the Marshall method (Table 7). The percentage density increase varies with gradation and also differs for different oil-stone ratios at the same gradation level. The density decreases with an increase in the asphalt content.

(2) Changes in the density, porosity, and mineral void ratio of GTM specimens with a change in the oil-stone ratio are similar to those observed with the Marshall method. When the oil-stone ratio is identical, the porosity and mineral aggregate clearance rate in asphalt concrete designed by GTM are much lower than in that designed with the Marshall method. This is advantageous for the stability and durability of the road.

(3) With a change in the oil-stone ratio, the change in GTM GSF is similar to that of Marshall, and there is a peak or abrupt change point. The GSF can be used as an indicator of shear strength. It can also be used to evaluate the sensitivity of gradation shear strength to variation in the mass ratio. When the GSF for asphalt changes slowly, it can be considered that the shear strength is less sensitive to the amount of asphalt. Asphalt has better high-temperature performance when the GSF for asphalt changes slowly.

(4) The asphalt content that would be selected on the basis of the Marshall method is higher than the maximum quantity determined with the GTM method at 0.7 MPa pressure. This may lead to rutting and the emergence of oil pan. Even when using the heavy traffic standard in the Marshall method, this problem is not fundamentally resolved. Additionally, the increased porosity would lead to poor water stability, and the degree of compaction would need to be increased to 99%.

3. Performance Test of an Antirutting Asphalt Mixture Designed with the GTM Method

3.1. High-Temperature Stability. To evaluate the high-temperature stability of the asphalt concrete, a rutting test was performed, applying JTJ052-2000 regulations [13]. Specimens measuring 300 mm × 300 mm × 50 mm and with 100% compaction were used. The density of the Marshall and GTM specimens were used as the standard density. The test results are shown in Table 8.

Table 7 indicates the following:

(1) The dynamic stability of different gradations does not increase monotonously with an increase in coarse aggregate content [2, 14]; indeed, extremely high gradation may have a negative impact on high-temperature stability. When using the heavy traffic variation of the Marshall method, the dynamic stability of the rutting test improves [15–17].

(2) A mixture designed using GTM shows better dynamic stability than the one designed using Marshall and in some cases meets the requirements of a modified asphalt mixture. In addition, GTM-designed mixtures show lower relative deformation than Marshall-designed mixtures.

(3) Asphalt mixtures of the same grade may have different high-temperature properties under the two methods. Of the four selected AC-16 mixes, gradation 4's high-temperature performance was the worst under the Marshall method but the best with GTM.

3.2. Water Stability. For evaluating the water stability of asphalt mixtures, the Lottman freeze-thaw splitting method currently shows the best correlation with the behavior of actual road surfaces [11, 18]. Therefore, GTM and Marshall mixtures were tested for water stability using this method. The results of this analysis indicate the following:

(1) The splitting strength indicates that a mixture designed using the GTM method is significantly more resilient than the one designed using the Marshall method: the splitting strength of AC16 was

TABLE 6: Summary table of the optimal oil-stone ratio for selected asphalt mixtures according to the GTM method.

Gradation number	Selected oil-stone ratio (%)	Theoretical maximum relative density (g/cm^3)	Measured density (g/cm^3)	Porosity (%)	Mineral material clearance rate (%)	Saturation (%)	Shear safety factor GSF	Rotation stability factor GSI
1	4.1	2.520	2.454	2.6	10.8	75.7	1.91	1.00
2	4.0	2.518	2.445	2.9	11.0	73.6	1.93	1.00
3	4.2	2.507	2.426	3.2	11.6	72.1	1.85	1.00
4	4.3	2.511	2.436	2.9	11.6	74.2	1.89	1.00

TABLE 7: Difference in bulk density for the two methods of asphalt mixture design.

	Percentage increase of GTM forming density relative to Marshall compaction in different asphalt ratios (%)				
Gradation number	Oil-stone ratio				
	3.5	4.0	4.5	5.0	Average
1	3.36	2.64	2.62	1.52	2.54
2	2.75	2.60	2.46	2.15	2.49
3	2.84	3.33	2.40	2.59	2.79
4	2.60	2.88	2.87	2.05	2.60
Average	2.89	2.86	2.59	2.08	2.60

TABLE 8: Results of a rutting test on the selected asphalt mixtures.

Gradation number	Design method		Test temperature (°C)	Test pressure (MPa)	Dynamic stability (mm/time)	Deformation rate (%)
1	Marshall method	Light load	60	0.7	1512	8.0
		Overload	60	0.7	2080	8.4
	GTM method	0.7 MPa	60	0.7	3315	5.2
2	Marshall method	Light load	60	0.7	1632	9.6
		Overload	60	0.7	2172	9.1
	GTM method	0.7 MPa	60	0.7	3480	5.7
3	Marshall method	Light load	60	0.7	1690	8.1
		Overload	60	0.7	2356	8.4
	GTM method	0.7 MPa	60	0.7	3320	5.0
4	Marshall method	Light load	60	0.7	828	10.8
		Overload	60	0.7	1940	9.6
	GTM method	0.7 MPa	60	0.7	3753	5.6

24.6% higher before freezing and 31.1% higher after freezing.

(2) The freeze-thaw splitting residual strength ratio also indicates that a mixture designed using GTM is an improvement on the one designed using the Marshall method: it is, on average, 5.1% higher for AC16-type mixtures.

In summary, because of the differences in oil-stone ratio and void ratio between the mixtures designed by the two methods, they have significantly different degrees of water stability. It is generally believed that the water stability of the asphalt mixture is better when the oil-stone ratio is higher or the asphalt film is thicker.

4. Conclusions

(1) For asphalt mixtures with the same proportions, the density of a GTM specimen is 1.52–3.57% higher than that of a Marshall specimen. The amount of density increase varies with gradation. For a given gradation, the density decreases with an increase in the oil-stone ratio. Therefore, simply reducing the oil-stone ratio to adapt to heavy traffic conditions in the Marshall method has limited usefulness, as one cannot rely on a consistent relationship between the two variables to replicate the GTM method.

(2) When using the different methods, asphalt mixtures with the same gradation may have completely different high-temperature performance. Especially for coarsely graded mixtures, special measures must be taken to prevent the selection of asphalt with poor high-temperature performance, such as an appropriate increase in porosity or the use of GTM design.

(3) The GTM method simulates conditions in the field, and its design performance indices (final density, GSF, GSI, etc.) are directly linked to the mechanical

parameters of the road. However, it abandons the use of the asphalt mixture volume index, which was the result of much practical experience. Its design performance indices are the result of theoretical reasoning, and there is still some debate as to the optimal values. Because of these issues, a valuable approach is to combine GTM mechanical design with traditional volume design and so benefit from the advantages of both. The antirutting performance of an asphalt mixture designed in this manner is improved over that of the one designed using the Marshall method, as is the water stability due to a reduction in the void fraction.

Conflicts of Interest

The authors declare that they have no conflicts of interest.

Acknowledgments

This publication was supported by the National Natural Science Foundation of China Project (51508109 and 51608085). The authors would also like to thank all those who contributed to the experimental part of this study. And thanks are due to Guangdong Hualu Traffic Technology Co. Ltd. for providing some experimental equipment and also for the guidance of engineers Li Shanqiang and Fang Yang in the experimental side.

References

[1] A. I. Zhangfa, *Structural Analysis and Performance Evaluation of Heavy-Duty Asphalt Pavement*, Southwest Jiaotong University, Chengdu, China, 2002.

[2] Z. Li and D. Wu, "Design and research on the warm-mix rubber-modified asphalt mixture based on the GTM method," *Advanced Materials Research*, vol. 598, pp. 543–551, 2012.

[3] W. Zhou, *Asphalt Pavement Engineering Handbook*, China Communications Press, Beijing, China, 2003.

[4] D. Han, L. Wei, and J. Zhang, "Experimental study on performance of asphalt mixture designed by different method," *Procedia Engineering*, vol. 138, pp. 407–414, 2016.

[5] X. Feng and Z. Gao, "Comparative study on GTM method and marshall mixes design method for large stone asphalt mixes," in *Proceedings of GeoHunan International Conference 2009*, vol. 193, pp. 106–111, Changsha, Hunan, China, August 2009.

[6] P. Chakroborty, A. Das, and P. Ghosh, "Determining reliability of an asphalt mix design: case of marshall method," *Journal of Transportation Engineering-ASCE*, vol. 136, no. 1, pp. 31–37, 2010.

[7] R. Namli and N. Kuloglu, "Experimental comparison of superpave and marshall methods," *Teknik Dergi*, vol. 18, no. 2, pp. 4103–4118, 2007.

[8] S. D. Hwang and B. Kim, "Method mix design marshall mix design and comparative evaluation with current marshall mix design method," *Journal of the Korean Society of Road Engineers*, vol. 6, no. 4, pp. 13–24, 2004.

[9] L. J. Ma, J. Y. Zhang, and Z. Li, "The road-test of high performance asphalt mixture with superpave," *Applied Mechanics and Materials*, vol. 361-363, pp. 1576–1579, 2013.

[10] T. Fan, "Influence of mix proportion of asphalt mixture on marshall index," in *Proceedings of International Conference on Electric Technology and Civil Engineering (ICETCE)*, pp. 5123–5127, Lushan, China, April 2011.

[11] Highway Research Institute of the Ministry of Communications, *JTG F40-2004 Technical Specifications for Asphalt Pavement Construction*, China Communications Press, Beijing, China, 2009.

[12] Y. Doh and K. Kim, "Volumetric property difference in mix design results by superpave and marshall method," *Journal of the Korean Society of Road Engineers*, vol. 6, no. 4, pp. 65–73, 2004.

[13] Ministry of Communications of the People's Republic of China, *JTG F40-2004 Technical Code for Construction of Highway Asphalt Pavement*, China Communications Press, Beijing, China, 2001.

[14] C. Wu and B. Tan, "Study on verification of evaluation indexes of shear resistance of asphalt mixtures based on GTM method," *Advances of Transportation: Infrastructure and Materials*, vol. 1, pp. 171–179, 2016.

[15] S. Saride, D. Avirneni, and S. C. P. Javvadi, "Utilization of reclaimed asphalt pavements in indian low-volume roads," *Journal of Materials in Civil Engineering*, vol. 28, no. 2, article 04015107, 2016.

[16] T. Xu and X. Huang, "Investigation into causes of in-place rutting in asphalt pavement," *Construction and Building Materials*, vol. 28, no. 1, pp. 525–530, 2012.

[17] P. Ayar, F. Moreno-Navarro, and M. C. Rubio-Gámez, "The healing capability of asphalt pavements: a state of the art review," *Journal of Cleaner Production*, vol. 113, pp. 28–40, 2016.

[18] J. Mills-Beale, Z. You, E. Fini et al., "Aging influence on rheology properties of petroleum-based asphalt modified with bio-binder," *Journal of Materials in Civil Engineering*, vol. 26, no. 2, pp. 358–366, 2014.

13

Simulation of Bending Fracture Process of Asphalt Mixture Semicircular Specimen with Extended Finite Element Method

Tian Xiaoge⑩, Ren Zhang, Zhen Yang, Yantian Chu, Shaohua Zhen, and Yichao Xv

School of Traffic & Transportation Engineering, Changsha University of Science & Technology, 960 Wanjiali Road, Tianxin District, Changsha, Hunan 410114, China

Correspondence should be addressed to Tian Xiaoge; tianxiaoge@126.com

Academic Editor: Wei Zhou

In order to numerically simulate the whole fracture process including the initiation and propagation of crack in asphalt concrete semicircular specimens under external force, the extended finite element method (XFEM) was adopted considering the shortcomings of the conventional finite element method (FEM). The fracture processes of the semicircular specimens under 5 kinds of loading modes, M_e, were analyzed, and the simulation results were compared to the actual fracture paths in the actual specimens. The results indicated that the critical effective stress intensity factor will decrease first and then increase with the increase of M_e, and the XFEM simulation results are similar to that of the actual specimens in crack initiation angle and propagation path in the 5 different loading modes. It is proved that the XFEM is very effective in simulating the fracture process and has obvious advantages compared with the FEM. According to the stress state at the crack tip, the initiation angle and its propagation paths were analyzed, and it was pointed out that the increase of the shear stress component caused the crack initial angle to increase with the increase of M_e.

1. Introduction

Cracking is one of the most important and common forms of damage in asphalt pavement. It will not only destroy the integrity of the pavement structure, which will affect the pavement structural bearing capacity, but also it will reduce smoothness of the pavement surface, which will affect the driving quality and travelling comfort. Besides, rain and snow water will also penetrate into the inner of the pavement structure, which will induce water damage, so accelerate destruction of the pavement structure [1]. So, research in the crack initiation and propagation process until fracture completely of asphalt concrete under external action is of great significance, which will help to improve the crack resistance of asphalt concrete in the selection of raw materials, gradation optimization design, toughening, and crack resistant design. The expansion of cracks in three-dimensional structures along curved or tortuous paths is a thorny problem in structural numerical analysis. The assumption in traditional fracture mechanics that cracks are straight is no longer valid. Therefore, various analytical methods for crack in traditional fracture mechanics are helpless. The application of computational mechanics to simulate the random propagation of cracks in three-dimensional solid has become a hot topic in this field [2].

Owing to its great flexibility and applicability, the FEM has been widely used in numerical simulation of various fracture mechanics. However, it uses continuous function as the geometric interpolation function and stipulates that the material properties cannot mutate within the unit. So, the FEM is a suitable numerical method for continuum analysis, but there are some limitations in analyzing discontinuities such as cracks [2].

At present, the commonly used analysis methods of fracture mechanics are traditional FEM and adaptive grid [3], nodal force release method [4], cohesive force model [5], and embedded discontinuous model [6]. However, these methods have some limitations when dealing with complex shape cracks; for example, the crack propagation path must be given in advance and the crack can only extend along the cell boundary.

To overcome the shortcomings of the FEM, the XFEM was proposed by Belytschko and Black in 1999. The XFEM is based on the conventional FEM and fracture mechanical theory. It is a new method to analyze complex fracture problem. Jump function and gradual displacement shape function at the crack tip were introduced to characterize the discontinuity between two surfaces of the crack. So, the inconvenience caused by remeshing the crack tip was avoided when the fracture problem is calculated using the conventional FEM. So, the XFEM has significant advantages in solving the effect of crack initiation and its propagation. So, it becomes a hot topic in mechanical research at home and abroad [2].

Pu and Manhan analyzed the extension of cracking in asphalt concrete overlay on cement concrete pavement using the XFEM [7]. Wang et al. analyzed crack propagation characteristics of asphalt mixture with the basalt fiber utilizing the XFEM [8]. Zhang et al. analyzed the two-dimensional expansion of surface cracks in asphalt pavement using the XFEM [9]. Xigang analyzed temperature cracks in asphalt pavement using the XFEM [10], and Pirmohammad and Ayatollahi analyzed the fracture toughness of asphalt concrete in different loading modes [11]. Zhang et al. simulated the propagation process of cracking in the cement concrete specimen with the XFEM [12]. Aliha et al. investigated the mixed mode I/II crack growth behavior of a sedimentary soft rock (Guiting limestone) experimentally and theoretically using both the centre cracked circular disc (CCCD) and the edge cracked semicircular bend (SCB) specimens [13]. Pook studied a collection of case on fatigue crack paths observed in metallic materials [14]. Shahani and Tabatabaei assessed the effect of specimen geometry on the T-stress value and effect of T-stress on different fracture criteria [15]. Sajjadi et al. proposed a novel mixed mode brittle fracture criterion for crack growth path prediction under static and fatigue loading [16]. Berto and Gomez investigated the possibility to use two fracture criteria in mixed mode fracture of notched components [17]. Matvienko and Morozov predicted the angle of surface crack growth under conditions of rolling-sliding using incremental method and integral (global) method [18]. Matvienko proposed the maximum average tangential stress (MATS) criterion for predicting the direction of fracture angle [19]. In this paper, the XFEM was utilized to analyze the fracture process and its path in asphalt concrete semicircular specimens under different loading modes.

2. Basic Principles of the XFEM

The analysis domain as shown in Figure 1 can be divided into three different regions: the continuous region, crack tip region, and crack fully penetrated region.

For the continuous region, the same method used in the ordinary FEM is used in the XFEM. For the crack and adjacent regions, due to the existence of discontinuity in displacement, stress and strain in the element, and stress concentration at the crack tip, an extended shape function should be introduced to deal with them.

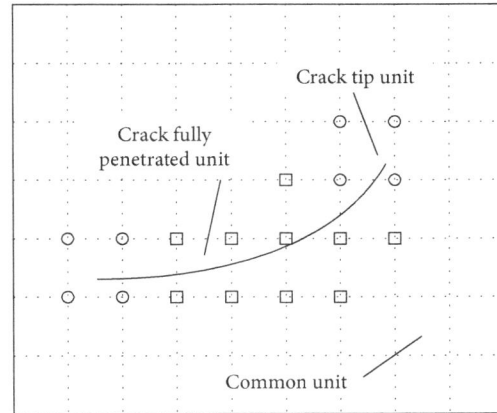

FIGURE 1: Analysis domain including a crack.

In the conventional FEM, the shape function is shown as follows:

$$u = \sum_I N_I(x)u_I, \qquad (1)$$

where N_I is the interpolation shape function of the node I and u_I is the displacement vector of the node I.

This kind of displacement mode is only applicable to units in the continuous region, but it is no longer suitable when the unit is discontinuous, such as unit containing cracks and/or holes.

Based on the conventional finite element method, an extended and discontinuous shape function was introduced in the XFEM to more accurately describe the discontinuities in the computational domain. In the XFEM, the unknown field u^h is composed of two parts:

$$u^h = \sum_I N_I(x)u_I + \psi(x), \qquad (2)$$

where $N_I(x)$ is the shape function of the node I in the conventional FEM and u_I is the degree of freedom at the node I. Both $N_I(x)$ and u_I are the same as in the FEM.

The expansion term $\psi(x)$ is used to improve the quality of characterizing of the unknown field. In the construction of the function $\psi(x)$, certain characteristics of the true solution of the unknown field, u^h, should be introduced to increase the convergence speed. In fact, it is often constructed based on the solution space of the real solution.

According to the properties of partition of unity, formula (2) can be further expressed as

$$u^h = \sum_I N_I(x)u_I + \sum_J N_J(x)\Phi(x)q_J. \qquad (3)$$

Let $\psi_J(x) = N_J(x)\Phi(x)$, so

$$u^h = \sum_I N_I(x)u_I + \sum_J \psi_J(x)q_J. \qquad (4)$$

Equation (3) or (4) is the approximate scheme of the shape function used in the XFEM. Compared with the conventional FEM, the biggest difference is the introduction of redundant degree of freedom at element nodes.

(1) For the fully penetrated crack unit, there is an abrupt change in the displacements on both sides of the crack, and the extended shape function $\psi_J(x)$ can be composed as

$$\psi_J(x) = N_J(x) \times H(f(x)), \qquad (5)$$

where $H(x)$ is the step function and $f(x)$ is the level set function. They are expressed as Equations (6) and (7), respectively:

$$H(x) = \begin{cases} 1, & x \geq 0, \\ -1, & x < 0, \end{cases} \qquad (6)$$

$$f(x) = \min\|x - \overline{x}\| \times \mathrm{sign}(n^+ \cdot (x - \overline{x})), \qquad (7)$$

where n^+ is the unit normal vector on a broken line, such as crack. For any node that is not on the broken line, $f(x)$ is the shortest distance from the node x to the broken line. And, if the location of a node x is coinciding with the direction of n^+, $f(x)$ takes a positive value. Or else, then $f(x)$ takes a negative value.

(2) The extended shape function $\psi_J(x)$ for nodes around the crack tip can be composed as follows:

$$\psi_J(x) = N_J(x) \times \Phi(x), \qquad (8)$$

where $\Phi(x)$ can be a linear combination of the following function basis:

$$\Phi(x) = \left[\sqrt{r}\, \sin\frac{\theta}{2}, \ \sqrt{r}\, \sin\frac{\theta}{2}\sin\theta, \ \sqrt{r}\, \cos\frac{\theta}{2}, \ \sqrt{r}\, \cos\frac{\theta}{2}\sin\theta \right]. \qquad (9)$$

In fact, the expansion function basis of $\Phi(x)$ is exactly the analytical displacement solution at the crack tip of the plane composite crack in linear elastic fracture mechanics [20]. The shape function of the crack tip composed with them can not only express the discontinuous property of the displacement at the crack but also accurately capture the displacement field at the crack tip. Actually, it is the reason that some information of the known solutions (such as physical properties of the obtained solution or partial analytical solution) were composed into the extended shape function. The XFEM can not only improve calculation accuracy but also significantly reduce the computation time.

So, the extended displacement mode, Equation (10), was introduced in the XFEM to describe the approximate displacement interpolation function for elements in the crack regions:

$$u = \sum_{i=1}^{N} N_I(x)\left[u_I + H(x)a_I + \sum_{\alpha=1}^{4} F_\alpha(x)b_I^\alpha \right]. \qquad (10)$$

Based on the conventional FEM, the process on crack analysis in the XFEM was divided into two parts:

(1) Meshing the whole domain while ignoring the inner boundary: cracks can pass across the unit, regardless of the boundary constraints of the unit.

(2) Extended shape function, formula (10), was used to improve the approximation space of the finite element. The other analytical processes are the same of the conventional FEM.

3. Bending Fracture Tests of Asphalt Mixture Semicircular Specimens

Semicircular specimens with prefabricated cracks were selected to conduct bending fracture tests at $-15°C$ to study the fracture resistance of asphalt concrete AC-13 under five different loading modes.

The gradation of asphalt mixture AC-13 is shown in Table 1. The materials used were basalt aggregates, limestone powder, and AH70 asphalt. The optimal asphalt-aggregate ratio of 5.3% was determined through Marshall tests [21].

The radius, R, of the specimens is 150 mm and their thickness, t, is 32 mm and the length of prefabricated crack, a, is 20 mm and its width, b, is 1.5 mm.

The different fracture loading modes at the tip of prefabricated crack are achieved through adjusting the locations (S_1 and S_2) of 2 bearings and the location of prefabricated crack (L), which are shown in Figure 2 and Table 2 [22]. The loading rate was controlled at 3 mm/min with MTS810.

The parameter, M_e, is the fracture loading mode with different positions of bearings, S_1 and S_2, and prefabricated crack L.

The bending fracture tests of semicircular specimens were conducted with a servohydraulic test machine, MTS at the temperature of $-15°C$. Before the test, specimens were kept at $-15°C$ for 4 hours in the environmental chamber to ensure that both the inside and outside of the specimen achieved the setting test temperature, $-15°C$. Then, the specimen was installed according to the locations of the bearing points in Table 2, and three point bending fracture tests were carried out. The loading rate was controlled with displacement increment at 3 mm/min. The data acquisition system collects force and displacement with time during the loading process, and the time interval of data acquisition is 0.01 s. After the test, the critical effective stress intensity factors (K_{eff}) are calculated using Equation (11) according to the maximum load P_{cr}, obtained from the tests:

$$K_{\mathrm{I}f} = Y_{\mathrm{I}}\frac{P_{\mathrm{cr}}}{2Rt}\sqrt{\pi a},$$

$$K_{\mathrm{II}f} = Y_{\mathrm{II}}\frac{P_{\mathrm{cr}}}{2Rt}\sqrt{\pi a}, \qquad (11)$$

$$K_{\mathrm{eff}} = \sqrt{K_{\mathrm{I}f}^2 + K_{\mathrm{II}f}^2},$$

where $K_{\mathrm{I}f}$, $K_{\mathrm{II}f}$, and K_{eff} are the critical effective stress intensity factors under the pure bending fracture mode, mode I, pure shear fracture mode, mode II, and mixed fracture mode, mixed mode I/II, respectively.

Sieve size (mm)	31.5	26.5	19	16	13.2	9.5	4.75	2.36	1.18	0.6	0.3	0.15	0.075
Passing rate (%)	100	100	100	100	95.2	71.8	51.6	32.9	23.6	18.3	12.1	9.3	5.8

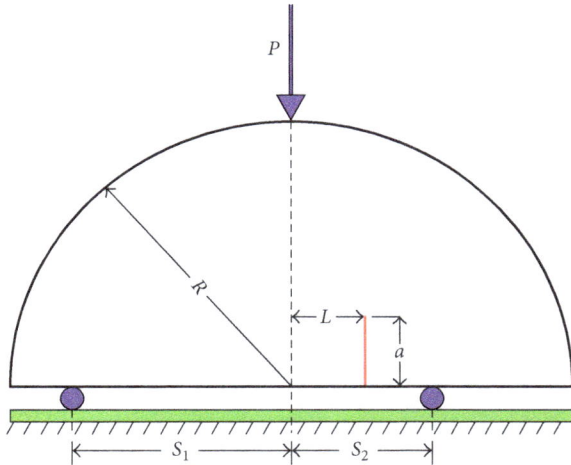

FIGURE 2: Positions of bearings and prefabricated crack in specimen.

TABLE 2: Positions of bearings and prefabricated crack at different loading modes.

Loading mode	S_1 (mm)	S_2 (mm)	L (mm)	M_e
Pure bending mode, I	50	50	0	1.0
	50	20	−2	0.8
Mixed loading mode, I/II	50	20	5	0.5
	50	20	11	0.2
Pure shear mode, II	50	20	16	0

Y_I and Y_{II} are geometric parameters reflecting the influence of the size and shape of the specimen, the position, length, and width of the prefabricated crack to the stress intensity factor. The values of Y_I and Y_{II} for different specimens and different fracture modes are listed in Table 3.

According to the above test plan, the critical effective stress intensity factor, K_{eff}, of asphalt concrete AC-13 at −15°C under different loading modes were obtained.

3.1. The Effect of Loading Mode on the Initial Angle and Propagation Path of the Crack.

The fracture propagation paths at the surface of the specimens under different loading modes are shown in Figure 3.

It can be seen from Figure 3 that the crack is initiating and propagating upward basically along the direction of the prefabricated crack in a straight line under the pure bending fracture mode, mode I (Figure 3(a)). But under other loading modes, pure shear mode, mode II, and 3 mixed modes I/II, the cracks initiate at a certain angle (which was named initiation angle) to the direction of the prefabricated crack, and then, they are propagating upward in a form of curves, finally arrived at the loading position (Figures 3(b)–3(e)).

With the decrease of the load mode parameter M_e, the values of initiation angle increases. The reason is that the horizontal tensile stress is the only action at the crack tip under the fracture mode I, the direction of the principal stress is the direction of horizontal tensile stress, so, the crack will initiate and propagate along the direction of the crack. But in the other modes (mode II or mixed mode I/II), there is the action of shear stress besides of horizontal tensile stress, so, the direction of principal stress is no longer perpendicular to the prefabricated crack surface. So, the crack will initiate at an angle to the prefabricated crack surface. The angle is named initiation angle. The value of the initiation angle is related to the relative value of the horizontal tensile stress and shear stress. The shear stress component increases with the decrease of the loading mode, M_e. And the relative values of shear stress to the horizontal tensile stress increase. So, the crack initial angle increases with the decrease of the loading mode, M_e.

3.2. Variation of Critical Stress Intensity Factor with the Fracture Mode.

The fracture load, P_{cr}, at different loading modes, M_e, is shown in Table 4, and the values of critical effective stress intensity factor (SIF) K_{eff} at different loading mode M_e were calculated and are shown in Table 4 and Figure 4.

It can be seen visually from Figure 4 that the critical effective stress intensity factor, K_{eff}, will decrease firstly and then increase with the increase of the load model parameter, M_e.

The critical effective stress intensity factor, K_{eff}, under mode I and mode II, is greater than the value of K_{eff} under all mixed loading modes. The minimum value of K_{eff} is obtained when the loading mode parameter M_e is 0.8.

4. Simulation of Bending Fracture Process in the Specimens under Different Loading Modes

The XFEM was used to analyze the initialization and propagation process of the crack in AC-13 specimens under different loading modes.

At the temperature −15°C of the bending fracture test conducted, asphalt mixture, AC-13, can be seen as an elastic material. Its elastic modulus E is 10400 MPa and Poisson's ratio is 0.25, measured via compressing modulus test at −15°C. The parameters used in the XFEM program are shown in Table 5.

The maximum displacement is the displacement when the specimen is fractured completely.

In the XFEM program, the maximum principal stress criterion was selected for fracture criterion, and the

TABLE 3: The values of Y_I and Y_{II} for different specimens.

Loading mode	S_1 (mm)	S_2 (mm)	L (mm)	M_e	Y_I (MPa·m$^{1/2}$)	Y_{II} (MPa·m$^{1/2}$)
I	50	50	0	1.0	3.734	0
I/II	50	20	−2	0.8	1.655	0.546
I/II	50	20	5	0.5	1.171	1.131
I/II	50	20	11	0.2	0.599	1.792
II	50	20	16	0	0	2.298

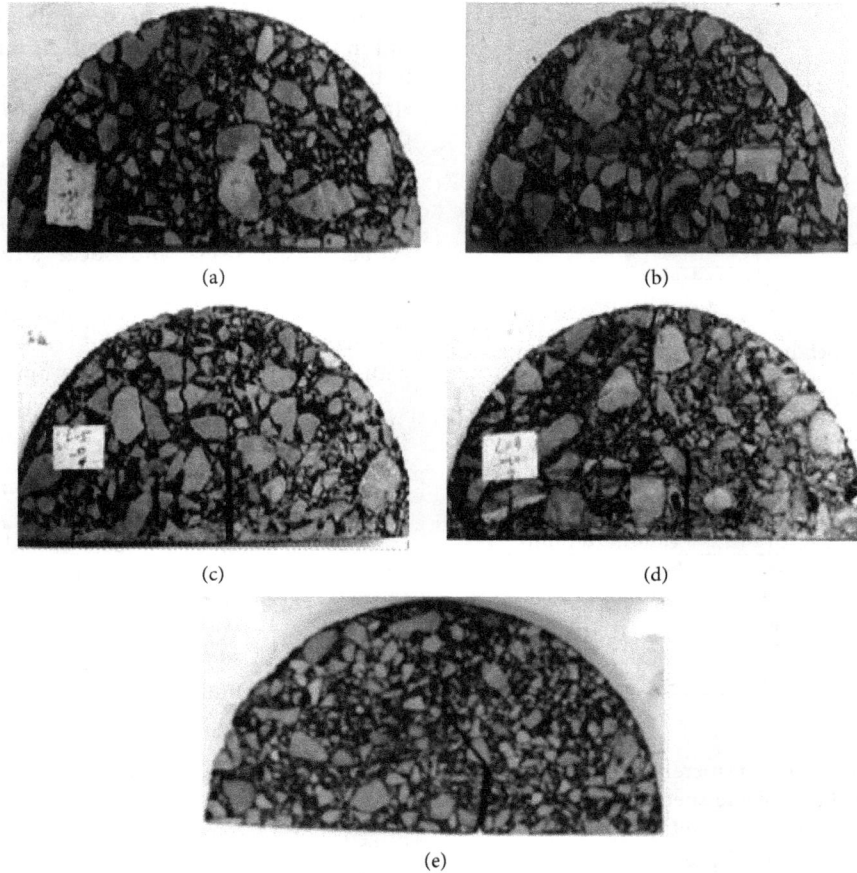

FIGURE 3: Crack routes at different loading modes. (a) M_e = 1.0. (b) M_e = 0.8. (c) M_e = 0.5. (d) M_e = 0.2. (e) M_e = 0.

displacement criterion was selected for damage evolution criterion.

4.1. Simulation of Fracture Process at Different Loading Modes

4.1.1. M_e = 1.0. The established plane strain model and its boundary conditions are shown in Figure 5. Its XFEM mesh is shown in Figure 6.

The simulation results of cracking in the XFEM are shown in Figure 7.

It can be seen from Figure 6 that the stress fields calculated with the XFEM are symmetrically distributed at the pure tension loading mode, Type I. During the crack growth, stress concentration occurs at the crack tip. It is not necessary to arrange a dense mesh at the crack tip or to

continuously redraw the mesh to satisfy the crack propagation during the analysis.

Comparing Figures 7 and 3(a), it can be found that the fracture propagation path simulated through the XFEM is basically the same as that of the actual specimen.

4.1.2. M_e = 0.8. The calculation results are shown in Figure 8.

Comparing Figures 8 and 3(b), it can be seen that in the mixed loading mode (M_e = 0.8), the initiation of the crack is no longer along with the prefabricated crack. There is a certain angle to the direction of the prefabricated crack. And then propagate upward in a curved form. It can also be seen from Figure 8 that the distributions of horizontal tensile stress and shear stress are asymmetrical.

TABLE 4: Fracture loads and critical effective SIFs at different loading modes.

Loading mode	Maximum displacement (mm)	Fracture load, P_{cr} (kN)	Effective SIF, K_{eff} (MPa·m$^{0.5}$)
$M_e = 1$	0.286	7.51	1.46
$M_e = 0.8$	0.585	10.59	0.96
$M_e = 0.5$	0.497	13.57	1.15
$M_e = 0.2$	0.801	14.40	1.42
$M_e = 0$	1.025	15.55	1.87

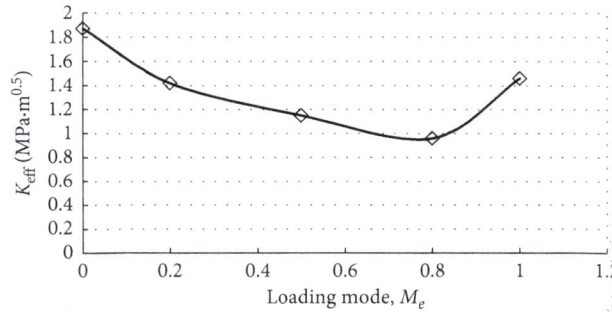

FIGURE 4: Variation of K_{eff} with loading mode M_e.

TABLE 5: Calculation parameters under different loading modes.

Loading mode, M_e	Maximum displacement (mm)	Loading speed	Elastic modulus, E (MPa)	Poisson's ratio
1	0.286			
0.8	0.585			
0.5	0.497	3 mm/min	10400	0.25
0.2	0.801			
0	1.025			

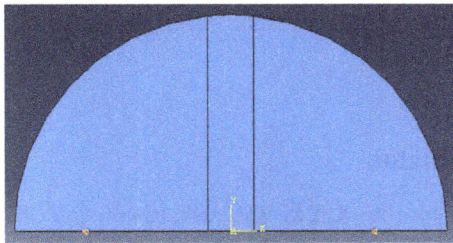

FIGURE 5: Geometric model and boundary conditions.

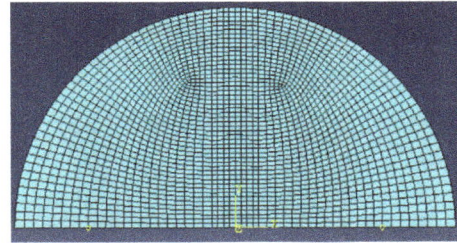

FIGURE 6: Mesh of XFEM.

Comparing Figures 8 and 3(b), the propagation paths of the crack are basically the same. The crack initiation angle (CIA) calculated with the XFEM is 29° and 30° in the actual specimen, which shows a good agreement.

4.1.3. $M_e = 0.5$. The calculation results are shown in Figure 9:

Comparing Figures 9 and 3(c), the expansion paths of the two are basically the same.

The CIA calculated with the XFEM is 33° and 32° in the actual fracture test, which shows a good agreement.

4.1.4. $M_e = 0.2$. The calculation results are shown in Figure 10:

Comparing Figures 10 and 3(d), the expansion paths of the two diagrams are basically the same.

The CIA calculated with the XFEM is 38° and 36° in the fracture test, which shows a good agreement.

4.1.5. $M_e = 0$. The calculation results are shown in Figure 11:

Comparing Figures 11 and 3(e), the crack expansion paths of the two diagrams are basically the same. The CIA calculated by the XFEM is 50° and 51° in the fracture test, which shows a good agreement.

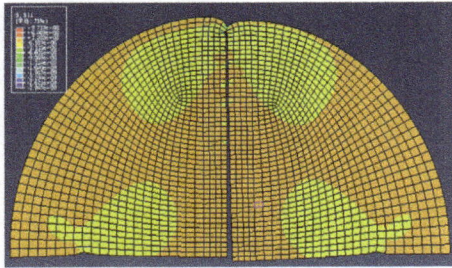

FIGURE 7: Crack propagation from XFEM.

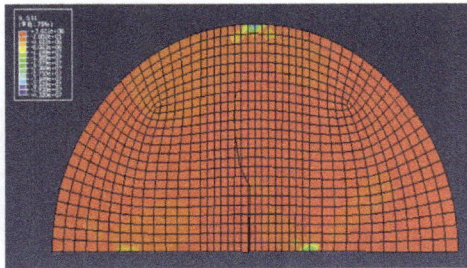

(a)

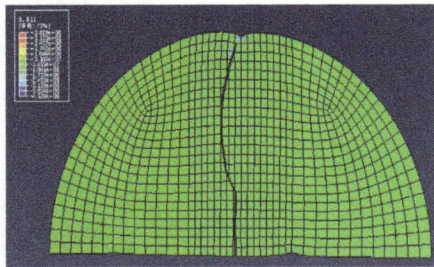

(b)

FIGURE 8: Distribution of stress and crack propagation process. (a) Horizontal tensile stress. (b) Shear stress.

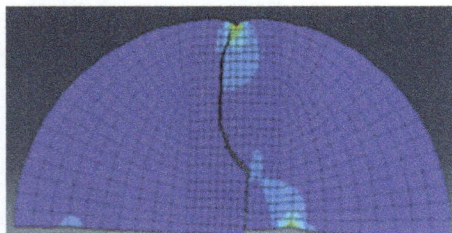

FIGURE 9: Distribution of stress and crack propagation path from XFEM.

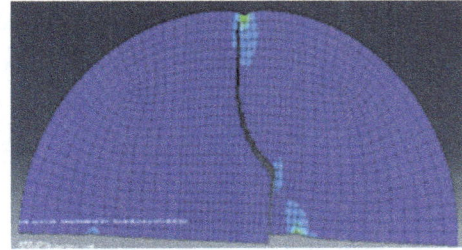

FIGURE 10: Distribution of stress and crack propagation path from XFEM.

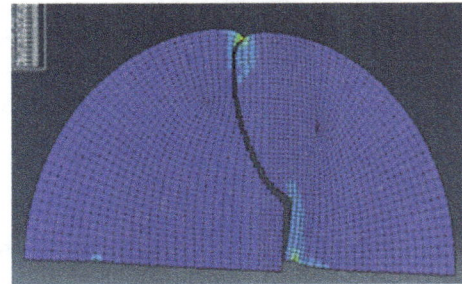

FIGURE 11: Distribution of stress and crack propagation path from XFEM.

TABLE 6: The CIA at different loading modes.

Loading mode, M_e	CIA (°)	
	Measured	Calculated with XFEM
1.0	0	0
0.8	30	29
0.5	32	33
0.2	36	38
0	51	50

4.2. The Variation of CIA to M_e. The calculated and measured crack initial angle (CIA) at crack tip under different loading modes are summarized in Table 6.

It can be seen from Table 6 that the calculated CIAs with the XFEM agree with the measured CIAs from the test. And the values of the CIA are increasing with the decrease of the loading mode. The reason is that the action of shear stress is stronger with the decrease of the loading mode, M_e.

5. Conclusion

In this paper, the XFEM was used to simulate the fracture process in asphalt concrete AC-13 semicircular specimen under −15°C with different loading modes. The following results were obtained:

(1) The critical effective stress intensity factor, K_{eff}, is decreasing firstly and then increasing with the decrease of the loading mode, M_e. And the minimum K_{eff} is achieved when the $M_e = 0.8$.

(2) The crack initiation angle (CIA) and propagation path varied with the loading mode. At the pure bending mode, the value of the CIA is 0° and it will propagate upward in a linear path, whereas under other loading modes, there is a certain angle between the cracking direction and the direction of the prefabricated crack. And then the crack propagates upward in a curved path because of the effect of shear stress at the crack tip.

(3) Under different loading modes, the crack shape, cracking propagation angle (CPA), crack propagation shape, and crack path curve of the XFEM simulation are in good agreement with those in the actual specimens, which indicated that the XFEM can be used well to calculate the crack effect and its propagation in the asphalt mixture specimen.

(4) The crack initiation angle is increased with the decrease of the value of the loading mode, M_e. This is due to the increasing effect of shear stress in the load.

Conflicts of Interest

The authors declare that there are no conflicts of interest regarding the publication of this paper.

Acknowledgments

The authors appreciate the support of the National Natural Science Foundation of China (50878032).

References

[1] K. Majidzadeh, E. M. Kauffmann, and D. V. Ramsamooj, "Application of fracture mechanics in the analysis of pavement fatigue," *Journal of Asphalt Pavement Technology*, vol. 40, pp. 227–246, 1971.

[2] Z. Zuo, Z. Liu, B. Cheng et al., *Extended Finite Element Method*, Tsinghua University Press, Beijing, China, 2012.

[3] C. Miehe and E. Gurses, "A robust algorithm for configurational force driven brittle crack propagation with R adaptive mesh alignment," *International Journal for Numerical Methods in Engineering*, vol. 72, no. 2, pp. 127–155, 2007.

[4] Z. Zhuang and P. E. O'Donoghue, "The recent development of analytical methodology for crack propagation and arrest in the gas pipelines," *International Journal of Fracture*, vol. 101, no. 3, pp. 269–290, 2000.

[5] X. P. Xu and A. Needleman, "Numerical simulations of fast crack growth in brittle solids," *Journal of the Mechanics and Physics of Solids*, vol. 42, no. 9, pp. 1397–1434, 1994.

[6] T. Belytschko, J. Fish, and B. E. Engelmann, "A finite element with embedded localization zones," *Computer Methods in Applied Mechanics and Engineering*, vol. 70, no. 1, pp. 59–89, 1988.

[7] P. Pu and S. Manhan, "Fracture propagation in asphalt overlay based on XFEM," *Highway Engineering*, vol. 37, no. 4, 2012.

[8] S. Wang, Y. Zhao, and L. Zhao, "Analysis of crack propagation characteristics of basalt mineral fiber asphalt mixture by using the extended finite element method (XFEM)," *China Science and Technology Information*, vol. 08, 2013.

[9] Z. Zhang, F. Li, Q. Yang et al., "Two-dimensional surface crack propagation in asphalt pavement using XFEM," *Chinese Journal of Civil Engineering and Management*, vol. 28, no. 2, 2011.

[10] Z. Xigang, *Finite Element Analysis of Temperature Crack in Asphalt Surface*, Zhengzhou University, Zhengzhou, China, 2012.

[11] S. Pirmohammad and M. R. Ayatollahi, "Fracture resistance of asphalt concrete under different loading modes," *Construction and Building Materials*, vol. 53, pp. 235–242, 2014.

[12] X. Zhang, Y. Ding, and X. Ren, "Extension finite element method simulation of crack propagation process in concrete," *Journal of Engineering Mechanics*, vol. 30, no. 7, 2013.

[13] M. R. M. Aliha, M. R. Ayatollahi, D. J. Smith, and M. J. Pavier, "Geometry and size effects on fracture trajectory in a limestone rock under mixed mode loading," *Engineering Fracture Mechanics*, vol. 77, no. 11, pp. 2200–2212, 2010.

[14] L. P. Pook, "The linear elastic analysis of cracked bodies, crack paths and some practical crack path examples," *Engineering Fracture Mechanics*, vol. 167, pp. 2–19, 2016.

[15] A. R. Shahani and S. A. Tabatabaei, "Effect of T-stress on the fracture of a four point bend specimen," *Materials & Design*, vol. 30, no. 7, pp. 2630–2635, 2008.

[16] S. H. Sajjadi, M. J. Ostad Ahmad Ghorabi, and D. Salimi-Majd, "A novel mixed-mode brittle fracture criterion for crack growth path prediction under static and fatigue loading," *Fatigue & Fracture of Engineering Materials & Structures*, vol. 38, no. 11, pp. 1372–1382, 2015.

[17] F. Berto and J. Gomez, "Notched plates in mixed mode loading (I + II): a review based on the local strain energy density and the cohesive zone model," *Engineering Solid Mechanics*, vol. 2017, pp. 1–8, 2017.

[18] Y. G. Matvienko and E. M. Morozov, "Two basic approaches in a search of the crack propagation angle," *Fatigue & Fracture of Engineering Materials & Structures*, vol. 40, no. 8, pp. 1191–1200, 2017.

[19] Y. G. Matvienko, "Maximum average tangential stress criterion for prediction of the crack path," *International Journal of Fracture*, vol. 176, no. 1, pp. 113–118, 2012.

[20] Z. Wang and S. Chen, *Advanced Fracture Mechanics*, Science Press, Beijing, China, 2009.

[21] JTGD50-2006, *Chinese Design Specifications of Highway Asphalt Pavement*, China Communications Press, Beijing, China, 2006.

[22] X. Tian, H. Han, X. Li et al., "Study on fracture characteristics of asphalt concrete under different loading modes," *Journal of Building Materials*, vol. 98, no. 4, pp. 150–153, 2016.

14

Emission Reduction Performance of Modified Hot Mix Asphalt Mixtures

Chaohui Wang,[1] **Qiang Li,**[2] **Kevin C. P. Wang,**[2] **Xiaolong Sun,**[1] **and Xuancang Wang**[1]

[1]*School of Highway, Chang'an University, Middle-Section of Nan'er Huan Road, Xi'an, CN 710064, China*
[2]*School of Civil and Environmental Engineering, Oklahoma State University, Stillwater, OK 74078, USA*

Correspondence should be addressed to Qiang Li; qiang.li@okstate.edu

Academic Editor: Luigi Nicolais

Three novel asphalt modifiers with pollutant emission reduction effects and new emissions measurement equipment compatible with several preexisting asphalt production systems are developed in this paper. The effects of various modifier, asphalt binder type, and gradation of hot mix asphalt (HMA) on pollutant emissions are evaluated in the lab through a comprehensive experimental design. Furthermore, road performances are monitored to evaluate the emissions reduction of modified HMA mixture for production. With increasing modifier content, the emissions reduction performance is improved markedly, with maximum reduction of 70.5%. However, the impact of modifier content on pollutant emissions reduction tends to be insignificant for dosages greater than 20% of the initial asphalt weight. Changes in asphalt type and asphalt mix gradation are found to moderately impact the emissions reduction effect. Finally, the mechanisms of emissions reduction are investigated, primarily attribute to their physical and chemical adsorption and pollutant reductive degradation characteristics.

1. Introduction

With the rapid development of civil infrastructure systems, industry, manufacturing, and many other sectors, environmental pollution has become one of the most serious problems in the world. Hot mix asphalt (HMA) is widely employed in the construction of road pavements due to its advantages during construction and field operation. However, HMA tends to emit large amounts of pollutant gases during both production and construction because of the high temperatures involved in these processes. These emissions not only pollute the air and exacerbate the greenhouse effect but are also harmful to the health of production and construction workers. This also violates the established targets for energy conservation and emission reduction [1–3]. Many existing studies have focused on reducing carbon emission during the application of HMA mixtures, including the use of warm mix asphalt (WMA), semiwarm mix asphalt (SWMA), and cold mix asphalt (CMA).

WMA releases less heat and emits less pollution during its production and application and greatly diminishes the environmental degradation associated with HMA [4–9]. The incorporation of a warm mixing agent also slows down the aging process of asphalt mixture. Shad verified that WMA develops an obvious indirect tensile strength with increased aging. Raghavendra et al. [10] conducted a comprehensive evaluation of the properties of WMA and compared a variety of these properties with the corresponding properties of HMA during asphalt production and application. Abdullah et al. [11] conducted laboratory testing to evaluate the quantity of pollutants emitted by HMA and WMA during the mixing process. They also employed a microtest method to determine the mechanism underlying the pollutant emission reductions of warm mix asphalt. Podolsky et al. [12] investigated the application of naturally occurring WMA modifiers and demonstrated that the hydrogenated glucose present in maize could reduce the mixing temperature of the asphalt mixture considerably, even to 30°C. Currently, WMA is still in the trial stage, and the additional cost associated with the preparation of warm mixing agents has limited its wider use in some ways [13]. In addition, many doubts remain regarding whether the quality and mechanical properties of WMA mixture meet the requirements of asphalt pavement construction [14, 15].

The principles underlying the production of SWMA are equivalent to those of WMA. Both employ technical methods to decrease the mixing temperature for achieving reduced pollutant emissions. The quantities of pollutants emitted by SWMA and HMA were compared by Del Carmen Rubio et al. [16] and the results showed that SWMA mixture provides for the reduced emissions of some specific pollutants at values ranging from 58% for CO_2 to 99.9% for SO_2. Botella et al. [17] conducted laboratory testing of a new semiwarm mix asphalt and demonstrated that the mechanical properties of the asphalt were a little different from those of HMA, indicating that semiwarm mix asphalt can perform well in actual engineering projects. While the use of semiwarm mix asphalt provides environmental benefits relative to HMA, its other road performance properties, such as stiffness, are not as good as those of HMA. Other considerations involve the additional cost associated with necessary equipment upgrading [18].

CMA has considerably lower heat requirements than either hot or warm mix asphalt and is therefore associated with a considerable reduction in pollution emissions, which offers considerable economic, social, and environmental benefits [19, 20]. However, cold mix asphalt is used far less often than HMA. Generally, it is used in road maintenance and requires a long setting time for good road performance and stability [21]. However, the relatively poor water stability of cold mix asphalt represents a significant challenge to its wide usage [22].

As discussed above, the use of the three alternatives (WMA modifier, SWMA modifier, and CMA modifier) to HMA has relatively profound environmental benefits and helps relieve the environmental problems associated with HMA. Nonetheless, producing those asphalt materials at lower temperatures while achieving the same high level of mechanical properties and field performance remains to be a challenge. Wang et al. [7, 8, 23] conducted a comprehensive study on the influence of various types and dosages of tourmaline on the asphalt fume density, the mechanism of hot mixing on the reduction of pollutant emissions, and the road performance of tourmaline-modified asphalt. In addition, the authors conducted a comparison of the asphalt fumes derived from ordinary modified asphalt and tourmaline-modified asphalt via emissions inspection in mixing plants and trial lots. Huang et al. [24] mixed graphite into HMA, assessed its reduction in emitted fumes, and determined the underlying emission-reduction mechanism via microexperimentation. However, the optimal experimental conditions for emissions testing and the influence of the asphalt type and the grade of the asphalt mixture on the reduction in the pollutant emissions of HMA require further study.

To reduce the pollutant emissions of HMA, this study selects three basic materials with various physical and chemical adsorption characteristics and pollutant reductive degradation principles. Subsequently asphalt modifiers are developed and employed in the preparation of modified asphalt mixtures to decrease the pollutant emissions (CO_x, NO_x, HC, and SO_2) during the HMA production and application process. In addition, new pollutant emission measurement equipment is developed for use with different asphalt production systems to determine the optimal experimental conditions for emissions testing. The effects of the modifier type and dosage, asphalt type, and grade of the asphalt mixture on the pollutant emissions of HMA are examined and quantified. Field performance tests are subsequently conducted to evaluate the pollutant emissions produced from the modified HMA mixtures during the process of spreading asphalt on the road surface.

2. Raw Experimental Materials

2.1. Selection of Raw Materials. Selection of basic materials is the first key step for the purpose of absorbing, reducing toxic substances, and reducing pollutant emission produced in the mixing and paving process of HMA. Based on physical characteristics, the chemical adsorption principle, and the reductive degradation principle, tourmaline powder, pyrite powder, and specularite powder, here denoted as T, P, and S, respectively, are selected as basic materials to develop new asphalt modifiers, which could reduce the pollutant emission of hot mix asphalt. During the selection process, the environmentally friendly principle is taken into account. Material stability, compatibility with asphalt, production conditions, and rate of repetition of use are fully considered. The characteristics of basic materials T, P, and S are shown in Table 1.

2.2. Asphalt Mixture. AC-13 asphalt mixture is selected for testing in this paper, and the asphalt is SBS I-C modified asphalt produced in Tianjin. The three indices of asphalt (penetration, softening point, and ductility), aging performance, and relative density could meet the requirements stipulated in the Technical Specifications for Construction of Highway Asphalt Pavements (JTGF40-2004) standards. Basalt aggregate and limestone filler are local materials in Tianjin and all technical indices satisfy the demands of the Standard "*Technical specification for Construction of Highway Asphalt Pavements (JTGF40-2004).*" The gradation used in this paper is AC-13, and its composite gradation is shown in Table 2. The optimal asphalt-aggregate ratio in this paper is 4.88% and volume of air voids (VV) is 4.2%.

3. Testing Method

3.1. Testing Equipment. The emission of pollutants from hot mix asphalt mainly occurs during the mixing process, which is generally conducted in mixing plants. However, most of the current studies of pollutant emission are performed through field tests [25]. However, many factors, such as open test environment, complex external influences, and poor test accuracy may result in imprecise results, which could not meet the requirements of testing and evaluation of pollutant emissions. Here, equipment for the testing of pollutant emission, as shown in Figure 1, is developed to achieve quantitative and precise elevation of pollutant emissions. The equipment consists of a sealing system, mixing system, power supply system for pollution emission transmission, test system, and waste gas treatment system.

TABLE 1: Characteristics of basic materials.

Basic materials	Crystal structure	Shape	Symmetry	Joint	Relative density (g/cm^3)	Emission reduction principle	Compatibility with asphalt	Side effect
T	Trigonal crystal	Columnar	Noncentrosymmetric	No joint	3.03–3.05	Physical adsorption, reductive degradation	Good	No toxic effect
P	Isometric system	Fine disseminated	Noncentrosymmetric	Not completely	4.9	Chemical adsorption, reductive degradation	Good	No toxic effect
S	Trigonal system	Schistose	Noncentrosymmetric	No joint	5.0–5.3	Physical adsorption, reductive degradation	Good	No toxic effect

TABLE 2: Gradation composition of AC-13.

Screen size	16	13.2	9.5	4.75	2.36	1.18	0.6	0.3	0.15	0.075
Passing rate (%)	100.0	94.8	72.3	48.2	33.5	23.3	18.0	14.4	8.7	6.6

The sealing system is mainly used to achieve the prevention of emission of pollutants in the mixing process of the asphalt mixture. The equipment includes a sealing groove and a stirring cover. The mixing system is composed of a stirring pot and a stirring paddle, and the power system of the pollution emission transmission allows pollutant emission with a steady speed from the mixing system to the test system. The test system includes the sealing test box and HA-856 gas analyzer. The purpose of establishing a waste gas treatment system is to prevent the experimental pollutants from emitting to the environment.

3.2. Performance Testing Method. The primary procedures employed in the experimental method used to assess the pollution emitted by HMA mixture are shown in Figure 2.

High-temperature stability testing is conducted at 60°C. Asphalt concrete slabs (30 cm × 30 cm × 5 cm) are rolled repeatedly along the same track using a load wheel with a loading of 0.7 MPa, simulating the wheels of a vehicle. Before the test, the asphalt should be incubated in an oven at 60°C for 6 h. The dynamic stability (DS) (i.e., the rolling times of every 1 mm rutting depth) is used to evaluate the high-temperature performance of the asphalt mixture.

The beam specimen for the bending test of the low temperature asphalt mixture is cut to the following specifications: 250 mm ± 2 mm in length, 30 mm ± 2 mm wide, and 35 mm ± 2 mm thick using an asphalt slab. Each group includes 4 specimens, which are placed in the incubator with −10°C for more than 4 h. The loading point is the midpoint of the specimen. The span is 200 mm and the experimental speed is 50 mm/min.

The residual stability of immersion Marshall test and residual strength of freezing and thawing split test are used to elevate the water stability of all asphalt mixtures. First,

standard Marshal specimens are made using a Marshall compactor, and then the stability of standard specimens is assessed. Then, one group of marshal specimens is placed in water at 60°C for 30 min and the other group is remained in the water at 60°C for 48 h. Finally, the stability of these specimens is tested. The specimen used in the residual strength of freezing and thawing split test is subjected to freezing and thawing split prior to obtaining the stability value.

4. Preparation of Modifiers

To make basic materials perform well in reducing pollutant emission, a high-energy ball mill is used to activate the potential of basic materials. After treatment with the high-energy ball mill, the specific surface area and surface energy of the three basic materials increased markedly. Moreover, the basic materials are in a nonthermodynamic stable state, and agglomeration takes place readily, which affected the performance of the materials described above. In this way, it is necessary to improve the dispersion of these three basic materials into the asphalt to prevent agglomeration and consistency problems. One or more dispersants and coupling agents are chosen for use with the three basic materials, and then the components are mixed thoroughly to make different kinds of modifiers. Finally, the best modifier would be selected after the comparison of pollutant emission reduction. The details of the preparation process are as follows.

(a) High-Energy Ball-Milling. Grind the basic material T, P, or S into micropowder though mechanical activation. Place the basic material T, P, or S into the tanks of the planetary ball mill. The selected mill ball's diameter is 10 mm, and the

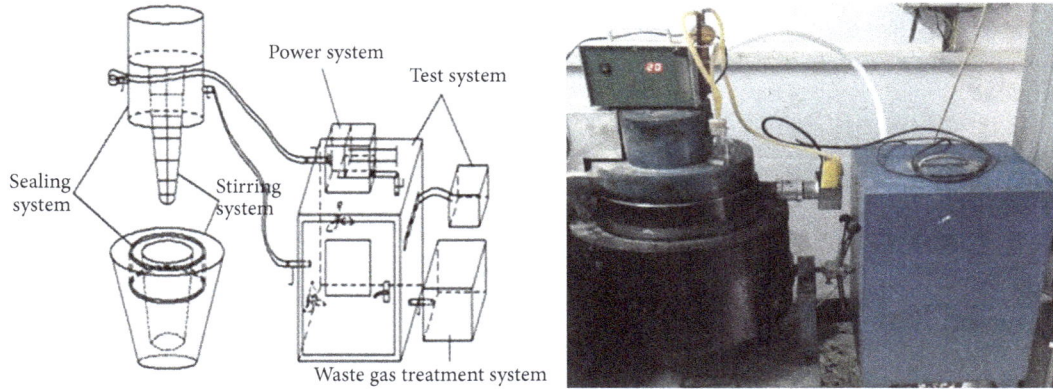

FIGURE 1: Test equipment for pollutant emission.

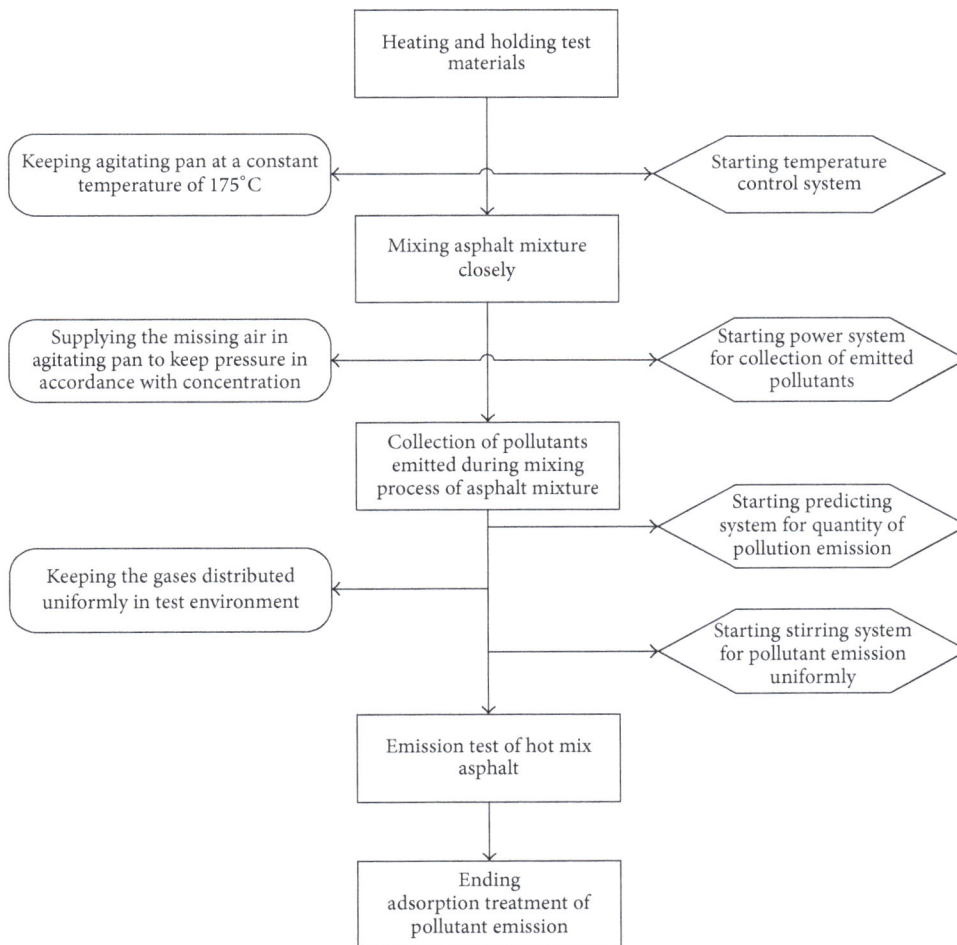

FIGURE 2: Experimental method.

weight ratio of basic materials and ball is about 1 : 20. Then seal the tanks and wash each one for about 3–5 minutes with highly pure inert N_2 gas. Finally, perform ball-milling for 30 min. The mechanical activation can enlarge the specific surface area and increase surface activity, which will improve the emission reduction performance of basic materials.

(b) Sift the Basic Materials. Sift materials T, P, and S into separate containers to remove impurities.

(c) Dry the Basic Materials in Vacuum Oven. Place the micropowder of basic material T, P, or S into a vacuum oven to dry for 5 h at 70°C and degree of vacuum 93.3–98.6 KPa. This way, vacuum drying can preserve the material's original properties well, reduce the loss of quality, and maintain the material in an easily dispersed state.

(d) Choose the Optimal Dispersant for Basic Materials. Determine the optimal dispersant for T, P, and S materials. Select one or more dispersants to improve the dispersion of basic materials T, P, and S into asphalt. Many reactions must be enhanced to improve dispersion, such as the repulsive effect and an obvious steric hindrance effect. A dispersion medium compatible with the surface wetting properties of the micropowder must be used to ensure stable dispersion of the system. Modified asphalt is made to be used for the dispersion test using SEM analysis method. The results show the optimal dispersant for basic materials T, P, and S to be sodium polyphosphate and sodium hexametaphosphate at a mass ratio of 1 : 2, sodium polyacrylate, sodium hexametaphosphate, and titanate coupling agent, respectively.

(e) Combine Basic Materials to Different Recipe. Use different proportions of dried basic materials T, P, and S to produce $(TP)_1$, $(TP)_2$, $(TS)_1$, $(TS)_2$, $(PS)_1$, $(PS)_2$, $(TPS)_1$, and $(TPS)_2$ asphalt modifiers. After blending the basic materials, the compound materials can be expected to emit less pollution gas during the process of HMA mixture production because of the synergistic effect.

(f) Determine the Dispersion of Modifiers in Asphalt. Choose the optimal dispersant and compound modifiers to prepare modified asphalt and modified asphalt mixture. The SEM analysis method is used to investigate the dispersion of basic materials in asphalt. A comparison of the SEM images taken at steps (d) and (f) shows that the dispersion of different compound modifiers is similar because they are all reasonably consistent with asphalt. No layering, cohesion, or segregation is observed in the asphalt.

(g) Determine the Emission Reduction Effect of Compound Modifiers. Choose the optimal dispersants and compound modifiers to prepare modified asphalt mixtures. Then assess the emissions associated with different modified asphalt mixtures. Here, the test results showed that, compared with base asphalt, the pollutant emission of modified asphalt decreased profoundly, and modifier $(TPS)_1$ is the most effective at reducing emissions, as indicated by the test results in step (f). The modifiers $(TP)_1$ and $(TPS)_2$ also have

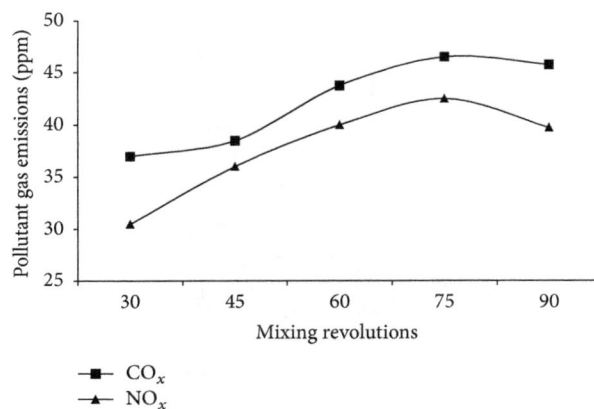

FIGURE 3: Effect of number of mixing revolutions on pollutant emissions.

better pollutant emission reduction performance than other compound modifiers, which are close to that of $(TPS)_1$.

Regarding emission reduction as the primary reference and considering the production conditions and the cost and dispersion of modifiers, the following is determined. After mechanical activation, purity, drying, compounding, and dispersal, compound modifier $(TPS)_1$ has the best comprehensive properties, followed by $(TP)_1$ and $(TPS)_2$. These compound modifiers are prepared to be used for further study, which are named as WEAM, WEP, and WES, respectively, in this paper.

5. Optimal Experimental Conditions

5.1. Mixing Revolutions. It is necessary to study the laws that govern changes in pollutant emissions from HMA mixed for different numbers of revolutions of the mixing pot. Samples mixed for 30, 45, 60, 75, and 90 revolutions are selected as the pollutant emission test conditions to identify the optimal mixing revolution. In particular, in this study, it was found that the modified asphalts with purifiers have decreased the amounts of CO_x and NO_x significantly. Due to the small amount of SO_2 emitted in this test, the emissions of NO_x and CO_x are set as the standard indicators to confirm the influence of mixing revolutions on the emission of pollution from HMA. The tests are carried out under the conditions of 175°C (mixing temperature) and 12,000 g (mixture weight). The tests mainly use the SBS I-C produced in Tianjin and AC-13 gradation. The results are shown in Figure 3.

As shown in Figure 3, the amount of pollution emitted increased gradually with the growth of mixing revolutions. After peaking at 75 revolutions, the amount of pollution emitted begins to decrease. This is because, during the process, asphalt is rapidly oxidized through mixing and the concentration of pollutants increases gradually until that of pollutant emitted from HMA mixture stops increasing. When the limited air runs up, the pollutant concentration peaks and begins to decrease. Considering the experimental error and accuracy, 75 is set as the optimal number of mixing revolutions for the pollutant emission testing.

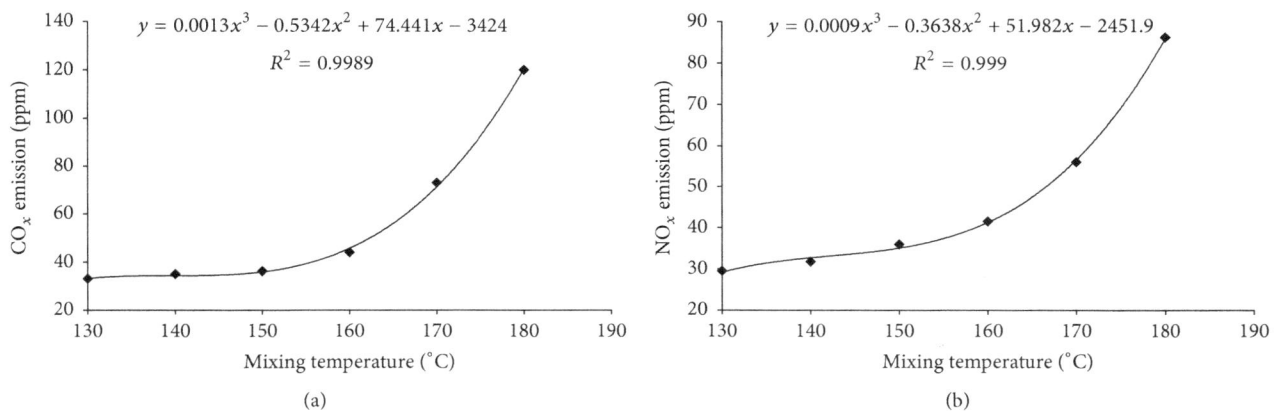

FIGURE 4: Effect of mixing temperature on pollutant emissions.

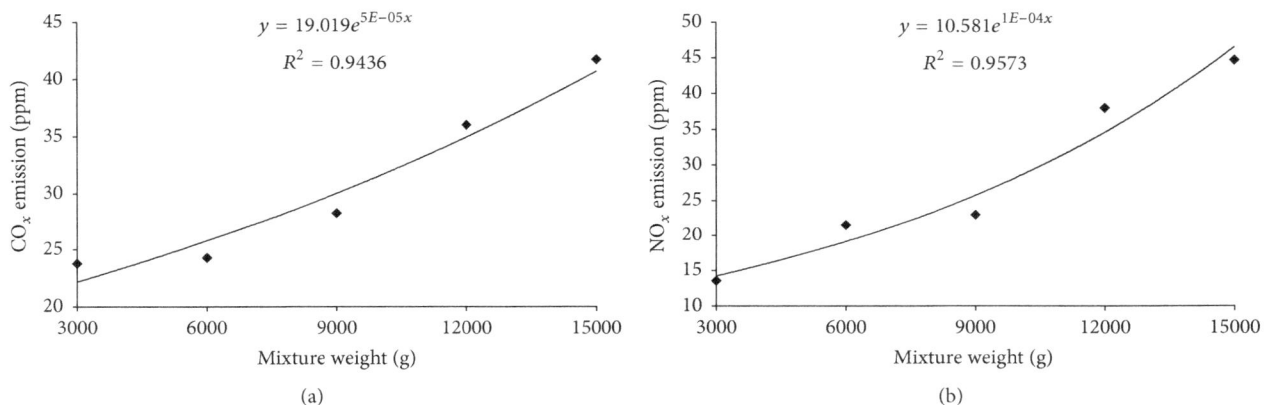

FIGURE 5: Effect of mixture weight on pollutant emissions.

5.2. Mixing Temperature. Mixing temperature is another important factor that affects the pollutant emission. 130°C, 140°C, 150°C, 160°C, 170°C, and 180°C are selected as test temperatures, in which the pollutant emissions are measured under the conditions of 12,000 g mixture weight and 75 mixing revolutions. The results are shown in Figure 4.

As shown in Figure 4, emissions gradually increase as mixing temperature increases. The pollutant emission and mixing temperature have a polynomial relationship, and the correlation coefficients of the regression curve are 0.9989 and 0.9990, respectively, which means that mixing temperature is an important factor to the pollutant emission of HMA mixture. In the relevant standard, the mixing temperature of petroleum asphalt is 140–160°C and that of modified asphalt is 160–175°C. In conclusion, 160°C is found to be the optimal mixing temperature based on the standard rules and engineering experience.

5.3. Mixture Weight. The different mixture weights are found to influence the pollutant emission because of different heated situations and the mixing degrees. Here, 3000 g, 6000 g, 9000 g, 12,000 g, and 15,000 g are selected as the test weights, in which the pollutant emission is measured at 160°C mixing

temperature and 75 mixing revolutions. The dosages of new modifiers are 15%, 20%, and 25% of asphalt weight and the percentage range of these three additives accounted for the weight of total mix is 0.72%~1.63%. The results are shown in Figure 5.

Figure 5 indicates that the pollutant emission increases gradually with the increase of mixture weight. The correlation between the emissions and mixture weight is readily visible, and the coefficients are 0.9436 and 0.9573, respectively. According to experimental observations, errors can be reduced by increasing mixture weight. To avoid the limitations of equipment and material utilization, 12,000 g is selected as the optimal mixture weight.

These results show 75 mixing revolutions, a mixing temperature of 160°C, and mixture weight of 12,000 g to be the optimal experimental conditions for measurement of pollutant emissions in the laboratory.

6. Results and Discussions

6.1. Impact of Modifier Types and Dosages on the Pollutant Emission. The effects of modifier types and dosages on the emission of NO_x and CO_x from asphalt mixtures are

TABLE 3: Emission reduction rates of different types of modified asphalt.

Dosage of modifier (%)	WEAM modifier			WES modifier			WEP modifier
	15	20	25	15	20	25	20
Emission reduction rate (%)							
NO_x	39.61	65.93	69.25	42.11	65.37	67.04	63.71
CO_x	40.45	70.50	78.09	37.08	64.89	70.20	61.80

Note: $\Delta Q = (Q1 - Q2)/Q1$;
ΔQ: emission reduction rate of HMA with emission reduction modifiers;
Q1: emissions of SBS modified asphalt mixture;
Q2: emissions of WEAM/WES/WEP modified asphalt mixture.

examined using the emission reduction rate. The WEAM, WES, and WEP modifiers are used in this test. The pollutant emission test is carried out under the optimal conditions of 75 mixing revolutions, 175°C mixing temperature, and 12,000 g mixture. The average reduction rates of the tests are calculated and shown in Table 3.

The information shown in Table 3 indicates the following:

(1) The NO_x and CO_x emissions of HMA mixture decrease visibly because of the use of WEAM, WES, and WEP modifiers, with the most pronounced reduction being 78%. This indicates that WEAM, WES, and WEP have good effects with respect to reducing emissions.

(2) Emission reduction becomes more pronounced at higher dosages of modifiers. As the dosage increases from 15% to 20%, the emission reduction rates of NO_x and CO_x increase by 55.24% and 74.29% with WEAM and 66.45% and 75% with WES. However, when the dosages increased from 20% to 25%, the reduction rates increase only slightly.

(3) There is no large difference in the emission reduction rates of CO_x among the three new modifiers when the dosage is 20%. With the increase of WES or WEAM, both emissions of NO_x and CO_x show the same variation trend. However, the emission reduction performance of WEAM is better than that of WES as indicated by the general downward trend of CO_x emissions.

6.2. Binder Type and the Pollutant Emissions. The constituents of asphalt are fairly different for the types and grades of asphalt mixture. It is necessary to study the influence of different types of asphalt on the emission reduction performance of HMA by emission tests. The tests are performed on various asphalts from different production areas. The tests are carried out under the optimal conditions, in which the dosages of new modifiers are 20% of asphalt weight and the percentage range of these three additives accounted for the weight of total mix is 0.96%~1.3%. The results are shown in Table 4.

The information presented in Table 4 indicates the following conclusions:

(1) The tests are performed using 70# matrix asphalt from different production areas with the WEAM,

TABLE 4: Type of asphalt and emission reduction rate.

Asphalt type	Modifier types	Emission reduction rate (%)	
		NO_x	CO_x
Shandong 70# matrix asphalt	WEAM	65.93	70.50
	WES	65.37	64.89
	WEP	63.71	61.80
Shanxi 70# matrix asphalt	WEAM	26.10	42.60
	WES	73.75	50.00
	WEP	62.37	39.00
Tianjin 70# matrix asphalt	WEAM	44.06	31.94
	WES	4.57	18.32
	WEP	12.33	12.04
SK-70# matrix asphalt	WEAM	45.41	45.41
	WES	26.53	19.39
	WEP	22.96	22.96
Shell 90# matrix asphalt	WEAM	42.06	32.51
	WES	27.90	22.17
	WEP	45.06	33.99
A area SBS asphalt	WEAM	39.38	39.38
	WES	15.03	17.10
	WEP	14.50	21.24
Shanxi SBS asphalt	WEAM	64.42	63.25
	WES	30.73	42.17
	WEP	28.03	48.43
Tianjin SBS asphalt	WEAM	31.28	29.75
	WES	16.11	18.25
	WEP	35.55	26.75
50# matrix asphalt	WEAM	20.57	14.09
	WES	20.83	19.17
	WEP	11.72	38.45
20# hard asphalt	WEAM	43.94	42.62
	WES	23.38	19.02
	WEP	32.39	27.54

WES, and WEP modifiers mixing into them and their emission reduction performances are found to be different. Modified asphalt from Shandong and Shanxi performs better than the others. The emission

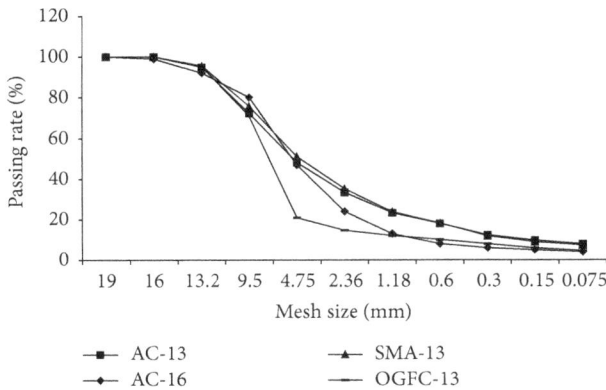

FIGURE 6: Different gradations of HMA mixture.

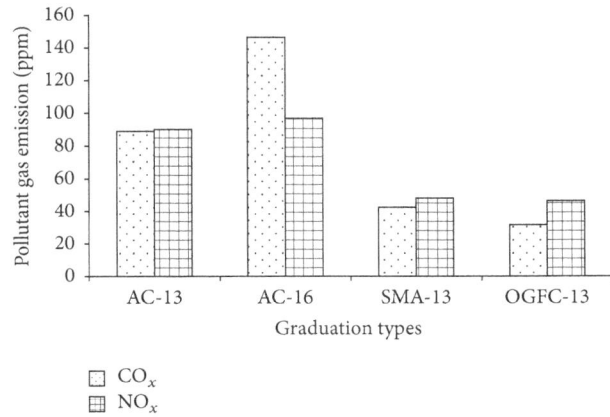

FIGURE 7: Gradation types and pollutant emissions.

reduction performance of WEAM modified asphalt is the best among the three modifiers.

(2) The SBS modified asphalt produced in Shanxi Province, modified with WEAM, WES, and WEP modifiers, shows the optimal emission reduction performance. When SBS asphalt is modified by WEAM, WES, and WEP, asphalt mixture modified by WEAM shows the best performance of emission reduction.

(3) WEAM, WES, and WEP have different emission reduction effects on matrix asphalt of different labels and production areas. The WEAM and WES have poor effect on the emission reduction performance of the 50# matrix asphalt and the 70# matrix asphalt, respectively. The improving effect of WEP modifier on matrix asphalt does not show any obvious rules.

(4) WEAM, WES, and WEP are found to have different effects on the emission reduction performances of different asphalts. The types and production areas are also associated with differences in asphalt components. The free asphalt from the mixture is adsorbed by WEAM, WES, and WEP to varying degrees, and it caused the differences in emission reduction performance.

6.3. Impact of Mix Gradation on the Pollutant Emission. The effect of gradation types on the pollutant emission of HMA mixture is assessed in this section. The AC-13, AC-16, SMA-13, and OGFC-13 asphalt mixtures are prepared for the emission reduction test, and the optimal asphalt-aggregate ratios are 4.88%, 4.8%, 6.5%, and 5.3%, respectively. The composite gradations of HMA mixtures used in this section are shown in Figure 6, and the pollutant emissions of asphalt mixture of different gradations are shown in Figure 7.

AC-16 and AC-13 asphalt mixtures emit more pollutant gas than SMA-13 and OGFC-13 mixtures. The NO_x emission of AC-16 asphalt mixture is about 6.72% higher than that of the AC-13 asphalt mixture. There are no large differences in the pollutant emissions of SMA-13 and OGFC-13 asphalt mixtures. The results indicate that the pollutant emissions of HMA are affected by the composite gradations of HMA mixture to an extent.

Based on the four composite gradations detailed above, the HMA mixture is prepared through addition of WEAM, WES, and WEP modifiers at 20% asphalt weight and the percentage range of these three additives accounted for the weight of total mix is 0.96%~1.3%. The results are shown in Figure 8. The information shown in Figure 8 indicates the following:

(1) The NO_x emission reduction performance of WEAM based on the AC-13 and AC-16 asphalt mixture is obvious, but, for the SMA-13 and OGFC-13 asphalt mixtures, WEAM shows poor NO_x emission reduction performance. The pollutant reduction effect of WEAM modifier on the AC-13 asphalt mixture is 96.86% and 90.22% more pronounced than on the SMA-13 and OGFC asphalt mixtures. In conclusion, the emission reduction performance of the new modifiers is affected by various gradations.

(2) The CO_x emission reduction performance of three new modifiers on the AC-13 and AC-16 asphalt mixture is significant. However, there are static differences in terms of CO_x emission reduction effects of WEAM, WES, and WEP modifiers on the SMA-13 and OGFC-13 asphalt mixture.

(3) The different gradations show different effects on the emission reduction performance of HMA mixtures. The asphalt membrane formed on the surface of coarse and fine aggregate of different asphalt mixtures show pronounced variations in thickness. The asphalt content is affected by the dosage of mineral powder and modifier. Due to the lasting effect on the mixture with various gradations by mixing machine, the inner stress of WEAM, WES, and WEP is always changed, which leads to differences in the emission reduction performance of HMA mixture.

6.4. Field Performance. The field performance tests of different modified asphalt mixtures are evaluated in this section to assess the effects of WEAM, WES, and WEP modifiers on the road performances of HMA mixtures. The road

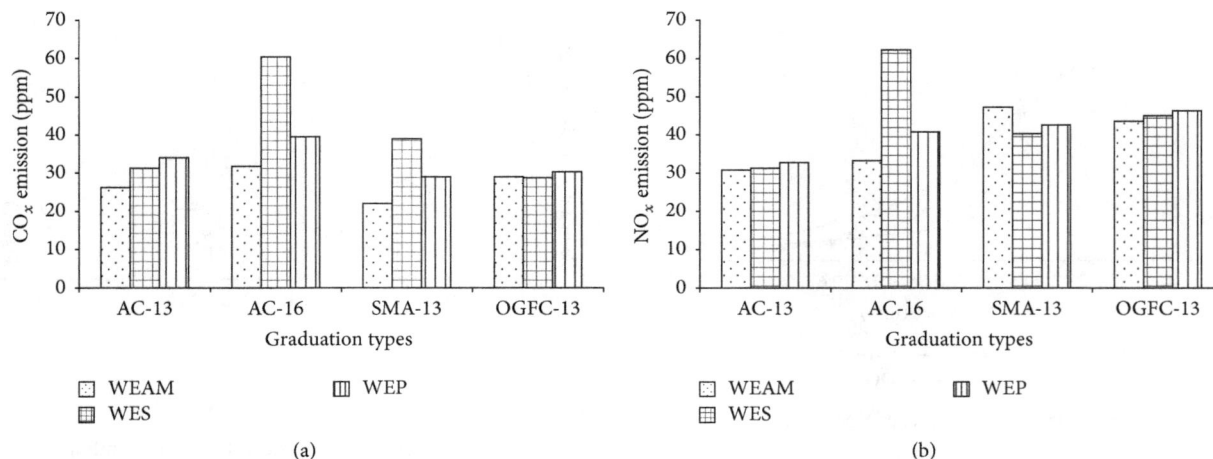

FIGURE 8: Different modifiers and gradations and the emission reduction rate.

TABLE 5: Road performance and various modified asphalt mixture.

Asphalt type	Dosage (%)	Dynamic stability (times/mm)	Split-tensile strain energy (kJ/m^3)	MS (%)	TSR (%)
SBS	—	8169	16,877.58	91.07	87.23
SBS -WEAM	15	10,329	21,215.53	92.97	91.32
	20	9673	18,212.17	96.73	95.86
	25	12,708	18,146.04	97.74	97.63
SBS -WES	20	9321	20,165.34	93.22	94.82
SBS -WEP	20	9012	17,246.27	94.00	94.09
Specification	—	≥2800	—	≥85	≥75

performance tests use the SBS I-C produced in TianJin and AC-13 gradation. The results are shown in Table 5.

As shown in Table 5, after adding the WEAM, WES, and WEP modifiers, the dynamic stability, split-tensile strain energy, residual stability, and splitting strength of asphalt mixture are increased significantly, which mean that the high-temperature performance, low-temperature performance, and water stability of asphalt mixture are improved by the addition of emission reduction modifier. Among different road performances, the high-temperature performance and water stability of WEAM modified asphalt mixture are superior to those of other modified asphalt mixtures, and WES modified asphalt mixture shows better low-temperature performance.

7. Mechanisms of Emission Reduction

7.1. Adsorption Mechanism. During the mixing of asphalt mixture at high temperatures, the changes in the polarization intensity of WEAM, WEP, and WES are caused by the changes in the temperature of asphalt mixture. Then, the bound charges are not completely shielded by the original free charge, and the free charges begin to appear on the surfaces of the modifiers. The particles with positive charges and pollutant gas molecules, derived from the mixing process, are adsorbed by the free charge, which lead to sedimentation.

However, the electric field forms around the modifiers through the secondary polarization reaction, which is caused by high temperature of HMA mixture. The charged particles from the emission of asphalt mixture are attracted and repelled by the electric field, and the charged particles with low weight are adsorbed. In this way, the dust and the pollutant gas of NO_x and CO_x inside asphalt fume are adsorbed, which produce the emission reduction effect on the pollutant gas and particles. The adsorption mechanism of emission reduction is shown in the following formula:

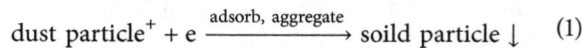

$$\text{dust particle}^+ + e \xrightarrow{\text{adsorb, aggregate}} \text{soild particle} \downarrow \quad (1)$$

7.2. Degradation Mechanism. After the WEAM, WEP, and WES added to the asphalt mixture, the inner polarization effect of the modifiers is activated due to the high-activity temperature field, which form a high-strength electrostatic field around the thickness of ten microns on the surface. The high-strength electrostatic field interacts with the atmosphere electrostatic field, which could generate DC static spontaneously and continuously [26, 27]. The DC static is capable of catalyzing and deoxidizing a part of pollutant particles into CO_2 and H_2O. Furthermore, when the high-strength electrostatic field comes into contact with the deoxidized H_2O, the instantaneous discharge transferred H_2O to HO^- and H^+ by electrolysis. Under the activating effect of the electric field, the negative ion will be produced by the reaction between HO^-, H_2O, and acid gas molecules from the air, which would consume a large amount of pollutant gas, such as CO_2 and CO. Infrared rays released from the modifiers could activate and degrade NO_x through catalysis of the reaction between NO_x and HO^-. The degradation mechanism of emission reduction is shown in the following formula.

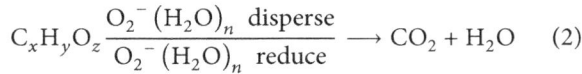

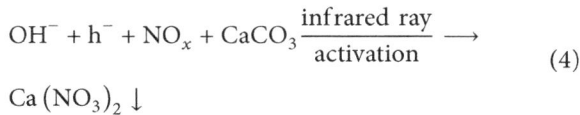

$$C_xH_yO_z \xrightarrow[\text{$O_2^- (H_2O)_n$ reduce}]{\text{$O_2^- (H_2O)_n$ disperse}} CO_2 + H_2O \quad (2)$$

$$CO_2 + H_2O + HO^- \longrightarrow CO_4^{4-}(H_2O)_n \quad (3)$$

$$OH^- + h^- + NO_x + CaCO_3 \xrightarrow[\text{activation}]{\text{infrared ray}} \longrightarrow$$
$$Ca(NO_3)_2 \downarrow \quad (4)$$

8. Conclusions

In this study, three new asphalt modifiers, denoted as WEAM, WEP, and WES, are developed and added into modified asphalt mixtures to reduce pollutant emissions. The effects of modifier type and dosage, binder type, and gradation of the asphalt mixture on the pollutant emission reductions of HMA mixtures are assessed via various experimental testing methods using newly developed equipment from this study. Road performance tests are subsequently conducted to evaluate the practical effect of modified HMA mixtures during application. On this basis, the mechanisms of emission reduction are investigated. The main conclusions of this study are as follows.

(i) Three basic materials are selected based on physical and chemical adsorption properties. New asphalt modifiers, named as WEAM, WEP, and WES, are developed after mechanical activation, purity, drying, compounding, and dispersal.

(ii) A new asphalt mixing system is independently designed for the study and refitted to measure the pollutant emissions from asphalt mixtures. The optimal experimental conditions are determined for pollutant emissions measurements. The optimal number of mixing revolutions, mixing temperature, and mixture weight are 75, 160°C, and 12,000 g, respectively.

(iii) Asphalt mixtures modified with WEAM, WES, and WEP can significantly reduce pollutant emissions relative to the unmodified asphalt. The extent of emissions reduction increases gradually with an increasing WEAM or WES dosage. When the dosage increases from 20% to 25% relative to the initial asphalt weight, both emission reductions of NO_x and CO_x tend to stabilize. It is observed that WEAM is outperforming WES in the reduction of CO_x emission.

(iv) The emission-reduction performance of HMA is affected to some extent by the asphalt binder type and mixture grade. For several particular types of binder, WEAM-modified asphalt mixtures achieve the best emission reduction performance.

(v) The road performance of HMA mixtures is significantly improved when the WEAM, WES, and WEP modifiers are used. During the production and application of the HMA mixture, the inner polarization effect of WEAM, WES, and WEP is enhanced with higher temperature of the asphalt mixture, enabling the absorption and degradation of different pollutants and thus reducing the emissions of the HMA mixture.

This paper provides novel asphalt modifiers to reduce the pollutant gas emitted from the production and construction of HMA mixtures. Future work includes further study on the application method and practical application effect in real engineering projects.

Conflicts of Interest

The authors declare that they have no conflicts of interest regarding the publication of this paper.

Acknowledgments

This paper describes research activities mainly requested and sponsored by the *Science and Technology projects of Ministry of Housing and Urban-Rural Development of the People's Republic of China* (Program 2014-R1-019), *Natural Science Basic Research Plan in Shaanxi Province of China* (Program no. 2014JM2-5045), *Key Scientific and Technological project in Henan Province of China* (Program no. 152102210113), and *Fundamental Research Funds for the Central Universities* (Program no. 310821162013). That sponsorship and interest are gratefully acknowledged.

References

[1] S. Swaroopa, A. Sravani, and P. K. Jain, "Comparison of mechanistic characteristics of cold, mild warm and half warm mixes for bituminous road construction," *Indian Journal of Engineering and Materials Sciences*, vol. 22, no. 1, pp. 85–92, 2015.

[2] X. Peng and Z. Li, "Study on regularity of fumes emitting from asphalt," *Intelligent Automation & Soft Computing*, vol. 16, no. 5, pp. 833–839, 2010.

[3] J. Zhang, E. R. Brown, P. S. Kandhal, and R. West, "An overview of fundamental and simulative performance tests for Hot Mix Asphalt," *Journal of ASTM International*, vol. 2, no. 5, pp. 205–219, 2005.

[4] P. Q. Cui, S. P. Wu, Y. Xiao, and H. H. Zhang, "Experimental study on the reduction of fumes emissions in asphalt by different additives," *Materials Research Innovations*, vol. 19, pp. S1158–S1161, 2015.

[5] M. C. Rubio, G. Martínez, L. Baena, and F. Moreno, "Warm mix asphalt: an overview," *Journal of Cleaner Production*, vol. 24, pp. 76–84, 2012.

[6] S. Sargand, M. D. Nazzal, A. Al-Rawashdeh, and D. Powers, "Field evaluation of warm-mix asphalt technologies," *Journal of Materials in Civil Engineering*, vol. 24, no. 11, pp. 1343–1349, 2012.

[7] C. H. Wang, T. T. Jiang, H. He, M. Y. Hou, and X. Q. Wang, "Function of warm-mixed flame retardant OGFC asphalt mixture," *Materials Review*, vol. 29, no. 4, pp. 122–128, 2015.

[8] C. Wang, P. Wang, Y. Li, and Y. Zhao, "Laboratory investigation of dynamic rheological properties of tourmaline modified bitumen," *Construction and Building Materials*, vol. 80, pp. 195–199, 2015.

[9] H. He, C. Wang, X. Sun, X. Wang, and X. Wang, "Preparation and road performance of new warm-mix modified asphalt," *Journal of Building Materials*, vol. 17, no. 5, pp. 927–932, 2014.

[10] A. Raghavendra, M. S. Medeiros, M. M. Hassan, L. N. Mohammad, and W. King, "Laboratory and construction evaluation of warm-mix asphalt," *Journal of Materials in Civil Engineering*, vol. 28, no. 7, 2016.

[11] M. E. Abdullah, M. R. Hainin, N. I. M. Yusoff, K. A. Zamhari, and N. Hassan, "Laboratory evaluation on the characteristics and pollutant emissions of nanoclay and chemical warm mix asphalt modified binders," *Construction and Building Materials*, vol. 113, pp. 488–497, 2016.

[12] J. H. Podolsky, A. Buss, R. C. Williams, and E. Cochran, "Comparative performance of bio-derived/chemical additives in warm mix asphalt at low temperature," *Materials and Structures*, vol. 49, no. 1, pp. 563–575, 2016.

[13] B. Prowell, G. Hurley, and E. Crews, "Field performance of warm-mix asphalt at national center for asphalt technology test track," *Transportation Research Record*, vol. 1998, 2007.

[14] A. Almeida-Costa and A. Benta, "Economic and environmental impact study of warm mix asphalt compared to hot mix asphalt," *Journal of Cleaner Production*, vol. 112, pp. 2308–2317, 2016.

[15] S. D. Capitão, L. G. Picado-Santos, and F. Martinho, "Pavement engineering materials: review on the use of warm-mix asphalt," *Construction and Building Materials*, vol. 36, pp. 1016–1024, 2012.

[16] M. Del Carmen Rubio, F. Moreno, M. J. Martínez-Echevarría, G. Martínez, and J. M. Vázquez, "Comparative analysis of emissions from the manufacture and use of hot and half-warm mix asphalt," *Journal of Cleaner Production*, vol. 41, pp. 1–6, 2013.

[17] R. Botella, F. Pérez-Jiménez, R. Miró, F. Guisado-Mateo, and A. Ramírez Rodríguez, "Characterization of half-warm—mix asphalt with high rates of reclaimed asphalt pavement," *Transportation Research Record: Journal of the Transportation Research Board*, vol. 2575, 2016.

[18] J. C. Nicholls and D. James, "Literature review of lower temperature asphalt systems," *Proceedings of Institution of Civil Engineers: Construction Materials*, vol. 166, no. 5, pp. 276–285, 2013.

[19] S. Al-Busaltan, H. Al Nageim, W. Atherton, and G. Sharples, "Mechanical properties of an upgrading cold-mix asphalt using waste materials," *Journal of Materials in Civil Engineering*, vol. 24, no. 12, pp. 1484–1491, 2012.

[20] S. Al-Busaltan, H. Al Nageim, W. Atherton, and G. Sharples, "Green Bituminous Asphalt relevant for highway and airfield pavement," *Construction and Building Materials*, vol. 31, pp. 243–250, 2012.

[21] B. Gómez-Meijide, I. Pérez, and A. R. Pasandín, "Recycled construction and demolition waste in cold asphalt mixtures: evolutionary properties," *Journal of Cleaner Production*, vol. 112, pp. 588–598, 2016.

[22] C. Ling, A. Hanz, and H. Bahia, "Measuring moisture susceptibility of cold mix asphalt with a modified boiling test based on digital imaging," *Construction and Building Materials*, vol. 105, pp. 391–399, 2016.

[23] C.-H. Wang, Y.-W. Li, R. Li, Y.-Z. Zhao, and P. Wang, "Preparation of low-carbon multi-function tourmaline modified asphalt and its performance evaluation," *China Journal of Highway and Transport*, vol. 26, no. 5, pp. 34–41, 2013.

[24] G. Huang, Z.-Y. He, C. Zhou, and T. Huang, "Suppression mechanism of expanded graphite for asphalt fume and dynamic performance of asphalt mixture of fume suppression," *China Journal of Highway and Transport*, vol. 28, no. 10, pp. 1–10, 2015.

[25] A. A. Cascione, R. C. Williams, W. G. Buttlar et al., "Laboratory evaluation of field produced hot mix asphalt containing post-consumer recycled asphalt shingles and fractionated recycled asphalt pavement," *Journal of the Association of Asphalt Paving Technologists*, vol. 80, pp. 377–418, 2011.

[26] D. Lee, S. H. Baek, T. H. Kim et al., "Polarity control of carrier injection at ferroelectric/metal interfaces for electrically switchable diode and photovoltaic effects," *Physical Review B—Condensed Matter and Materials Physics*, vol. 84, no. 12, Article ID 125305, 2011.

[27] I. Akihiro and K. Kay, "Pyroelectric effect and possible ferroelectric transition of helimagnetic," *Journal of Physics: Condensed Matter*, vol. 8, no. 15, pp. 2673–2678, 1996.

15

Evaluation and Analysis of Variance of Storage Stability of Asphalt Binder Modified by Nanotitanium Dioxide

Xiaolong Zou,[1,2,3] **Aimin Sha,**[2] **Biao Ding,**[4] **Yuqiao Tan,**[2] **and Xiaonan Huang**[2]

[1]*School of Architecture and Civil Engineering, Xi'an University of Science and Technology, Xi'an, Shaanxi, China*
[2]*Key Laboratory for Special Area Highway Engineering of Ministry of Education, Chang'an University, Xi'an, Shaanxi, China*
[3]*Guangxi Key Lab of Road Structure and Materials, Guangxi Transportation Research & Consulting Co., Ltd., Nanning, Guangxi, China*
[4]*CCCC First Highway Consultants Co., Ltd., Xi'an, Shaanxi, China*

Correspondence should be addressed to Xiaolong Zou; zouxiaolong_1234@163.com

Academic Editor: Qingli Dai

To investigate the effects of nanoparticle content, storage time, and storage temperature on the storage stability of asphalt binders modified by nanoparticles, hot tube storage tests, softening point tests, and dynamic-shearing rheometer (DSR) tests were adopted to evaluate the properties of two kinds of nanotitanium dioxide (TiO_2) modified asphalt binders. A statistical one-way analysis of variance (ANOVA) test was employed to analyze the effects of those variations on the storage stability of the nano-TiO_2 modified asphalt binders. The results indicated that the softening point, the failure temperature, the dynamic-shear viscosity, and $|G^*|/\sin(\delta)$ of the binders increased with nanoparticle content. The storage stability of the binders decreased with nanoparticle content. The impact of storage time on the storage stability of the binders was remarkable when the storage time was more than 48 h. Moreover, the storage stability of the binders at low temperatures was better than that at high temperatures. Based on the one-way ANOVA, the size of nanoparticle had little influence on the storage stability of the nano-TiO_2 modified asphalt binders in this study. Reducing the nanoparticle size cannot effectively enhance the storage stability of the nanoparticle modified asphalt binder due to the agglomeration of nanoparticle.

1. Introduction

Nanomaterials have application in asphalt mixtures with increase of performance requirement of pavement [1]. The one dimension of nanomaterials is usually less than 100 nm which has a larger surface area-to-volume ratio than that of conventional materials, so that nanomaterials have some special properties to improve the performance of asphalt mixture [2]. The nanomaterials using in asphalt construction include nanoclay, nano-SiO_2, and nanotitanium dioxide (TiO_2) [3, 4].

Nanoclay has a good compatibility with asphalt because of its special composite. The addition of nanoclay into asphalt is useful to improve the short- and long-term aging resistance of the modified asphalt binder. At the same time, it contributes to the physical, mechanical, and rheological properties of modified asphalt binder [5–9]. In addition, nanoclay

can improve the rutting resistance of styrene-butadiene-styrene (SBS) modified asphalt binder [10].

Nano-SiO_2 also is added to an SBS modified asphalt binder in order to improve the performance of asphalt mixture. The research of Mojtaba et al. shows that the asphalt mixture using asphalt binder modified by nano-SiO_2 and SBS has better performance than the asphalt mixture using unmodified asphalt binder [11].

TiO_2 has the special crystal structure which can absorb or catalytic decomposition part of automobile exhaust; thus it can be applied in road construction and play a significant role in promoting for environmental protection [12]. Due to the increasing awareness of environmental protection, more and more attention has been paid to the environment-friendly roads. With the continuous increase in the number of car ownerships, automobile exhaust is one of the main

(a) TiO$_2$-50   (b) TiO$_2$-100

FIGURE 1: Electron microscope scanning images.

sources of pollution [13]. The study of TiO$_2$ use in construction of pavement has gradually become a hot topic. Researchers in Japan, China, Italy, and France tried the photocatalytic decomposition of TiO$_2$ in the construction of roads [14].

Nano-TiO$_2$ has the advantages in both photocatalytic decomposition and the scale effect, so that nano-TiO$_2$ attracted the attention. In addition, nano-TiO$_2$ has a good shielding effect on ultraviolet (UV) light, which can be used as a UV absorbent to mitigate UV aging of asphalt [15, 16]. Xiao and Li evaluated the performance of the mixture used nano-TiO$_2$ as a modifier in SBS modified asphalt binder. Their research showed that the mixture using nano-TiO$_2$ and SBS composite modified asphalt binder had better high-temperature stability, water stability, and low temperature cracking resistance than those of the mixture using unmodified asphalt binder [17]. Hassan et al. studied the UV aging resistance of nano-TiO$_2$ modified asphalt binder. Nano-TiO$_2$ modified asphalt binder had the lower penetration loss rate, the lower softening point increase, and the lower ductility loss rate after UV aging than those of the ordinary asphalt binder, which indicated that nano-TiO$_2$ improved the UV aging resistance of asphalt binder [18].

Those studies mainly focus on the effects of nanoparticles on the performance of the modified asphalt binders; however, there is still a lack of research on the effects of nanoparticles on the storage stability of modified asphalt binders.

When a modifier is added to asphalt binder, the variation in the storage stability of the binder should be changed [19]. Moreover, the storage stability could have an influence on the physical parameters of asphalt binder, such as density, softening point, viscosity, and rheological properties [20].

Because the nano-TiO$_2$ has large surface area and high surface free energy, the agglomeration phenomenon is easy to occur. Deepening the understanding of the influences of the storage stability on the properties of asphalt binder is helpful to take corresponding measures to improve the storage stability, so it is necessary to study the storage stability of the nano-TiO$_2$ modified asphalt. This study adopted hot tube storage test, softening point test, and dynamic-shearing rheometer (DSR) test to investigate the storage stability of

TABLE 1: Physical properties and chemical components of the base asphalt.

Items	Test results	Test specification
Softening point (°C)	43.9	ASTM D36
Ductility (15°C) (cm)	>150	ASTM D113
Penetration (25°C) (0.1 mm)	103	ASTM D5
Saturates (%)	31	
Aromatics (%)	40	ASTM D4124
Resins (%)	22	
Asphaltene (%)	7	

TABLE 2: Characteristics of nano-TiO$_2$.

Items	TiO$_2$-50	TiO$_2$-100
Crystal type	Anatase	Anatase
Average particle size (nm)	50	100
Purity (%)	99.8	99.5
Color	White	White

the nano-TiO$_2$ modified asphalt binders with the variation of nanoparticle content, storage time, and storage temperature.

2. Materials and Test Methods

2.1. Materials and Sample Preparation

2.1.1. Materials. A base asphalt was used in this research, which was produced by Karamay Petrochemical Company, China. The physical and chemical parameters of the base asphalt are shown in Table 1.

Two types of nano-TiO$_2$ from Nanjing Emperor Nano Material Co., Ltd., China, were selected for this investigation: the one with the average particle size of 50 nm was named TiO$_2$-50 and the other with the average particle size of 100 nm was named TiO$_2$-100. The characteristics of the two types of nano-TiO$_2$ are shown in Table 2. The morphology of the nano-TiO$_2$ was observed by scanning electron microscope (SEM). The images of SEM are shown as Figure 1. From

Figure 1, according to the scale in the images, the particle of TiO_2-50 is pie-shaped with the diameter of about 100 nm and the particle of TiO_2-100 is pie-shaped with the diameter of about 400 nm. In addition, the agglomeration of TiO_2-50 is more obvious than that of TiO_2-100.

2.1.2. Preparation of Modified Asphalt Binder.
The base asphalt was heated to 160°C in a small container until it flowed fully, and the temperature was kept constant. A certain amount of nanoparticle material was slowly added and mixed into asphalt under 3000 rpm for 30 min. Mixing was then continued at 160°C with 500 rpm for 30 min and the blend became essentially homogenous. After completion, the homogeneous modified asphalt binders were cooled to room temperature for further testing.

Two series of modified asphalt binders were prepared with the two types of nano-TiO_2. The nanoparticle contents used in the tests were 0.50%, 1.00%, 1.50%, 2.00%, and 2.50% by weight of the corresponding blends. The base asphalt was also tested as a reference.

2.2. Test Methods

2.2.1. Hot Tube Storage Test.
In order to study effects of nanoparticle content on the storage stability of the modified asphalt binders, two types of modified asphalt binders with nanoparticle contents of 0.50%, 1.00%, 1.50%, 2.00%, and 2.50% were applied in this study for hot tube storage test according to the conventional procedure [21, 22].

The modified asphalt binders were melted in an oven at 163°C and then poured into an aluminum foil tube (25 mm in diameter and 140 mm in height). The tubes containing the modified asphalt binders were sealed and stored vertically in an oven at 163°C for 48 h. The tubes were removed from the oven and cooled in a freezer at −20°C for 4 h to solidify the modified asphalt binders completely. The cooled modified asphalt binders were cut transversely into three equal sections. The modified asphalt binders from the top and the bottom sections of the same tube were applied for further softening point tests and dynamic-shear rheometer (DSR) tests.

To study the effects of storage time on the storage stability, the two types of the modified asphalt binders with a nanoparticle content of 2.50% were applied for hot tube storage test at the storage time of 6, 12, 24, 48, 72, and 120 h and at storage temperature of 163°C.

To study the effects of storage time on the storage stability, the two types of modified asphalt binders with a nanoparticle content of 2.50% were applied for hot tube storage test at the storage time of 48 h and at storage temperature of 120°C, 163°C, and 175°C.

2.2.2. Softening Point Test.
The modified asphalt binders from the top and the bottom sections of the same tube were applied for softening point tests according to the standard test method ASTM D36. The difference of the softening points between the top and the bottom sections of the binders was calculated to evaluate the high-temperature storage stability of the modified asphalt binders.

FIGURE 2: The dynamic-shear rheometer.

2.2.3. Dynamic-Shear Rheometer (DSR) Test.
The accuracy of softening point test is relatively low, which is hard to evaluate the different of storage stability of modified asphalt binders, especially with low nanoparticle content. However, DSR test has a high accuracy and is commonly used for the fundamental rheological characterization of asphalt binder. Asphalt binder has different rheological characterization due to different composition, so DSR test can be used to investigate the storage stability of asphalt binder.

The DSR test was performed on the binders from the top and the bottom sections of the same tube using a TA rheometer, shown as Figure 2. The test was performed under controlled strain loading conditions using stain 12% at frequency 10 rad/s at temperature 58°C. The test samples were diameter of 25 mm and thickness of 1000 μm. Test procedure followed AASHTO standard TP 5.

The viscoelastic properties of the asphalt binder were evaluated by measuring the complex shear modulus G^* and phase angle δ. Superpave specification uses rutting factor $G^*/\sin(\delta)$ to characterize the high-temperature resistance to permanent deformation of asphalt binder. The principal viscoelastic parameters obtained from the DSR were the magnitude of the complex shear modulus (G^*) and the phase angle (δ) and $|G^*|/\sin(\delta)$ was calculated. The failure temperature and the dynamic-shear viscosity ($|n^*|$) at 135°C were determined following the Superpave mix design specifications. $|G^*|/\sin(\delta)$, the failure temperature, and the dynamic-shear viscosity were used as indexes to evaluate the storage stability of modified asphalt binders.

3. Results and Discussion

3.1. Effects of Nanoparticle Content.
To study the effects of nanoparticle content on the storage stability of the nanoparticle modified asphalt binders, the asphalt binder samples with nanoparticle contents of 0.50%, 1.00%, 1.50%, 2.00%, and 2.50% after hot tube storage tests were applied to investigate the values of softening point, failure temperature, dynamic-shear viscosity, and $|G^*|/\sin(\delta)$.

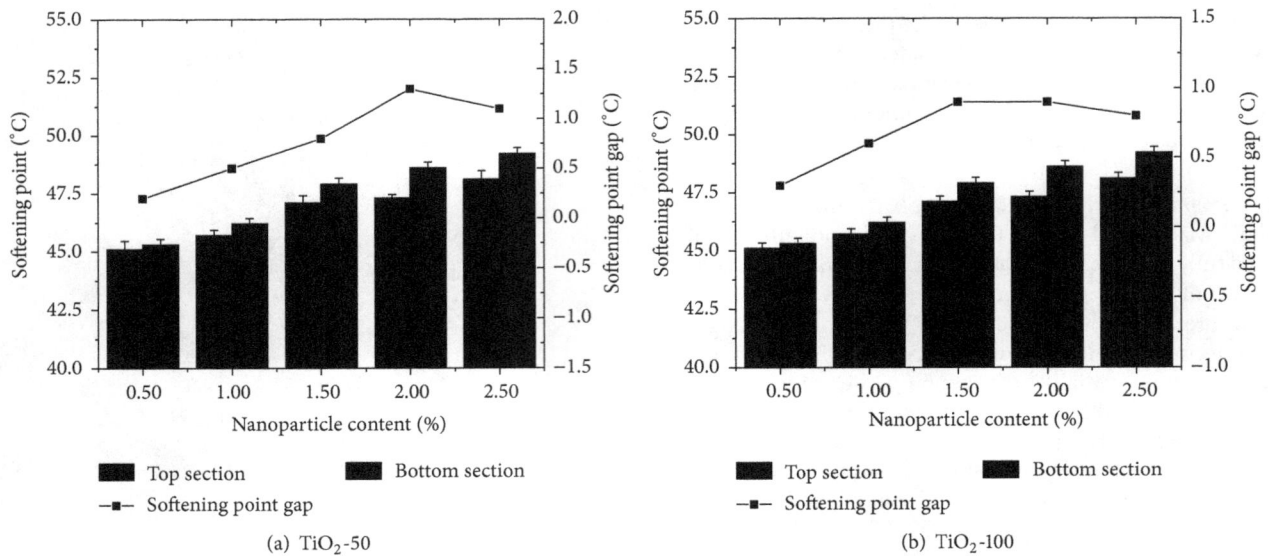

FIGURE 3: The effects of nanoparticle content on the softening point.

(1) The Effects of Nanoparticle Content on the Softening Point. The results of the softening point test using the two types of asphalt binders modified with different nanoparticle contents were shown as Figure 3. It can be observed that, with the increase of the nanoparticle content, the softening point of the asphalt binders modified by TiO_2-50 and TiO_2-100 increased gradually. Comparing the differences between the two modified asphalt binders, it can be found that the softening point of TiO_2-50 modified asphalt binder was relatively large.

A softening point gap was the difference of the softening point of the nano-TiO_2 modified asphalt binders between the top and the bottom sections in the aluminum foil tubes after the hot tube storage test. To evaluate the storage stability of the modified asphalt binders, the softening point gaps under the conditions of different nanoparticle content were calculated and the results are presented in Figure 3. With the increase of the nanoparticle content, the softening point gap of TiO_2-50 modified asphalt binder increased gradually and reached the peak at the nanoparticle content of 2.00%. Similarly, the softening point gap of TiO_2-100 modified asphalt binder reached the peak at the nanoparticle content of 1.50% and decreased lightly after that.

It was considered that the dispersion of nanoparticle decreased with the increasing nanoparticle content and the agglomeration became more significant leading to the decrease of the storage stability of modified asphalt binders.

The softening point test is a common test for storage stability of modified asphalt binder. When the difference of the softening points between the top and the bottom sections of the sample treated for 48 h at 163°C was less than 2.5°C, the sample was considered to have good high-temperature storage stability [2]. Based on this principle, the two types of nano-TiO_2 modified asphalt binders had good high-temperature storage stability.

(2) The Effects of Nanoparticle Content on the Dynamic-Shear Viscosity. The effects of nanoparticle content on the

dynamic-shear viscosity are shown as Figure 4. The dynamic-shear viscosity of two kinds of modified asphalt binders correspondingly increased with the increasing nanoparticle content, which indicated that the content of nanoparticle affected the dynamic-shear viscosity of asphalt binder modified by TiO_2-50 and TiO_2-100.

The Krieger-Dougherty equation gives a description of the relationship between the particle volume fraction and the relative viscosity in a monodisperse system [23, 24]:

$$\eta_r = \left(1 - \frac{\phi}{\phi_{max}}\right)^{-[\eta]\phi_{max}}, \qquad (1)$$

where η_r is the relative viscosity, $[\eta]$ is the intrinsic viscosity, ϕ is the volume fraction, and ϕ_{max} is the maximum volume fraction of the latex. According to (1), it can be found that the larger the volume fraction, the more the relative viscosity, correspondingly, the higher the nanoparticle content, the more the dynamic-shear viscosity.

The dynamic-shear viscosity gaps under the conditions of different nanoparticle content are shown as Figure 4. For the asphalt binders modified by TiO_2-50 and TiO_2-100, the dynamic-shear viscosity gap increased as the nanoparticle content increased. The results indicated that the difference of the shear viscosity between the top section and the bottom section in the aluminum foil tube increased with the nanoparticle content.

With the increase of nanoparticle content, the agglomeration of nanoparticle increased due to the effect of large surface energy. Besides, the interaction between the agglomerated nanoparticles and asphaltene molecules by Van der Waals forces transfers the dispersed system of modified asphalt binders from a stable state to a metastable state, attribute to decreasing the storage stability.

(3) The Effects of Nanoparticle Content on the Failure Temperature. Although the softening point test is a common test

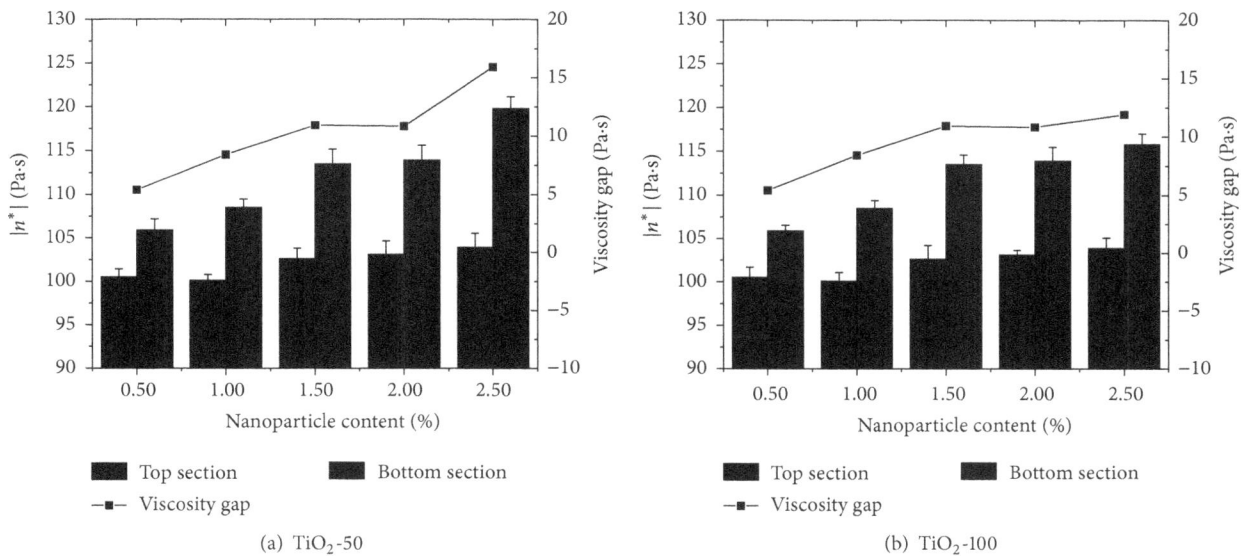

(a) TiO$_2$-50   (b) TiO$_2$-100

FIGURE 4: The effects of nanoparticle content on the dynamic-shear viscosity.

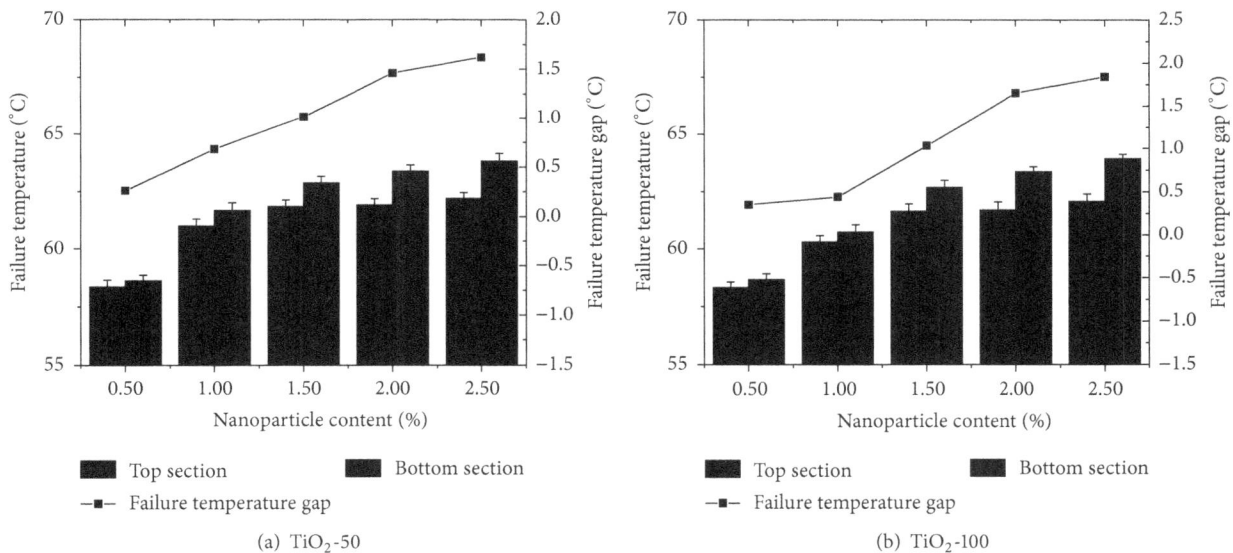

(a) TiO$_2$-50   (b) TiO$_2$-100

FIGURE 5: The effects of nanoparticle content on the failure temperature.

method, the accuracy of the test is not high, especially for the modified asphalt with a small amount of nanoparticles. It was necessary to apply a high accuracy test method to investigate the storage stability, so DSR test was applied in this study.

The effects of nanoparticle content on the failure temperature are shown as Figure 5. With the increase of nanoparticle content, the failure temperatures of the asphalt binder modified by TiO$_2$-50 and TiO$_2$-100 presented a continued slight growth. When the nanoparticle content increased from 0.50% to 2.50%, the failure temperature of TiO$_2$-50 modified asphalt binder increased from 58.35°C to 62.21°C and that of

TiO$_2$-100 modified asphalt binder increased from 58.31°C to 62.09°C.

Failure temperature gaps under the conditions of different nanoparticle content were also calculated and the results are presented in Figure 5. The failure temperature gap of TiO$_2$-50 modified asphalt binder steadily increased with the nanoparticle content and that of TiO$_2$-100 modified asphalt binder increased slowly before nanoparticle content of 1.00% and then presented a larger rise.

Overall, the TiO$_2$-50 modified asphalt binder and the TiO$_2$-100 modified asphalt binder had similar change regulation of failure temperature over the nanoparticle content.

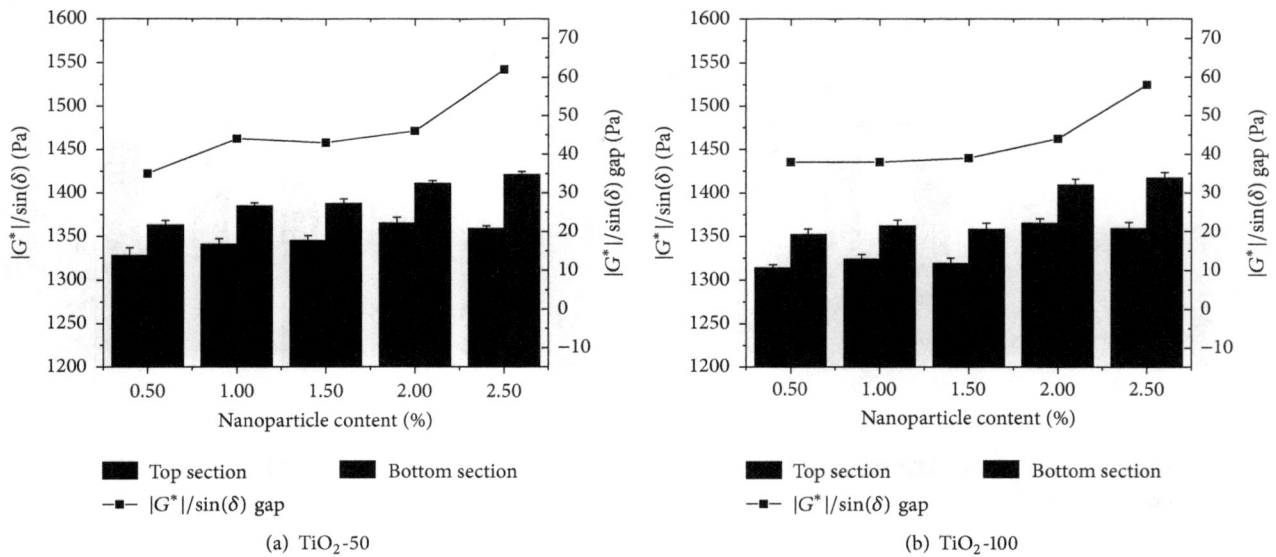

(a) TiO$_2$-50

(b) TiO$_2$-100

FIGURE 6: The effects of nanoparticle content on $|G^*|/\sin(\delta)$.

(4) The Effects of Nanoparticle Content on the Rutting Factor. The effects of nanoparticle content on $|G^*|/\sin(\delta)$ are shown as Figure 6. For the two kinds of modified asphalt binders, $|G^*|/\sin(\delta)$ increased slowly as the nanoparticle content increased.

$|G^*|/\sin(\delta)$ gaps under the conditions of different nanoparticle content were calculated and are presented in Figure 6. According to the curve of the $|G^*|/\sin(\delta)$ gap of TiO$_2$-50 modified asphalt binder, the $|G^*|/\sin(\delta)$ gap increased slowly when the amount of TiO$_2$-50 was less than 2.00%, while the $|G^*|/\sin(\delta)$ gap increased significantly in the amount of 2.50%. By contrast, the $|G^*|/\sin(\delta)$ gap of TiO$_2$-100 modified asphalt binder showed a slow increasing trend in the range of test content of 0.50%~2.50%. The results indicated that the nanoparticle content had effects on the $|G^*|/\sin(\delta)$.

In general, the gap of softening point, the gap of failure temperature, the gap of dynamic-shear viscosity, and the gap of $|G^*|/\sin(\delta)$ increased with the nanoparticle content. For the asphalt binders modified by TiO$_2$-50 and TiO$_2$-100, the storage stability of nanoparticle modified asphalt binders decreased with the increase of the nanoparticle content, which could be mainly due to the agglomeration of nanoparticle increased with the increase of nanoparticle content based on the microscale analysis.

3.2. Effects of Storage Time.

To study effects of storage time on the storage stability of the nanoparticle modified asphalt binders, the two types of the modified asphalt binder samples with a nanoparticle content of 2.50% after hot tube storage tests at the storage time of 6, 12, 24, 48, 72, and 120 h and at storage temperature of 163°C were applied to investigate the values of $|G^*|/\sin(\delta)$.

Figure 7 shows the effects of storage time on $|G^*|/\sin(\delta)$. For both TiO$_2$-50 modified asphalt binder and TiO$_2$-100 modified asphalt binder, the values of $|G^*|/\sin(\delta)$ were

relatively small at the storage time before 48 hours and dramatically increased after 48 hours.

To analyze the effects of storage time on storage stability of nanoparticle modified asphalt binders, the $|G^*|/\sin(\delta)$ gaps under different conditions of storage time were calculated and are presented in Figure 7. For TiO$_2$-50 modified asphalt binder, the $|G^*|/\sin(\delta)$ gap had a trend of increase with the storage time before 48 hours and changed little after 48 hours. For TiO$_2$-100 modified asphalt binder, the variation of $|G^*|/\sin(\delta)$ gap was like that of TiO$_2$-50. The results indicated that the storage time had an impact on storage stability of nanoparticle modified asphalt binders and the impact was remarkable when the storage time was more than 48 hours. Moreover, it can be found that the $|G^*|/\sin(\delta)$ gap of TiO$_2$-50 at 72 hours and that of TiO$_2$-100 at 120 hours decreased lightly, that it was considered that both of the top and the bottom section of binder specimens were aged heavily, affecting the $|G^*|/\sin(\delta)$ gap when the storage time was more than 48 hours.

Nanoparticles tend toward agglomeration to depress the surface energy based on the lowest energy principle [24]. With increasing storage time, the agglomerated nanoparticles will redisperse or escalate agglomeration resulting in reducing the storage stability of the modified asphalt binders.

3.3. Effects of Storage Temperature.

To study the effects of storage time on the storage stability of the nanoparticle modified asphalt binders, the two types of the modified asphalt binder samples with a nanoparticle content of 2.50% after hot tube storage tests at the storage time of 48 h and at storage temperature of 120°C, 163°C, and 175°C were applied to investigate the values of $|G^*|/\sin(\delta)$.

Figure 8 shows the effects of storage temperature on $|G^*|/\sin(\delta)$. For both the TiO$_2$-50 modified asphalt binder

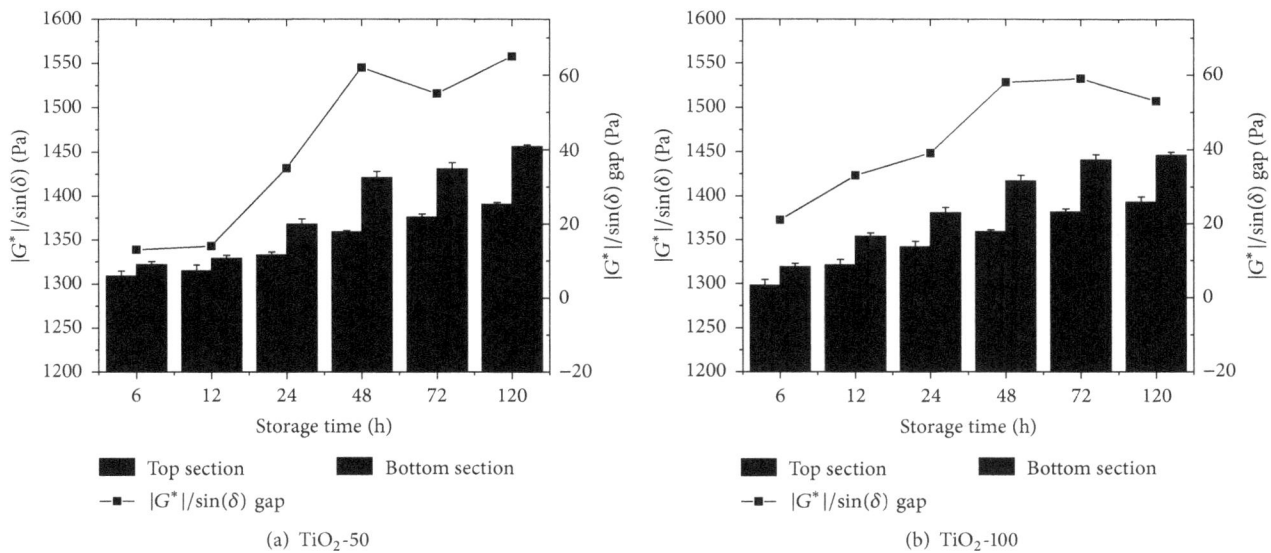

(a) TiO$_2$-50

(b) TiO$_2$-100

FIGURE 7: The effects of storage time on $|G^*|/\sin(\delta)$.

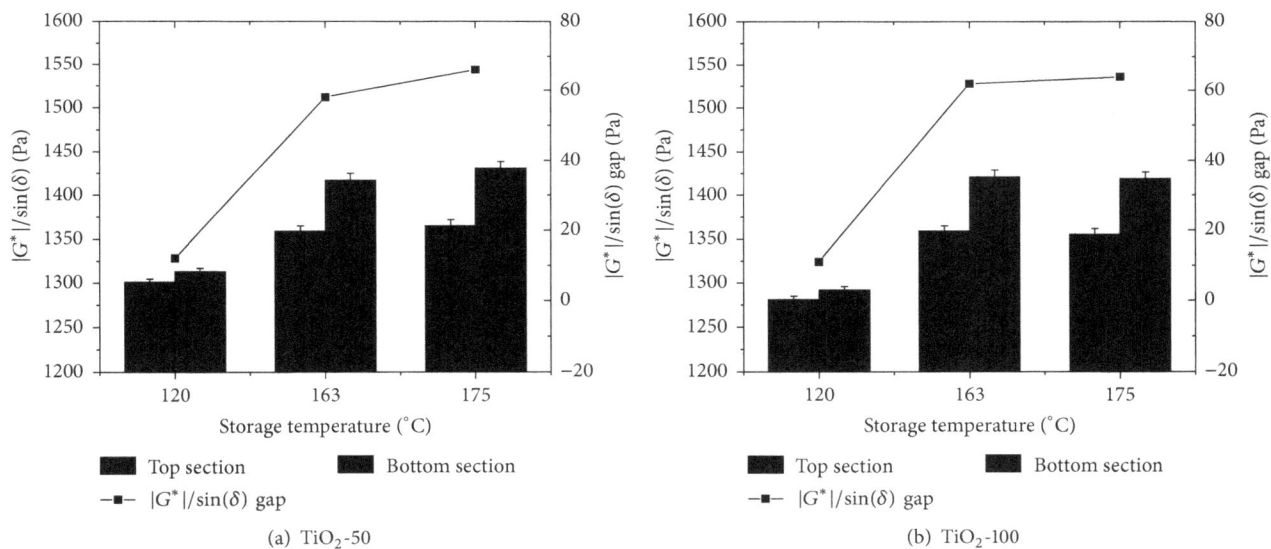

(a) TiO$_2$-50

(b) TiO$_2$-100

FIGURE 8: The effects of storage temperature on $|G^*|/\sin(\delta)$.

and the TiO$_2$-100 modified asphalt binder, $|G^*|/\sin(\delta)$ increased with the storage temperature.

The $|G^*|/\sin(\delta)$ gaps at different storage temperatures were calculated to analyze the effects of storage temperature on the storage stability and the calculation is presented as curves of $|G^*|/\sin(\delta)$ gap in Figure 8. It can be seen from Figure 8 that the two kinds of modified asphalt binders had similar $|G^*|/\sin(\delta)$ gap variation with the storage temperature: the $|G^*|/\sin(\delta)$ gap between the top and bottom section was smaller at low storage temperature (120°C), the $|G^*|/\sin(\delta)$ gap was larger at 163°C, and the $|G^*|/\sin(\delta)$ gap

slightly rose when the storage temperature rose to 175°C. It can be found that storage temperature had an influence on the storage stability of nanoparticle modified asphalt binders. The storage stability of nanoparticle modified asphalt binders was better at low temperatures than that at high temperatures.

A nanoparticle modified asphalt binder is a dispersed system. The molecular heat motion in the dispersed system of the modified asphalt binders increased with temperature. The agglomerated nanoparticle and micelle composition of heavy asphaltene trended to sink at high storage temperature resulting in $|G^*|/\sin(\delta)$ gap increasing. Consequently, the storage stability of nanoparticle modified asphalt binders decreased with storage temperatures.

4. Analysis of Variance (ANOVA)

In statistics, analysis of variance (ANOVA) is an important technique to determine whether there is a significant difference between two or more sample means of populations [25, 26]

To statistically evaluate the effects of nanoparticle size, nanoparticle content, storage time, and storage temperature on the storage stability of nanoparticle modified asphalt binders, one-way ANOVA (single factor) was carried out in this research.

(1) ANOVA of Nanoparticle Size. The one-way ANOVA (single factor) was carried out in this research with 95% confidence interval ($\alpha = 0.05$) to determine whether there was a significant difference of the storage stability between TiO_2-50 and TiO_2-100 groups. The primary variables included softening point, failure temperature, dynamic-shear viscosity, and rutting factor ($|G^*|/\sin(\delta)$).

The null hypothesis tested by ANOVA is that the population means for all conditions are the same. The hypotheses for the two-tailed test were as follows.

The Null Hypothesis, H_0. The storage stability of the TiO_2-50 modified asphalt binder was similar to that of the TiO_2-100 modified asphalt binder.

The Proposed Hypothesis, H_1. The storage stability of TiO_2-50 modified asphalt binder was different from that of the TiO_2-100 modified asphalt binder.

If the null hypothesis was rejected as the significance value was smaller than the significance level, α (0.05), then it can be concluded that at least one of the population means was different from at least one other population mean. Therefore, the two populations were found to be statistically significantly different. Table 3 presents the ANOVA between the TiO_2-50 and TiO_2-100 groups.

From Table 3, the significance value of 0.736 in softening point was larger than α (0.05); thus the proposed hypothesis was rejected and it indicated that there was no significant difference of the storage stability between TiO_2-50 and TiO_2-100 groups according to the softening point.

Correspondingly, it can be seen from Table 3 that the significance value of 0.890 in failure temperature, the significance value of 0.711 in dynamic-shear viscosity, and the significance value of 0.668 in rutting factor were larger than α (0.05). The one-way ANOVA with single factor showed that there was no statistically significant difference of the storage stability between TiO_2-50 and TiO_2-100 groups based on the indexes of failure temperature, viscosity, and rutting factor.

The test of one-way ANOVA indicated that the nanoparticle size had no significant influence on the storage stability of nanoparticle modified asphalt binders, which could be mainly due to the fact that the agglomeration of nanoparticle reduced the dispersion of nanoparticle in the dispersed

system of modified asphalt binder and the difference of size effects for nanoparticle on the storage stability. The Stokes-Einstein equation provides an estimate of the terminal velocity of a particle in a dispersed system [27]:

$$v_s = \frac{d^2 \Delta \rho g}{18\eta_c},\qquad(2)$$

where v_s is the terminal velocity, d is the particle diameter, $\Delta \rho$ is the density difference between dispersed phase and continuous phase, g is the gravitational acceleration, and η_c is the continuous phase viscosity.

From Figure 1, the agglomeration of TiO_2-50 is more obvious than that of TiO_2-100. The agglomeration of TiO_2-50 will escalate in the dispersed system of modified asphalt binder to further increase the terminal velocity of the agglomerated particles to sink. Therefore, reducing the nanoparticle size cannot effectively enhance the storage stability of nanoparticle modified asphalt binders, but improving the dispersion of nanoparticle was thought to be a positive method.

(2) ANOVA of Nanoparticle Content. To analyze the effects of nanoparticle content on the storage stability of nanoparticle modified asphalt binders, the one-way ANOVA was also employed. The hypotheses for the two-tailed test were as follows.

The Null Hypothesis, H_0. The storage stability of the modified asphalt binders was the same in the conditions of different nanoparticle content.

The Proposed Hypothesis, H_1. The storage stability of modified asphalt binder was different in the conditions of different nanoparticle content.

Tables 4–7 present the variance analysis of softening point, failure temperature, viscosity, and rutting factor between the groups of different nanoparticle contents.

As shown in Table 4, the results of the ANOVA indicated that the significance value of 0.020 in softening point was less than the significance level, $\alpha = 0.05$; therefore, it can be concluded that there was enough evidence to suggest a difference between the groups of different nanoparticle contents. It can be found that there is a significant difference between the groups of different nanoparticle contents in the softening point value.

According to one-way ANOVA in Tables 5–7, it suggested that there were significant differences in failure temperature, viscosity, and rutting factor between the groups of different nanoparticle contents, which was further verification of the effects of nanoparticle content on the storage stability of nanoparticle modified asphalt binders.

The test of one-way ANOVA with single factor showed that there was a statistically significant difference between the groups of different nanoparticle contents. It could be learnt that the nanoparticle content of 0.50%~2.50% had a significant influence on the storage stability of asphalt binders based on the ANOVA.

TABLE 3: ANOVA: single factor between the TiO_2-50 and TiO_2-100 groups.

Properties	Source of variation		Sum of squares	Degree of freedom	Mean Square	F	Significance
Softening point	Between groups	(Combined)	0.016	1	0.016	0.122	0.736
		Linear term	0.016	1	0.016	0.122	0.736
	Within groups		1.048	8	0.131		
	Total		1.064	9			
Failure temperature	Between groups	(Combined)	0.008	1	0.008	0.020	0.890
		Linear term	0.008	1	0.008	0.020	0.890
	Within groups		3.092	8	0.386		
	Total		3.100	9			
Dynamic shear viscosity	Between groups	(Combined)	1.600	1	1.600	0.147	0.711
		Linear term	1.600	1	1.600	0.147	0.711
	Within groups		87.016	8	10.877		
	Total		88.616	9			
Rutting factor	Between groups	(Combined)	16.900	1	16.900	0.198	0.668
		Linear term	16.900	1	16.900	0.198	0.668
	Within groups		681.200	8	85.150		
	Total		698.100	9			

TABLE 4: ANOVA: single factor between the groups of different particle contents on softening point value.

Source of variation			Sum of squares	Degree of freedom	Mean square	F	Significance
Between groups	Linear term	(Combined)	0.924	4	0.231	8.250	0.020
		Contrast	0.761	1	0.761	27.161	0.003
		Deviation	0.163	3	0.054	1.946	0.240
Within groups			0.140	5	0.028		
Total			1.064	9			

TABLE 5: ANOVA: single factor between the groups of different particle contents on failure temperature.

Source of variation			Sum of squares	Degree of freedom	Mean square	F	Significance
Between groups	Linear term	(Combined)	3.024	4	0.756	50.205	0
		Contrast	2.957	1	2.957	196.335	0
		Deviation	0.068	3	0.023	1.495	0.323
Within groups			0.075	5	0.015		
Total			3.100	9			

TABLE 6: ANOVA: single factor between the groups of different particle contents on viscosity.

Source of variation			Sum of squares	Degree of freedom	Mean square	F	Significance
Between groups	Linear term	(Combined)	80.616	4	20.154	12.596	0.008
		Contrast	75.272	1	75.272	47.045	0.001
		Deviation	5.344	3	1.781	1.113	0.426
Within groups			8.000	5	1.600		
Total			88.616	9			

TABLE 7: ANOVA: single factor between the groups of different particle contents on rutting factor.

Source of variation			Sum of squares	Degree of freedom	Mean square	F	Significance
Between groups	Linear term	(Combined)	657.600	4	164.400	20.296	0.003
		Contrast	520.200	1	520.200	64.222	0
		Deviation	137.400	3	45.800	5.654	0.046
Within groups			40.500	5	8.100		
Total			698.100	9			

TABLE 8: ANOVA: single factor between the groups of different storage time on rutting factor.

Source of variation			Sum of squares	Degree of freedom	Mean square	F	Significance
Between groups	Linear term	(Combined)	3659.750	5	731.950	14.236	0.003
		Contrast	3177.779	1	3177.779	61.804	0
		Deviation	481.971	4	120.493	2.343	0.168
Within groups			308.500	6	51.417		
Total			3968.250	11			

TABLE 9: ANOVA: single factor between the groups of different storage temperatures on rutting factor.

Source of variation			Sum of squares	Degree of freedom	Mean square	F	Significance
Between groups	Linear term	(Combined)	3493.000	2	1746.500	499.000	0
		Contrast	2862.250	1	2862.250	817.786	0
		Deviation	630.750	1	630.750	180.214	0.001
Within groups			10.500	3	3.500		
Total			3503.500	5			

(3) ANOVA of Storage Time. The test of ANOVA of the rutting factors between the groups of different storage time was applied to investigate the effect of storage time on the storage stability of nanoparticle modified asphalt binders. The hypotheses for the two-tailed test are as follows.

The Null Hypothesis, H_0. The storage stability of the modified asphalt binders was the same in the conditions of different storage time.

The Proposed Hypothesis, H_1. The storage stability of modified asphalt binder was different in the conditions of different storage time.

The results of ANOVA between the groups of different storage time on rutting factor are shown in Table 8. The results showed that the significance value of 0.03 was less than α (0.05). The null hypothesis was rejected; thus, it can be concluded that there was a significant difference between the groups of different storage time concerning the rutting factors.

The ANOVA indicated that storage time had a significant influence on storage stability of nanoparticle modified asphalt binders.

(4) ANOVA of Storage Temperature. The test of ANOVA on rutting factor between the groups of different storage temperature was applied to investigate the effect of storage temperature on the storage stability of nanoparticle modified asphalt binder. The hypotheses for the two-tailed test were as follows.

The Null Hypothesis, H_0. The storage stability of the modified asphalt binders was the same in the conditions of different storage temperature.

The Proposed Hypothesis, H_1. The storage stability of modified asphalt binder was different in the conditions of different storage temperature.

The results of ANOVA between the groups of different storage temperatures on rutting factor are shown in Table 9. The significance value was less than α (0.05). The null hypothesis was rejected, so it can be concluded that there was a significant difference between the groups of different storage time. The results indicated that storage temperature had a significant influence on the storage stability of nanoparticle modified asphalt binders.

The results of ANOVA further verified the previous analysis of the tests on the modified asphalt binders. In order to improve the storage stability of nanoparticle modified asphalt binders, proper nanoparticle contents, lower storage time, and temperatures can be chosen. At the same time, it is important to improve the dispersion of nanoparticle modified asphalt binders.

5. Conclusions

This study adopted the hot tube storage test, the softening point test, and the DSR test to investigate the storage stability of the nano-TiO_2 modified asphalt with the variation of nanoparticle content, storage time, and storage temperature. Based on the results and the ANOVA, the following conclusions can be drawn:

(a) The storage stability of nanoparticle modified asphalt binders decreased with the nanoparticle content.

(b) The storage time had an impact on storage stability of nanoparticle modified asphalt binders and the impact was remarkable when the storage time was more than 48 h. It is necessary to take measures to improve the storage stability of nanoparticle modified asphalt binders when the storage time was more than 48 h.

(c) The storage stability of nanoparticle modified asphalt binders at low temperatures was better than that at high temperatures.

(d) Reducing the nanoparticle size cannot effectively enhance the storage stability of the nanoparticle modified asphalt binder due to the agglomeration of nanoparticle.

Disclosure

The results and opinions presented are those of the authors and do not necessarily reflect those of the sponsoring agencies.

Conflicts of Interest

The authors declare that they have no conflicts of interest.

Authors' Contributions

Xiaolong Zou and Aimin Sha conceived and designed the experiments; Xiaolong Zou, Yuqiao Tan, and Xiaonan Huang performed the experiments; Xiaolong Zou wrote the paper; Xiaolong Zou and Biao Ding analyzed the data.

Acknowledgments

This research was sponsored by the Ministry of Science and Technology of China (2014BAG05B04), the Ph.D. Research Startup Foundation of Xi'an University of Science and Technology (2017QDJ024), the opening fund of Guangxi Key Lab of Road Structure and Materials (2017gxjgclkf-001), and the Natural Science Foundation of Shaanxi Province (2016JQ5115). The authors would also like to thank Mr. Hao Liu for his contribution to this research.

References

[1] A. N. Amirkhanian, F. Xiao, and S. N. Amirkhanian, "Characterization of unaged asphalt binder modified with carbon nano particles," *International Journal of Pavement Research and Technology*, vol. 4, no. 5, pp. 281–286, 2011.

[2] S. W. Goh, M. Akin, Z. You, and X. Shi, "Effect of deicing solutions on the tensile strength of micro- or nano-modified asphalt mixture," *Construction and Building Materials*, vol. 25, no. 1, pp. 195–200, 2011.

[3] J. Yang and S. Tighe, "A review of advances of nanotechnology in asphalt mixtures," *Procedia - Social and Behavioral Sciences*, vol. 96, pp. 1269–1276, 2013.

[4] T. F. Pamplona, B. De C. Amoni, A. E. V. De Alencar et al., "Asphalt binders modified by SBS and SBS/nanoclays: Effect on rheological properties," *Journal of the Brazilian Chemical Society*, vol. 23, no. 4, pp. 639–647, 2012.

[5] M. V. De Ven, A. Molenaar, J. Besamusca, and J. Noordergraaf, "Nanotechnology for binders of asphalt mixtures," in *Proceedings of the 4th Eurasphalt and Eurobitume Congress*, Copenhagen, Denmark, 2008.

[6] B. Golestani, F. Moghadas Nejad, and S. Sadeghpour Galooyak, "Performance evaluation of linear and nonlinear nanocomposite modified asphalts," *Construction and Building Materials*, vol. 35, pp. 197–203, 2012.

[7] B. Golestani, B. H. Nam, F. Moghadas Nejad, and S. Fallah, "Nanoclay application to asphalt concrete: Characterization of polymer and linear nanocomposite-modified asphalt binder

and mixture," *Construction and Building Materials*, vol. 91, pp. 32–38, 2015.

[8] M. Ameri, S. Nobakht, K. Bemana, H. Rooholamini, and M. Vamegh, "Effect of nanoclay on fatigue life of hot mix asphalt," *Petroleum Science and Technology*, vol. 34, no. 11-12, pp. 1021–1025, 2016.

[9] S. G. Jahromi and A. Khodaii, "Effects of nanoclay on rheological properties of bitumen binder," *Construction and Building Materials*, vol. 23, no. 8, pp. 2894–2904, 2009.

[10] L. G. A. T. Farias, J. L. Leitinho, B. D. C. Amoni et al., "Effects of nanoclay and nanocomposites on bitumen rheological properties," *Construction and Building Materials*, vol. 125, pp. 873–883, 2016.

[11] G. Mojtaba, M. S. Morteza, T. Majid, K. R. Jalal, and T. Reza, "Modification of stone matrix asphalt with nano-SiO_2," *Journal of Basic and Applied Scientific Research*, vol. 2, no. 2, pp. 1338–1344, 2012.

[12] L. Mohammad, M. Hassan, and S. Cooper III, "Mechanical characteristics of asphaltic mixtures containing titanium-dioxide photocatalyst," *Journal of Testing and Evaluation*, vol. 40, no. 6, pp. 1–8, 2012.

[13] A. Beeldens, "An environmental friendly solution for air purification and self-cleaning effect: the application of TiO_2 as photocatalyst in concrete," in *Proceedings of the Transport Research Arena Europe–TRA*, Göteborg, Sweden, 2006.

[14] M. M. Hassan, H. Dylla, L. N. Mohammad, and T. Rupnow, "Evaluation of the durability of titanium dioxide photocatalyst coating for concrete pavement," *Construction and Building Materials*, vol. 24, no. 8, pp. 1456–1461, 2010.

[15] J. Tanzadeh, F. Vahedi, T. K. Pezhouhan, and R. Tanzadeh, "Laboratory study on the effect of nano Tio2 on rutting performance of asphalt pavements," *Advanced Materials Research*, vol. 622, pp. 990–994, 2013.

[16] G. Shafabakhsh, S. M. Mirabdolazimi, and M. Sadeghnejad, "Evaluation the effect of nano-TiO2 on the rutting and fatigue behavior of asphalt mixtures," *Construction and Building Materials*, vol. 54, pp. 566–571, 2014.

[17] P. Xiao and X. Li, "Research on the performance and mechanism of nanometer ZnO/SBS modified asphalt," *Journal of Highway and Transportation Research and Development*, vol. 24, no. 6, pp. 12–16, 2007.

[18] M. M. Hassan, L. N. Mohammad, S. B. Cooper, and H. Dylla, "Evaluation of nano-titanium dioxide additive on asphalt binder aging properties," *Transportation Research Record*, no. 2207, pp. 11–15, 2011.

[19] C. Ouyang, S. Wang, Y. Zhang, and Y. Zhang, "Thermorheological properties and storage stability of SEBS/kaolinite clay compound modified asphalts," *European Polymer Journal*, vol. 42, no. 2, pp. 446–457, 2006.

[20] H. U. Bahia and H. Zhai, "Storage stability of modified binders using the newly developed LAST procedure," *Road Materials and Pavement Design*, vol. 1, no. 1-2, pp. 53–73, 2000.

[21] Y. Niu, Z. Zhu, J. Xiao, Z. Liu, and B. Liang, "Evaluation of storage stability of styrene-butadiene-styrene block copolymer-modified asphalt via electrochemical analysis," *Construction and Building Materials*, vol. 107, pp. 38–43, 2016.

[22] H. Fu, L. Xie, D. Dou, L. Li, M. Yu, and S. Yao, "Storage stability and compatibility of asphalt binder modified by SBS graft copolymer," *Construction and Building Materials*, vol. 21, no. 7, pp. 1528–1533, 2007.

[23] P. F. Luckham and M. A. Ukeje, "Effect of particle size distribution on the rheology of dispersed systems," *Journal of Colloid and Interface Science*, vol. 220, no. 2, pp. 347–356, 1999.

[24] X. Zou, A. Sha, W. Jiang, and X. Huang, "Modification mechanism of high modulus asphalt binders and mixtures performance evaluation," *Construction and Building Materials*, vol. 90, pp. 53–58, 2015.

[25] B. G. Tabachnick and L. S. Fidell, *Using Multivariate Statistics*, Allyn & Bacon, 5th edition, 2006.

[26] H. Asli, E. Ahmadinia, M. Zargar, and M. R. Karim, "Investigation on physical properties of waste cooking oil - Rejuvenated bitumen binder," *Construction and Building Materials*, vol. 37, pp. 398–405, 2012.

[27] D. D. Li and M. L. Greenfield, "Viscosity, relaxation time, and dynamics within a model asphalt of larger molecules," *The Journal of Chemical Physics*, vol. 140, no. 3, article 034507, 2014.

16

Preparation Parameter Analysis and Optimization of Sustainable Asphalt Binder Modified by Waste Rubber and Diatomite

Hanbing Liu, Mengsu Zhang, Yubo Jiao ⓘ, and Liuxu Fu

College of Transportation, Jilin University, Changchun 130025, China

Correspondence should be addressed to Yubo Jiao; jiaoyb@jlu.edu.cn

Academic Editor: Estokova Adriana

In this study, crumb rubber and diatomite were used to modify asphalt binder. Wet process was adopted as a preparation method, and the corresponding preparation process was determined firstly. The effects of six preparation parameters (crumb rubber concentration, diatomite concentration, shear time, shear speed, shear temperature, and storing time) on properties of modified asphalt binder (penetration at 25°C, softening point, ductility, viscosity at 135°C, elastic recovery, and penetration index) were investigated, and multiresponse optimization was conducted using the response surface method. The results revealed that softening points, viscosity, elastic recovery, and penetration index increase, while penetration and ductility decrease with the increase of crumb rubber concentration. Softening points, viscosity, and penetration index increase, while penetration and ductility decrease with the increase of diatomite concentration, which presents little influence on elastic recovery of binder. Shear temperature presented significant effects on penetration, softening point, viscosity, and ductility. Shear speed, shear time, and storing time have similar effects on binder properties because of their similar mechanism of action. Based on the model obtained from the response surface method, optimized preparation parameters corresponding to specific criteria can be determined, which possess favorable accuracy compared with experimental results.

1. Introduction

Disposal of industrial wastes has become a critical problem in the world under the promotion of environmental management and mission of sustainable development [1, 2]. Millions of end-of-life tyres (ELTs) are generated each year due to the increasing number of vehicles, which have been regarded as one of the largest and most problematic waste polymeric materials with large quantity and durability [3]. Current researches have shown that crumb rubber (CR) derived from grinding of ELTs into asphalt modification can be an effective modifier for construction and maintenance of asphalt pavement [4]. For one thing, CR-modified asphalt and corresponding mixture possess favorable environmental benefits. Farina et al. [5] demonstrated the environmental results of recycled materials in bituminous mixtures for pavement wearing courses using life cycle assessment (LCA) methodology. Results reveal that the use of wearing course containing asphalt rubber obtained by the wet process presents significant benefits for energy saving, environmental impact, human

health, preservation of ecosystems, and minimization of resource depletion. The reductions of the gross energy requirement and global warming potential range between 36% and 45% compared with standard paving solutions. Bartolozzi et al. [6] compared the environmental performances between rubberized asphalt pavement and the conventional one. Results show that the global environmental performances of the rubberized asphalt road are improved by 30–40%. La Rosa et al. [7] evaluated the environmental benefit of using recycled waste tyres in mixture with virgin rubber. It can be concluded that addition of ground tyre rubber (GTR) in styrene-isoprene-styrene (SIS) formulation considerably reduces the influences in terms of global warming potential, human toxicity, energy consumption, and so on. Yu et al. [8] evaluated the environmental burdens of plastic-rubber asphalt (PRA) mixture and styrene-butadiene-styrene (SBS) asphalt mixture by using a cradle-to-gate LCA model. The results revealed that PRA mixture is more environmental-friendly than SBS asphalt mixture. CR has been proved to be an effective modifier for virgin asphalt. Adding CR into asphalt binder can improve its viscosity and elastic properties at high temperature, which can enhance the resistance ability to permanent deformation under traffic load. It can also increase the flexibility of asphalt binder at low temperature, thus making it more capable of resisting low temperature cracking [9–13]. Therefore, recycled ELTs have been widely used as a road-paving material in the world.

Diatomite is a naturally occurring, soft, siliceous sedimentary rock that can be easily crumbled into fine powder. The typical chemical components of diatomite are silica, alumina, and iron oxide [14]. Diatomite has been widely used as a asphalt modifier due to its low cost, large storage, and high absorptive ability [14–18]. The addition of diatomite into asphalt can improve its high temperature performance, storage stability, rutting resistance, and long-term aging resistance. Asphalt mastic modified by diatomite possesses more favorable high and medium temperature performances than that modified by limestone, hydrated lime, and fly ash [16]. However, diatomite-modified asphalt presents some drawbacks such as reducing the thermal conductivity and low temperature deformation ability of asphalt mixture [17]. In order to overcome this limitation, crumb rubber and diatomite can be used as a compound modifier for asphalt binder. Liu et al. [19] investigated the short-term aging properties of diatomite and crumb rubber compound–modified asphalt (DRA). The results revealed that DRA could combine the advantages of diatomite and crumb rubber and achieve better performance in short-term aging resistance than diatomite-modified asphalt (DA) and crumb rubber–modified asphalt (RA).

It is worth noting that different procedures and parameters can be applied to prepare crumb rubber and diatomite compound–modified asphalt binders [20]. The different preparation procedures and parameters can result in different interactions among crumb rubber, asphalt, and diatomite, thus affecting the properties of modified asphalt. But previous studies have rarely investigated the effect of preparation factors on the property of crumb rubber and diatomite compound–modified asphalt, and even the influences of preparation factors on properties of crumb rubber–modified asphalt and diatomite-modified asphalt are limited [20, 21].

In general, there are two kinds of production methods for crumb rubber–modified asphalt binder: the dry process and wet process [3, 22]. In the dry process, coarse crumb rubber is used to substitute a part of the mineral aggregate, which is suitable for producing asphalt mixture with favorable elasticity and anti-icing properties. In the wet process, crumb rubber and asphalt are mixed and allowed to react for a period that can take advantage of the benefits of both base ingredients. Crumb rubber–modified asphalt obtained through the wet process has spread worldwide and got much success by roads built in the past 30 years [3]. For the production of diatomite-modified asphalt, the wet process is the most widely used method. In the preparation of both crumb rubber–modified asphalt binder and diatomite-modified one, the device of high-speed mixing and shear stirring is usually adopted to obtain a homogeneous asphalt binder. The processing parameters including temperature, time, and speed determine the interaction

condition, which is important for the properties of modified asphalt [23]. However, different preparation parameters were applied in recent researches. For the preparation of crumb rubber–modified asphalt, shear temperature 175°C, shear speed 5000 rpm, and shear time 40 min were adopted by Wang et al. [24]. Shear temperature 170–180°C, shear speed 5000 rpm, and shear time 5 min were used by Kedarisetty et al. [23]. In the research by Peralta et al. [25], asphalt rubber binder was heated at 180°C for 60 min and stirred at 230 rpm. For the preparation of diatomite-modified asphalt, the shear temperature is 150°C, shear speed is 3000 rpm, and shear time is 120 min in the research by Cong et al. [14]. The shear temperature is 150°C, shear speed is 4000 rpm, and shear time is 40 min in the research by Guo et al. [17]. The shear temperature is 160°C, shear speed is 5000 rpm, and shear time is 40 min in the research by Tan et al. [15]. As can be concluded from current researches, different preparation parameters were adopted. The effects of preparation parameters on properties of modified asphalt have not been demonstrated.

In this study, the preparation process for crumb rubber and diatomite compound–modified asphalt binder was determined. The effects of six kinds of preparation parameters (crumb rubber concentration, diatomite concentration, shear time, shear speed, shear temperature, and storing time) on properties of binder (penetration at 25°C, softening point, ductility, viscosity at 135°C, elastic recovery, and penetration index (PI)) were investigated using the response surface method (RSM). Relationships between preparation parameters and asphalt properties were obtained, and multiresponse optimization was conducted to determine the optimum preparation parameters corresponding to the satisfactory properties.

2. Materials and Methods

2.1. Raw Materials. Base asphalt AH-90 from Panjin Petrochemical Industry, crumb rubber particle from Changchun Yuxing Rubber Materials Co., Ltd., and diatomite produced by Changchun Diatomite Products Co., Ltd. were used in this study. The detailed properties for raw materials were given in [19].

2.2. Preparation of Crumb Rubber and Diatomite Compound–Modified Asphalt. The wet process was adopted to prepare the crumb rubber and diatomite compound–modified asphalt. A high-speed shear homogenizer (KRH-I, Shanghai Konmix Mechanical & Electrical Equipment Technology Co. Ltd., China) was employed, which can control its internal temperature through a heating device and oil bath. The preparation process mainly included three steps: heating of raw materials; shearing and mixing of asphalt, crumb rubber, and diatomite; and storing of modified asphalt binder. In this study, the influences of six important preparation parameters on compound-modified asphalt binder were investigated. They were content of crumb rubber by weight of the neat asphalt, content of diatomite by weight of the neat asphalt, shear temperature, shear speed, shear time, and storing time. Contents of crumb rubber and diatomite are essential parameters affecting the properties of compound-modified asphalt. Shear speed, shear time, shear temperature, and storing time are key factors in the interaction process among asphalt, crumb rubber, and diatomite to achieve a homogeneous and stable modified binder. The adopted six preparation parameters have not been standardized in current researches, and the researchers have used various parameters. Based on previous research results and considering the workability of binder, the range for content of crumb rubber was 5–15%; it was 5–15% for content of diatomite, 160–190°C for shear temperature, 3000–6000 rpm for shear speed, 30–60 min for shear time, and 30–60 min for storing time in this study. Therefore, the detailed preparation process was determined as follows: firstly, neat asphalt, crumb rubber, and diatomite were separately heated in an oven at 140°C, and asphalt was at the fluid state that can be easily stirred. Secondly, the crumb rubber and diatomite were added into asphalt, which was placed into a shear homogenizer after preliminary mixing. Then, shear temperature and speed were set to the specific values; the blend of asphalt, crumb rubber, and diatomite was mixed and sheared for the specified shear time after the temperature of the blend reached the specified shear temperature. Finally, the blend was contained in the shear homogenizer to swell for the specified storing time. The crumb rubber and diatomite compound–modified asphalt binder was prepared for the tests.

2.3. Characterization Method. In order to evaluate the effects of content of crumb rubber, content of diatomite, shear temperature, shear speed, shear time, and storing time on asphalt properties, penetration, softening point, ductility, viscosity, elastic recovery, and penetration index (PI) were tested and investigated for compound-modified asphalt binder. Detailed illustrations for penetration, softening point, ductility, viscosity, and elastic recovery were conducted in [19].

The results obtained from penetration and softening point tests can be used to calculate the penetration index (PI). PI is an important property which is related to Van der Poel's monograph and useful to determine the stiffness of asphalt. PI was usually used for the classification of pure asphalt. However, it has been verified to be useful for polymer-modified ones [26]. PI can be calculated by the following classical approach [27]:

$$PI = \frac{30}{1 + 50\left(\lg 800 - \lg P_{(25°C, 100\,g, 5\,s)}/T_{R\&B} - 25\right)} - 10,$$

$$(1)$$

where $P_{(25°C, 100\,g, 5\,s)}$ is the penetration at 25°C and $T_{R\&B}$ is the softening point temperature.

2.4. Response Surface Method. In order to analyze the effects of six preparation parameters (content of crumb rubber, content of diatomite, shear temperature, shear speed, shear time, and storing time) on the asphalt property, the response

surface method (RSM) was utilized. The RSM is a group of mathematical and statistical techniques useful to design experiments. It can save cost and time by reducing the overall number of tests required [28, 29]. The RSM has been widely used in many areas to identify the effect of individual variables and interaction of different independent variables on the response [30–32]. An optimal quadratic model was used to determine the optimal condition of response:

$$y = \beta_0 + \sum_{i=1}^{k} \beta_i x_i + \sum_{i=1}^{k} \beta_{ii} x_i^2 + \sum_{i<j}^{k} \beta_{ij} x_i x_j + \varepsilon, \quad (2)$$

where y is the predicted response; x_i and x_j are the coded values of the preparation parameters; k is the number of independent variables; β_i is the linear effect of x_i; β_{ij} is the linear interaction between x_i and x_j; β_{ii} is the secondary effects of x_i; and ε is the random error.

In this study, the three level Box-Behnken design with six factors (content of crumb rubber, X_1; content of diatomite, X_2; shear time, X_3; shear speed, X_4; shear temperature, X_5; and storing time, X_6) was carried out. Actual and coded values of the independent variables are shown in Table 1. Combinations of X_1 (5%, 10%, 15%), X_2 (5%, 10%, 15%), X_3 (30, 45, 60 min), X_4 (3000, 4500, 6000 rpm), X_5 (160°C, 175°C, 190°C), and X_6 (30, 45, 60 min) were selected as independent variables. Penetration at 25°C, Y_1; softening point, Y_2; ductility at 15°C, Y_3; viscosity at 135°C, Y_4; elastic recovery at 25°C, Y_5; and penetration index (PI), Y_6 were adopted as the responses. The software Design-Expert 7.0 was adopted for the scheme design, model generating, statistical analysis, and optimization of the preparation parameters.

3. Results and Discussion

3.1. Experimental Scheme and ANOVA Results

3.1.1. Experimental Scheme. Experimental scheme for property analysis of compound-modified asphalt binder under the effect of preparation parameters was established based on the RSM. This scheme is composed of 54 experimental groups including 6 replicates at the central point, which is much less than a full test plan $3^6 = 729$. Three specimens are included in each group, and the corresponding average value is regarded as a representative one. The detailed experimental scheme and test results of responses are listed in Table 2. The relationships between independent variables and responses were analyzed by 3D response surface plots. A 3D plot is used to represent the dependent variables in function of two independent parameters, when other four variables are kept constant at level 0 in the code value. The interaction between independent variables can also be demonstrated in a 3D plot.

As can be seen from Table 2, the penetration ranged from 34.8 to 71.0 (0.1 mm). Softening point ranged from 46.8°C to 62.4°C. Ductility ranged from 141.0 to 379.7 mm. Viscosity ranged from 497.3 to 5752.0 mPa·s. Elastic recovery ranged from 21% to 63%. PI ranged from −1.5 to 0.83. According to these experimental ranges, the minimum and maximum values of penetration were obtained at Group

TABLE 1: Actual and coded values of the independent variables.

Coded value	Actual value					
	X_1 (%)	X_2 (%)	X_3 (min)	X_4 (rpm)	X_5 (°C)	X_6 (min)
−1	5	5	30	3000	160	30
0	10	10	45	4500	175	45
+1	15	15	60	6000	190	60

nos. 50 (X_1: 15%; X_2: 15%; X_3: 45 min; X_4: 3000 rpm; X_5: 175°C; and X_6: 45 min) and 46 (X_1: 5%; X_2: 5%; X_3: 45 min; X_4: 3000 rpm; X_5: 175°C; and X_6: 45 min), respectively. The minimum and maximum values of softening point were obtained at Group nos. 46 (X_1: 5%; X_2: 5%; X_3: 45 min; X_4: 3000 rpm; X_5: 175°C; and X_6: 45 min) and 50 (X_1: 15%; X_2: 15%; X_3: 45 min; X_4: 3000 rpm; X_5: 175°C; and X_6: 45 min), respectively. The minimum and maximum values of ductility were obtained at Group nos. 54 (X_1: 15%; X_2: 10%; X_3: 45 min; X_4: 3000 rpm; X_5: 160°C; and X_6: 45 min) and 14 (X_1: 15%; X_2: 5%; X_3: 45 min; X_4: 4500 rpm; X_5: 190°C; and X_6: 60 min), respectively. The minimum and maximum values of viscosity were obtained at Group nos. 46 (X_1: 5%; X_2: 5%; X_3: 45 min; X_4: 3000 rpm; X_5: 175°C; and X_6: 45 min) and 50 (X_1: 15%; X_2: 15%; X_3: 45 min; X_4: 3000 rpm; X_5: 175°C; and X_6: 45 min), respectively. The minimum and maximum values of elastic recovery were obtained at Group nos. 21 (X_1: 10%; X_2: 10%; X_3: 45 min; X_4: 4500 rpm; X_5: 190°C; and X_6: 60 min) and 50 (X_1: 15%; X_2: 15%; X_3: 45 min; X_4: 3000 rpm; X_5: 175°C; and X_6: 45 min), respectively. The minimum and maximum values of PI were obtained at Group nos. 49 (X_1: 15%; X_2: 5%; X_3: 45 min; X_4: 3000 rpm; X_5: 175°C; and X_6: 45 min) and 7 (X_1: 5%; X_2: 10%; X_3: 30 min; X_4: 4500 rpm; X_5: 175°C; and X_6: 60 min), respectively.

3.1.2. Analysis of Variance (ANOVA) Results for Quadratic Models and Independent Variables.

Analysis of variance (ANOVA) was used to evaluate the statistical significance of independent parameters and interactions among them. The adequacy of the models constructed for penetration at 25°C, softening point, ductility at 15°C, viscosity at 135°C, elastic recovery at 25°C, and penetration index (PI) was checked by coefficient of determination (R^2), adjusted coefficient of determination (Adj. R^2), Adeq. precision, and Fisher's test value (F value). Models and factors were considered significant when $p < 0.01$. ANOVA results for quadratic models and independent variables are obtained and listed in Tables 3 and 4, respectively. The detailed explanations of results are given in Sections 3.2–3.7.

3.2. Penetration

3.2.1. Analysis of the Model for Penetration.

The results of ANOVA for the penetration model indicated that the model possessed satisfactory levels of R^2 (0.90), Adj. R^2 (0.80), and Adeq. precision (11.014), which were significant at $p < 0.0001$. R^2 and Adj. R^2 values of the model were higher than 0.8, which revealed a close agreement between experimental results and predicted ones. Adeq. precision is the measure

of the signal-to-noise ratio, and greater ratio than 4 is desirable. An Adeq. precision of 11.014 indicated that the model can be used to navigate the design space.

The second-order quadratic polynomial models were established based on the regression coefficients that were determined with the least squares method. Stepwise regression was used to identify the statistically significant variables. As can be seen from the ANOVA results of independent variables for the penetration model, the significant variables ($p < 0.05$) for the penetration model included X_1, X_2, X_5, X_{23}, X_{46}, X_{22}, X_{33}, and X_{44}. The reduced second-order model in terms of coded factors for penetration was obtained by removing the insignificant variables:

$$Y_1 = 43.81 - 6.69 \times X_1 - 4.97 \times X_2 + 5.17 \times X_5$$
$$+ 3.10 \times X_{23} - 3.01 \times X_{46} + 2.87 \times X_{22} + 6.24 \times X_{33}$$
$$+ 6.22 \times X_{44}.$$

(3)

3.2.2. Effect of Preparation Parameters on Penetration.

As shown in Table 4, the most significant linear variables on penetration include content of crumb rubber, content of diatomite, and shear temperature with $p < 0.0001$. Meanwhile, quadratic terms of shear time and shear speed present the most significant effect with $p < 0.0001$. The quadratic term of diatomite content has a significant effect with $p < 0.05$. Furthermore, the interaction terms of (diatomite content) ∗ (shear time) and (shear speed) ∗ (storing time) have significant effects on penetration with $p < 0.05$.

The relationships between preparation parameters and penetration were demonstrated in 3D response surface plots, which are shown in Figure 1. As can be seen from the figure, the penetration decreases with the increase of diatomite concentration and crumb rubber concentration. However, it increases with the increase of shear temperature. For the relationships of penetration with shear time, shear speed, and storing time, they present the trends of decreasing firstly and then increasing.

The interaction of crumb rubber and asphalt binder is essential for the performance of modified asphalt binder. Rubber particle swelling and significant absorption take place when rubber particles are added into heated asphalt binder. The polymer chains of rubber particles can absorb the lighter and solvating fractions of binder. Rubber particle size and shape are changed, and a gel-like structure is generated [3], which causes the reduction of penetration for asphalt binder.

Diatomite has a large specific area and good adsorption performance, which can effectively absorb the lower molecular group and lower polar aromatic molecular group for its mesoporous structure. The anchorage structure forms

TABLE 2: Experimental scheme with test results for property analysis of compound-modified asphalt binder.

Group number	Preparation parameters						Responses					
	X_1 (%)	X_2 (%)	X_3 (min)	X_4 (rpm)	X_5 (°C)	X_6 (min)	Y_1 (0.1 mm)	Y_2 (°C)	Y_3 (mm)	Y_4 (mPa·s)	Y_5 (%)	Y_6
1	5	10	45	6000	160	45	47.9	52.5	185.0	1343.3	33	−0.69
2	15	10	30	4500	175	30	47.5	56.2	218.3	2682.3	49	0.12
3	10	5	30	4500	190	45	67.7	50.5	339.5	1280.0	36	−0.33
4	15	5	45	6000	175	45	51.0	57.2	259.0	2865.0	58	0.50
5	10	15	45	4500	160	30	42.6	55.4	158.3	2098.0	36	−0.31
6	10	10	45	4500	175	45	44.8	53.3	203.3	1907.0	43	−0.66
7	5	10	30	4500	175	60	60.8	47.1	295.0	692.4	25	−1.50
8	10	10	45	4500	175	45	42.4	54.5	205.6	1919.0	44	−0.52
9	10	15	30	4500	190	45	50.6	54.8	294.3	1792.7	42	−0.04
10	10	10	45	4500	175	45	44.3	53.9	203.3	1907.0	43	−0.55
11	15	10	30	4500	175	60	51.4	55.1	244.0	1778.7	44	0.07
12	10	10	45	4500	175	45	43.6	53.9	204.5	1906.0	44	−0.58
13	10	5	45	4500	160	60	45.3	53.1	181.5	1833.3	42	−0.68
14	10	5	45	4500	190	60	55.0	51.0	379.7	1676.0	41	−0.73
15	10	10	30	3000	175	30	55.5	54.1	215.5	1844.7	34	0.03
16	5	10	30	4500	175	30	54.6	51.2	255.7	1140.7	25	−0.70
17	10	10	45	4500	175	45	43.1	54.1	204.9	1906.5	44	−0.57
18	10	15	60	4500	190	45	56.9	55.3	197.0	2017.0	51	0.36
19	10	5	30	4500	160	45	53.1	53.6	175.0	1402.0	40	−0.20
20	10	15	30	4500	160	45	41.8	55.9	171.0	2513.0	36	−0.24
21	10	10	30	3000	175	60	58.8	51.8	316.5	1301.7	21	−0.37
22	10	10	45	4500	175	45	44.8	53.3	203.3	1907.0	43	−0.66
23	10	5	60	4500	190	45	61.6	50.5	301.0	953.1	41	−0.58
24	5	5	45	6000	175	45	68.6	50.6	284.0	770.1	31	−0.27
25	15	10	45	6000	160	45	36.9	59.6	172.7	3821.0	54	0.24
26	10	5	60	4500	160	45	47.8	52.7	184.0	1034.0	46	−0.65
27	5	10	45	3000	160	45	54.6	50.7	219.3	880.4	27	−0.82
28	10	10	30	6000	175	60	53.8	54.2	231.0	2008.3	38	−0.02
29	10	15	45	4500	190	60	53.1	55.4	243.5	2109.0	53	0.21
30	15	10	60	4500	175	30	43.6	59.2	214.3	3675.7	57	0.54
31	5	15	45	6000	175	45	48.8	53.2	216.3	1356.0	29	−0.50
32	10	5	45	4500	190	30	56.8	51.4	341.3	1627.0	42	−0.55
33	10	10	60	6000	175	60	56.6	52.4	308.5	1930.3	47	−0.32
34	10	10	30	6000	175	30	57.9	52.2	354.3	1508.0	33	−0.31
35	5	10	45	3000	190	45	63.0	49.4	315.0	1188.0	32	−0.80
36	10	10	60	3000	175	30	53.3	54.0	269.5	2229.7	44	−0.09
37	10	15	60	4500	160	45	49.0	56.0	174.0	2430.0	42	0.15
38	10	15	45	4500	190	30	49.7	54.5	260.3	2054.0	46	−0.15
39	5	10	60	4500	175	60	49.9	50.9	249.0	913.4	34	−0.99
40	10	5	45	4500	160	30	49.2	53.3	197.7	1860.0	41	−0.44
41	5	15	45	3000	175	45	55.2	51.5	357.7	1202.7	15	−0.60
42	10	10	60	3000	175	60	60.3	50.9	329.0	774.4	25	−0.53
43	5	10	45	6000	190	45	59.3	49.9	314.5	704.8	30	−0.82
44	15	15	45	6000	175	45	37.1	60.4	191.7	4847.3	61	0.40
45	10	10	60	6000	175	30	66.3	49.5	348.0	1735.0	41	−0.65
46	5	5	45	3000	175	45	71.0	46.8	307.3	497.3	23	−1.21
47	15	10	45	3000	190	45	50.6	57.6	292.3	3509.0	54	0.57
48	15	10	45	6000	190	45	47.0	57.0	278.3	3539.0	60	0.26
49	15	5	45	3000	175	45	50.8	58.8	224.5	3724.3	55	0.83
50	15	15	45	3000	175	45	34.8	62.4	160.0	5752.0	63	0.64
51	10	15	45	4500	160	60	39.1	55.6	170.0	2198.7	32	−0.45
52	5	10	60	4500	175	30	58.4	49.1	309.5	765.9	23	−1.07
53	15	10	60	4500	175	60	40.6	59.5	198.7	4491.7	59	0.43
54	15	10	45	3000	160	45	40.2	59.3	141.0	3916.3	45	0.37

and the stiffening effect presents [33], which decrease the penetration of binder.

As for shear temperature, the interactions among asphalt, crumb rubber, and diatomite are strengthened with the increase of shear temperature. Rubber particle dissolution happens because of depolymerization and devulcanization, which break not only the polymer chains of rubber but also the crosslink bonds. Then, the swelling reaction is reduced,

TABLE 3: ANOVA results for quadratic models of modified asphalt properties.

Sources	R^2	Adj. R^2	Adeq. precision	F value	p value	Significant
Y_1	0.90	0.80	11.014	8.68	<0.0001	Yes
Y_2	0.95	0.90	17.854	18.37	<0.0001	Yes
Y_3	0.90	0.79	10.875	8.46	<0.0001	Yes
Y_4	0.94	0.87	13.392	13.94	<0.0001	Yes
Y_5	0.92	0.84	14.374	11.01	<0.0001	Yes
Y_6	0.93	0.85	14.466	12.00	<0.0001	Yes

TABLE 4: ANOVA results for independent variables.

Sources	Y_1 F	Y_1 p	Y_2 F	Y_2 p	Y_3 F	Y_3 p	Y_4 F	Y_4 p	Y_5 F	Y_5 p	Y_6 F	Y_6 p
X_1	74.86	<0.0001****	358.54	<0.0001****	26.1	<0.0001****	220.32	<0.0001****	279.97	<0.0001****	236.0	<0.0001****
X_2	41.34	<0.0001****	60.43	<0.0001****	17.3	0.0003***	23.60	<0.0001****	0.19	0.6701	15.2	0.0006***
X_3	0.26	0.6144	0.38	0.5416	0.0	0.8445	1.81	0.1899	19.69	0.0001****	0.0	0.9208
X_4	0.84	0.3668	0.07	0.7993	0.0	0.9755	0.03	0.8620	15.23	0.0006***	0.0	0.8415
X_5	44.66	<0.0001****	15.07	0.0006***	104.4	<0.0001****	1.66	0.2085	6.66	0.0159*	1.4	0.2500
X_6	0.33	0.5691	0.36	0.5541	0.0	0.9802	0.46	0.5041	0.24	0.6299	1.8	0.1901
X_{12}	0.27	0.6055	0.02	0.8832	2.0	0.1675	4.45	0.0448*	2.78	0.1077	1.4	0.2395
X_{13}	0.49	0.4923	3.59	0.0692	0.5	0.4847	8.96	0.0060**	2.33	0.1388	1.3	0.2606
X_{14}	0.98	0.3306	7.37	0.0116*	6.2	0.0197**	1.50	0.2312	0.35	0.5609	7.2	0.0125*
X_{15}	0.00	0.9454	0.02	0.8961	0.2	0.6962	0.08	0.7832	1.16	0.2905	0.4	0.5503
X_{16}	0.09	0.7717	0.26	0.6137	0.1	0.7023	0.03	0.8700	1.59	0.2184	1.1	0.3138
X_{23}	5.37	0.0286*	0.24	0.6252	0.6	0.4287	0.42	0.5223	0.14	0.7099	7.0	0.0134*
X_{24}	0.03	0.8562	0.71	0.4083	2.2	0.1461	0.02	0.8992	0.00	0.9543	1.8	0.1916
X_{25}	0.27	0.6110	2.23	0.1478	7.0	0.0137**	0.14	0.7147	11.70	0.0021**	3.4	0.0758
X_{26}	0.28	0.6017	0.31	0.5800	0.1	0.7374	0.01	0.9184	0.07	0.7989	1.3	0.2724
X_{34}	1.23	0.2783	1.33	0.2591	0.0	0.9543	0.05	0.8231	0.11	0.7442	0.4	0.5388
X_{35}	0.03	0.8635	0.18	0.6718	3.4	0.0782	0.07	0.7891	0.02	0.8861	0.2	0.6993
X_{36}	2.40	0.1332	2.93	0.0986	0.8	0.3941	0.36	0.5522	0.48	0.4967	1.1	0.3038
X_{45}	0.07	0.8001	0.53	0.4748	0.0	0.8845	0.41	0.5299	0.68	0.4156	0.3	0.5728
X_{46}	5.05	0.0334*	11.53	0.0022**	16.1	0.0005***	4.37	0.0466*	12.78	0.0014**	6.7	0.0158*
X_{56}	0.70	0.4095	0.03	0.8704	0.1	0.7497	0.00	0.9816	0.46	0.5027	1.0	0.3236
X_{11}	0.39	0.5398	7.87	0.0094**	0.4	0.5287	10.11	0.0038**	0.84	0.3692	3.6	0.0693
X_{22}	5.90	0.0223*	2.61	0.1183	0.9	0.3585	1.37	0.2530	0.10	0.7563	13.0	0.0013**
X_{33}	27.96	<0.0001****	3.72	0.0647	4.3	0.0471*	4.40	0.0458*	3.92	0.0585	5.1	0.0318*
X_{44}	27.76	<0.0001****	0.38	0.5442	23.2	<0.0001****	0.50	0.4840	5.90	0.0224*	13.0	0.0013**
X_{55}	0.29	0.5974	0.04	0.8452	0.0	0.8253	0.47	0.4968	0.09	0.7687	0.3	0.5725
X_{66}	1.71	0.2025	3.17	0.0869	11.8	0.0020**	0.10	0.7596	3.99	0.0562	0.8	0.3675

Note. $^*p \leq 0.05$, $^{**}p \leq 0.01$, $^{***}p \leq 0.001$, and $^{****}p \leq 0.0001$.

and the lighter fractions of binder increase. Therefore, the penetration increases with the increase of shear temperature.

Shear speed, shear time, and storing time present similar effects on penetration. At the early stage, rubber and diatomite particles are distributed more evenly in asphalt binder with the increase of shear speed, shear time, and storing time. The swelling effect is significant, and distances between particles are shortened. A gel-like structure and an anchorage structure are formed and strengthened. The penetration decreases. However, the crosslink bonds between particles are destroyed if the shear time and storing time are longer and the shear speed becomes higher, which increase the penetration of binder.

3.3. Softening Point

3.3.1. Analysis of the Model for Softening Point. The softening point model possessed satisfactory levels of R^2 (0.95),

Adj. R^2 (0.90), and Adeq. precision (17.854), which were significant at $p < 0.0001$. R^2 and Adj. R^2 values of the model were higher than 0.8, which revealed a close agreement between experimental results and predicted ones. An Adeq. precision of 17.854 indicated that the model can be used to navigate the design space.

As can be seen from the ANOVA results of independent variables for the softening point model, the significant variables ($p < 0.05$) included X_1, X_2, X_5, X_{14}, X_{46}, and X_{11}. The reduced second-order model in terms of coded factors for softening point was obtained by removing the insignificant variables:

$$Y_2 = 53.83 + 4.15 \times X_1 + 1.70 \times X_2 - 0.85 \times X_5$$
$$- 0.73 \times X_{14} + 1.29 \times X_{46} + 0.94 \times X_{11}. \quad (4)$$

3.3.2. Effect of Preparation Parameters on Softening Point. As shown in Table 4, the most significant linear

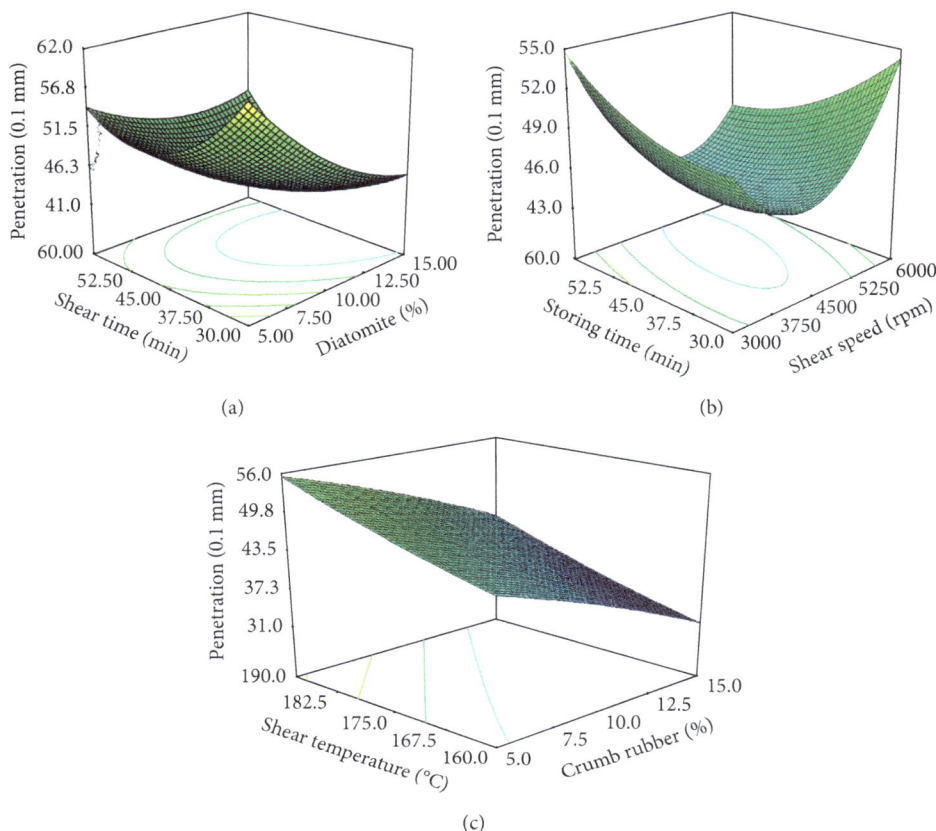

FIGURE 1: Response surface plots for the effect of preparation parameters on penetration. (a) Diatomite content ∗ shear time, (b) shear speed ∗ storing time, and (c) crumb rubber content ∗ shear temperature.

variables on softening point include content of crumb rubber and content of diatomite with $p < 0.0001$, followed by shear temperature with $p < 0.001$. Meanwhile, the quadratic term of crumb rubber content has a significant effect with $p < 0.01$. Furthermore, the interaction terms of (shear speed) ∗ (storing time) have significant effects on softening point with $p < 0.01$, followed by (crumb rubber content) ∗ (shear speed) with $p < 0.05$.

The softening points are applied to evaluate the high stability of modified asphalt binder. The higher the softening point is, the better the high susceptibility of asphalt is. The relationships between preparation parameters and softening point were demonstrated in 3D response surface plots, which are shown in Figure 2. As can be seen from the figure, the softening point increases with the increase of diatomite concentration and crumb rubber concentration. However, it decreases with the increase of shear temperature. For the relationships of the softening point with shear time, shear speed, and storing time, they present the trends of increasing firstly and then decreasing.

As discussed above, the polymer chains of rubber particles can absorb the lighter and solvating fractions of binder, and a gel-like structure is generated, which shortens the distances between particles and increases the softening point of binder. Diatomite can effectively absorb the lower molecular group and lower polar aromatic molecular group.

The anchorage structure can improve the high temperature stability of asphalt binder with the increase of diatomite content. As for shear temperature, higher temperature will lead to the breakdown of polymer chains of rubber and crosslink bonds of binder. Therefore, the softening point increases with the increase of shear temperature. Shear speed, shear time, and storing time can strength the gel-like structure and anchorage structure of asphalt binder at the early stage, which make the softening point increase. However, the crosslink bonds between particles are destroyed if the shear time and storing time are longer and the shear speed becomes higher, which decrease the softening point of binder.

3.4. Ductility

3.4.1. Analysis of the Model for Ductility. The ductility model possessed satisfactory levels of R^2 (0.90), Adj. R^2 (0.79), and Adeq. precision (10.875), which were significant at $p < 0.0001$. R^2 value of the model was higher than 0.8 and Adj. R^2 value was highly close to 0.8, which revealed a close agreement between experimental results and predicted ones. An Adeq. precision of 10.875 indicated that the model can be used to navigate the design space.

As can be seen from the ANOVA results of independent variables for the ductility model, the significant variables

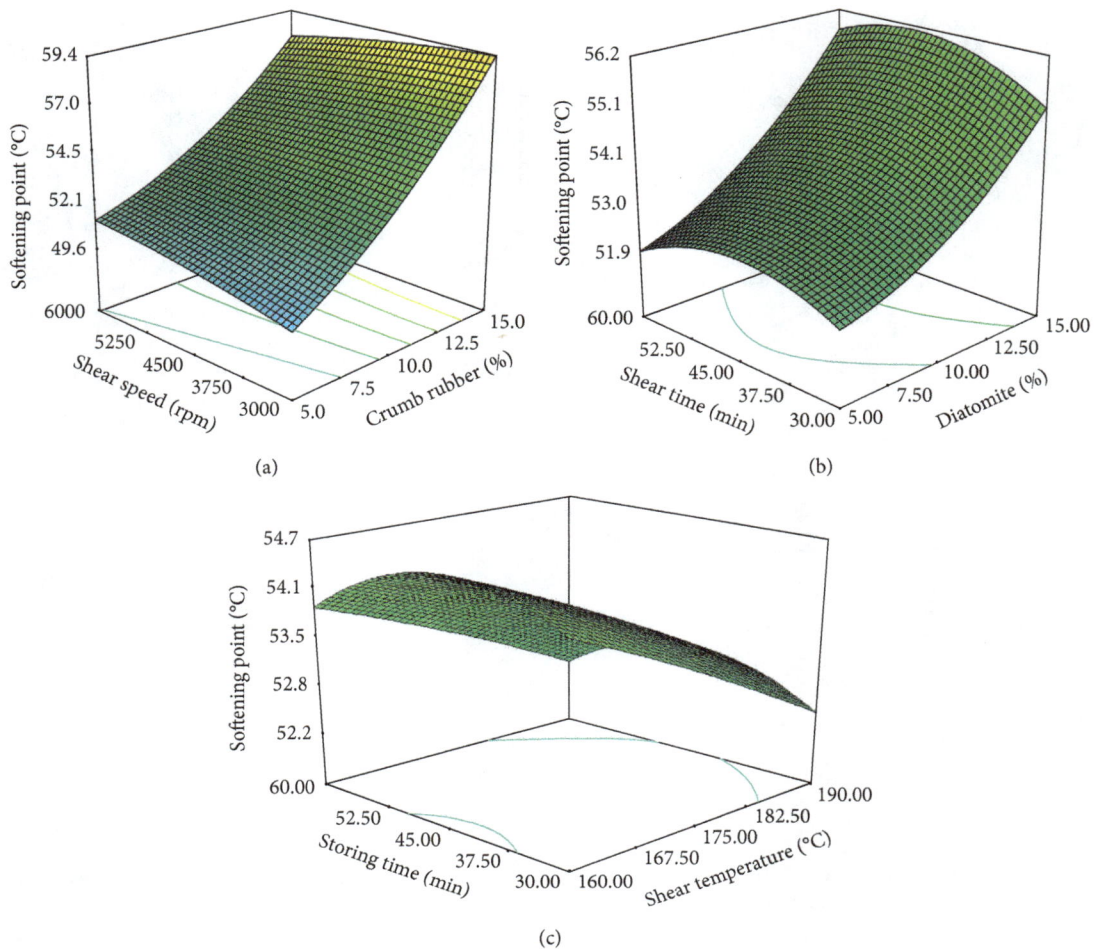

FIGURE 2: Response surface plots for the effect of preparation parameters on softening point. (a) Crumb rubber content ∗ shear speed, (b) diatomite content ∗ shear time, and (c) shear temperature ∗ storing time.

($p < 0.05$) included X_1, X_2, X_5, X_{14}, X_{25}, X_{46}, X_{33}, X_{44}, and X_{66}. The reduced second-order model in terms of coded factors for ductility was obtained by removing the insignificant variables:

$$Y_3 = 204.63 - 29.73 \times X_1 - 24.18 \times X_2 + 59.47 \times X_5$$
$$+ 17.71 \times X_1 \times X_4 - 18.84 \times X_{25} - 40.42 \times X_{46} \quad (5)$$
$$+ 18.53 \times X_{33} + 42.80 \times X_{44} + 30.59 \times X_{66}.$$

3.4.2. Effect of Preparation Parameters on Ductility. As shown in Table 4, the most significant linear variables on ductility include content of crumb rubber and shear temperature with $p < 0.0001$, followed by content of diatomite with $p < 0.001$. Meanwhile, the quadratic term of shear speed has a significant effect with $p < 0.0001$, followed by storing time with $p < 0.001$ and shear time with $p < 0.05$. Furthermore, the interaction terms of (shear speed) ∗ (storing time) have significant effects on ductility with $p < 0.001$, followed by (crumb rubber content) ∗ (shear speed) and (diatomite content) ∗ (shear temperature) with $p < 0.01$.

Ductility can be used to reflect the low temperature performance of asphalt binder. The relationships between preparation parameters and ductility were demonstrated in 3D response surface plots, which are shown in Figure 3. As can be seen from the figure, the ductility increases with the increase of shear temperature. However, it decreases with the increase of diatomite concentration and crumb rubber concentration. For the relationships of ductility with shear time, shear speed, and storing time, they present the trends of decreasing firstly and then increasing.

The reasons lie in that addition of diatomite and crumb rubber can absorb the lighter fractions of the asphalt and result in large progressive increase of stiffness. The higher the stiffness is, the faster the failure stress reaches [16, 34]. Therefore, the elongation of compound-modified asphalt binder decreases. Another important reason is the stress concentration phenomenon occurring at the interface between the particle and asphalt. Therefore, the reduction of ductility does not indicate that the low temperature performance for crumb rubber and diatomite compound–modified asphalt is unsatisfactory. With the increase of shear temperature, viscosity toughness is reduced, resulting in the increase of

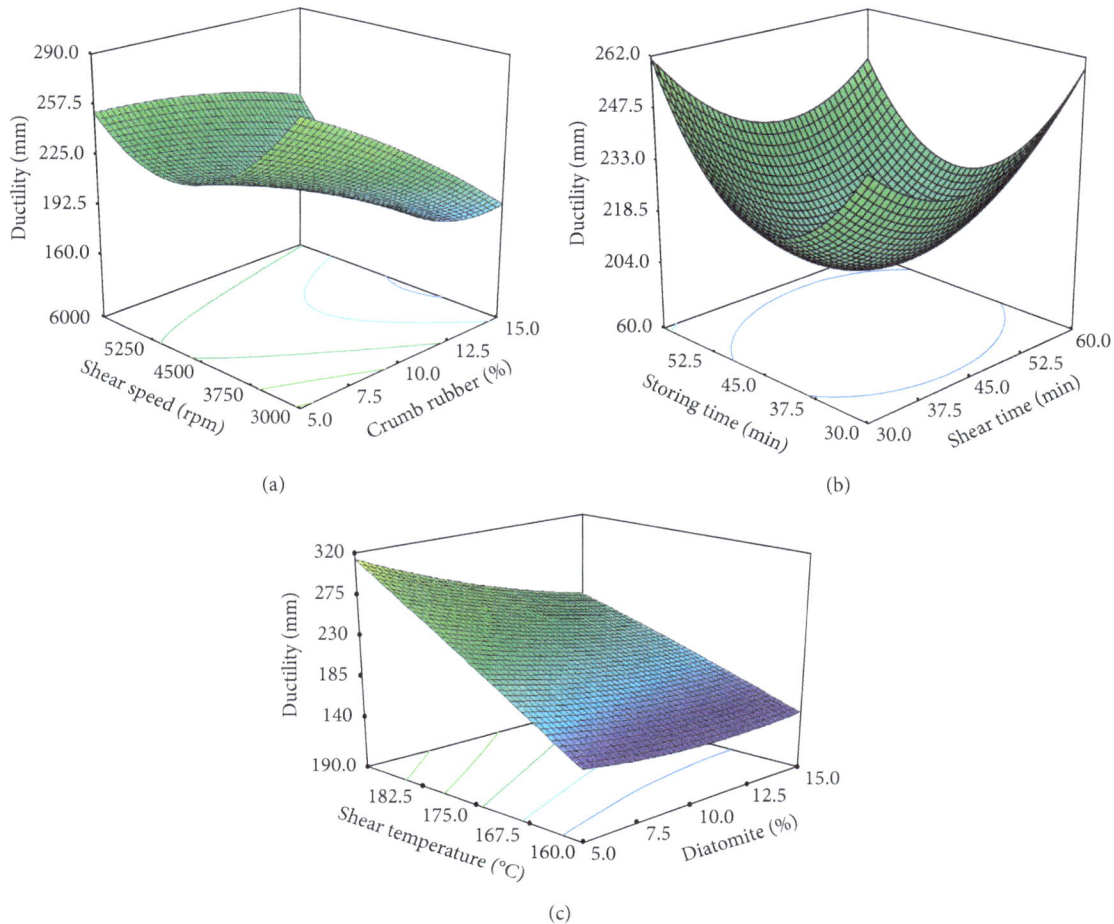

FIGURE 3: Response surface plots for the effect of preparation parameters on ductility. (a) Crumb rubber content ∗ shear speed, (b) storing time ∗ shear time, and (c) shear temperature ∗ diatomite content.

ductility. As for shear speed, shear time, and storing time, they can strength the gel-like structure and anchorage structure of asphalt and increase the stiffness of binder when their values are relatively lower. However, the stable structure will be destroyed when their values reach the critical points. Therefore, ductility decreases firstly and then increases with the increase of shear time, shear speed, and storing time.

3.5. Viscosity

3.5.1. Analysis of the Model for Viscosity.
The viscosity model possessed satisfactory levels of R^2 (0.94), Adj. R^2 (0.87), and Adeq. precision (13.392), which were significant at $p < 0.0001$. R^2 and Adj. R^2 values of the model were higher than 0.8, which revealed a close agreement between experimental results and predicted ones. An Adeq. precision of 13.392 indicated that the model can be used to navigate the design space.

As can be seen from the ANOVA results of independent variables for the viscosity model, the significant variables ($p < 0.05$) included X_1, X_3, X_4, X_5, X_{25}, X_{46}, and X_{44}. The reduced second-order model in terms of coded factors for

viscosity was obtained by removing the insignificant variables:

$$Y_4 = 1907.75 + 1381.14 \times X_1 + 125.24 \times X_3$$
$$- 16.34 \times X_4 - 120.2 \times X_5 - 42.11 \times X_{25} \quad (6)$$
$$+ 336.74 \times X_{46} + 100.93 \times X_{44}.$$

3.5.2. Effect of Preparation Parameters on Viscosity.
As shown in Table 4, the most significant linear variables on viscosity include content of crumb rubber and shear time with $p \leq 0.0001$, followed by shear speed with $p < 0.001$ and shear temperature with $p < 0.05$. Meanwhile, the quadratic term of shear speed has a significant effect with $p < 0.05$. Furthermore, the interaction terms of (shear speed) ∗ (storing time) and (diatomite content) ∗ (shear temperature) have significant effects on viscosity with $p < 0.01$.

The relationships between preparation parameters and viscosity were demonstrated in 3D response surface plots, which are shown in Figure 4. As can be seen from the figure, the viscosities increase with the increase of diatomite concentration and crumb rubber concentration. However, they

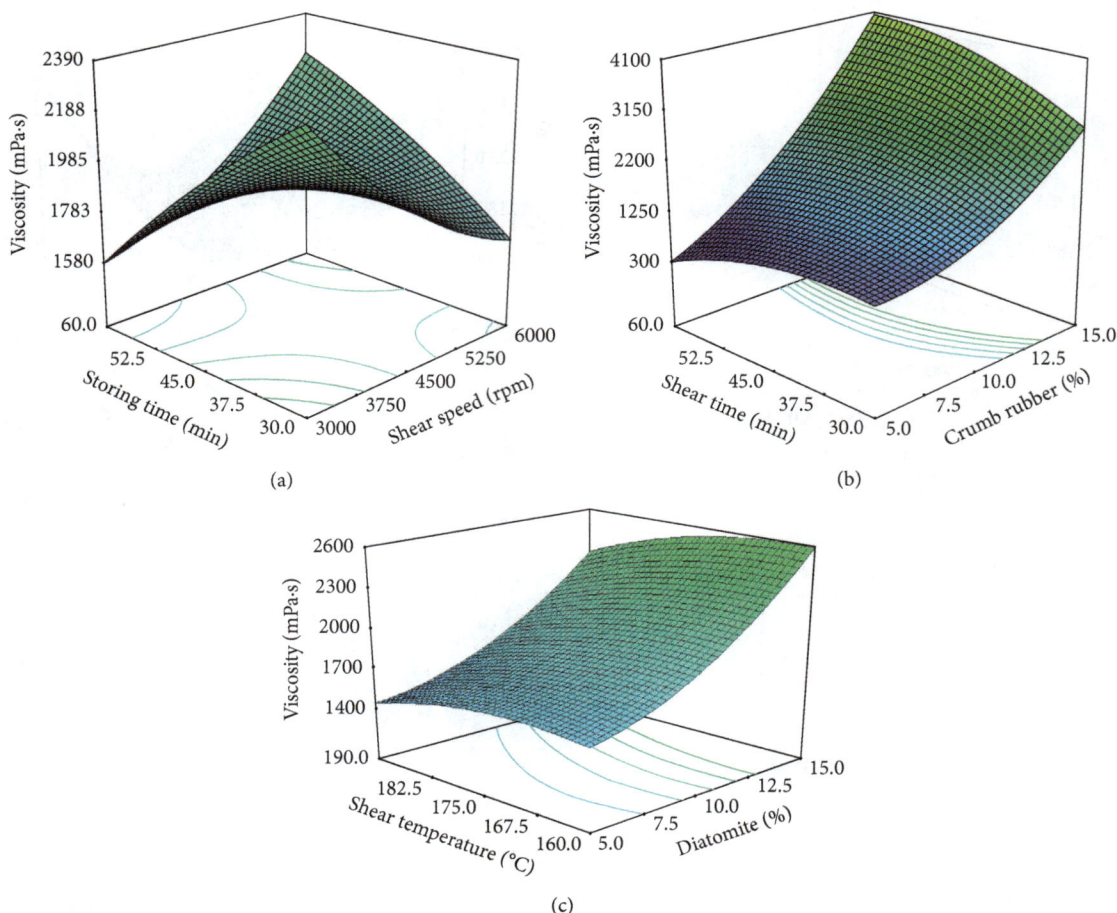

FIGURE 4: Response surface plots for the effect of preparation parameters on viscosity. (a) Storing time * shear speed, (b) crumb rubber content * shear time, and (c) shear temperature * diatomite content.

present a declining trend with the increase of shear time, shear speed, shear temperature, and storing time.

This is because addition of diatomite and crumb rubber into asphalt can form a stable gel-like structure and an anchorage structure. The change in particle size and shape shortens the distances between particles, causing an increase in binder viscosity. If the values of shear temperature, shear speed, shear time, and storing time are too large, rubber particles will be broken down and dissolved into the liquid phase of binder. The crosslink bonds between particles will be destroyed, which will cause a reduction in binder viscosity.

3.6. Elastic Recovery

3.6.1. Analysis of the Model for Elastic Recovery. The elastic recovery model possessed satisfactory levels of R^2 (0.92), Adj. R^2 (0.84), and Adeq. precision (14.374), which were significant at $p < 0.0001$. R^2 and Adj. R^2 values of the model were higher than 0.8, which revealed a close agreement between experimental results and predicted ones. An Adeq. precision of 14.374 indicated that the model can be used to navigate the design space.

As can be seen from the ANOVA results of independent variables for the elastic recovery model, the significant variables ($p < 0.05$) included X_1, X_2, X_{12}, X_{13}, X_{46}, X_{11}, and X_{33}. The reduced second-order model in terms of coded factors for elastic recovery was obtained by removing the insignificant variables:

$$Y_5 = 43.40 + 13.92 \times X_1 + 0.36 \times X_2 + 2.40 \times X_{12}$$
$$+ 2.20 \times X_{13} + 5.15 \times X_{46} + 1.16 \times X_{11} - 2.51 \times X_{33}.$$
$$(7)$$

3.6.2. Effect of Preparation Parameters on Elastic Recovery. As shown in Table 4, the most significant linear variables on elastic recovery include content of crumb rubber and content of diatomite with $p < 0.0001$. Meanwhile, the quadratic term of crumb rubber concentration has a significant effect with $p < 0.01$, followed by shear speed with $p < 0.05$. Furthermore, the interaction terms of (crumb rubber concentration) * (shear time) have significant effects on elastic recovery with $p < 0.01$, followed by (crumb rubber content) * (diatomite content) and (shear speed) * (storing time) with $p < 0.05$.

(a)

(b)

(c)

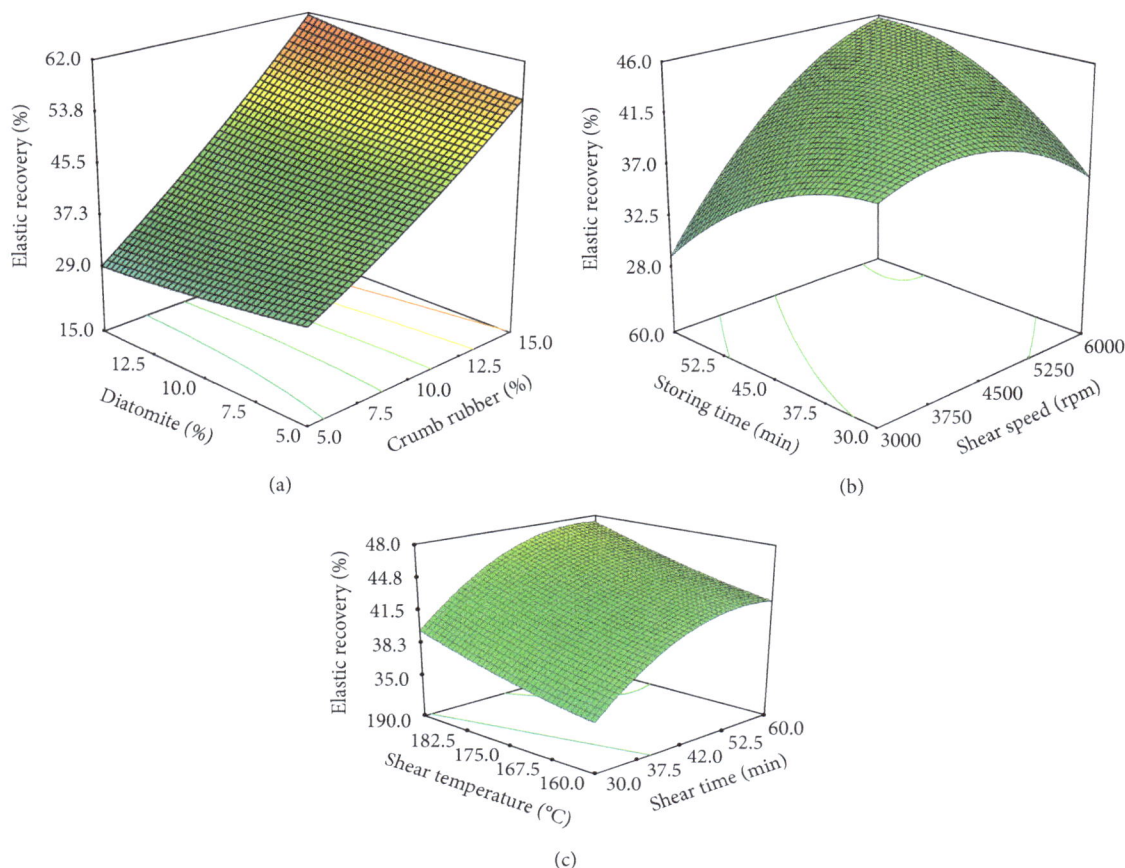

FIGURE 5: Response surface plots for the effect of preparation parameters on elastic recovery. (a) Crumb rubber content ∗ diatomite content, (b) storing time ∗ shear speed, and (c) shear temperature ∗ shear time.

The relationships between preparation parameters and elastic recovery were demonstrated in 3D response surface plots, which are shown in Figure 5. As can be seen from the figure, the elastic recovery increases with the increase of crumb rubber concentration and shear temperature. It increases with the increase of shear time and shear speed but at a decreasing rate. For the relationships of elastic recovery with storing time, they present the trend of increasing firstly and then decreasing slightly. Diatomite concentration can slightly increase the elastic recovery of binder. However, its effect is limited.

As can be seen from the results, the elastic recovery performance of the modified asphalt binder is mainly determined by crumb rubber concentration. The reasons lie in that crumb rubber possesses favorable elastic performance. Additions of crumb rubber into asphalt can strengthen the bonding effect of binder after swelling. The molecular force between particles is increased. Therefore, the ability of elastic recovery is improved. As for diatomite, it is a porous material without good elastic property. Its anchorage structure can enhance the elastic performance of binder to a certain extent. However, the effect is little.

3.7. Penetration Index (PI)

3.7.1. Analysis of the Model for PI.
The PI model possessed satisfactory levels of R^2 (0.93), Adj. R^2 (0.85), and Adeq.

precision (14.466), which were significant at $p < 0.0001$. R^2 and Adj. R^2 values of the model were higher than 0.8, which revealed a close agreement between experimental results and predicted ones. An Adeq. precision of 14.466 indicated that the model can be used to navigate the design space.

As can be seen from the ANOVA results of independent variables for the PI model, the significant variables ($p < 0.05$) included X_1, X_2, X_{14}, X_{23}, X_{46}, X_{22}, X_{33}, and X_{44}. The reduced second-order model in terms of coded factors for PI was obtained by removing the insignificant variables:

$$Y_6 = -0.59 + 0.62 \times X_1 + 0.16 \times X_2 - 0.13 \times X_{14}$$
$$+ 0.19 \times X_{23} + 0.18 \times X_{46} + 0.22 \times X_{22} \qquad (8)$$
$$+ 0.14 \times X_{33} + 0.22 \times X_{44}.$$

3.7.2. Effect of Preparation Parameters on Penetration Index (PI).
As shown in Table 4, the most significant linear variables on PI are content of crumb rubber with $p < 0.0001$, followed by content of diatomite with $p < 0.001$. Meanwhile, the quadratic term of diatomite concentration and shear speed has a significant effect with $p < 0.01$, followed by shear time with $p < 0.05$. Furthermore, the interaction terms of (crumb rubber concentration) ∗ (shear speed), (shear time) ∗ (diatomite content), and (shear speed) ∗ (storing time) have significant effects on PI with $p < 0.05$.

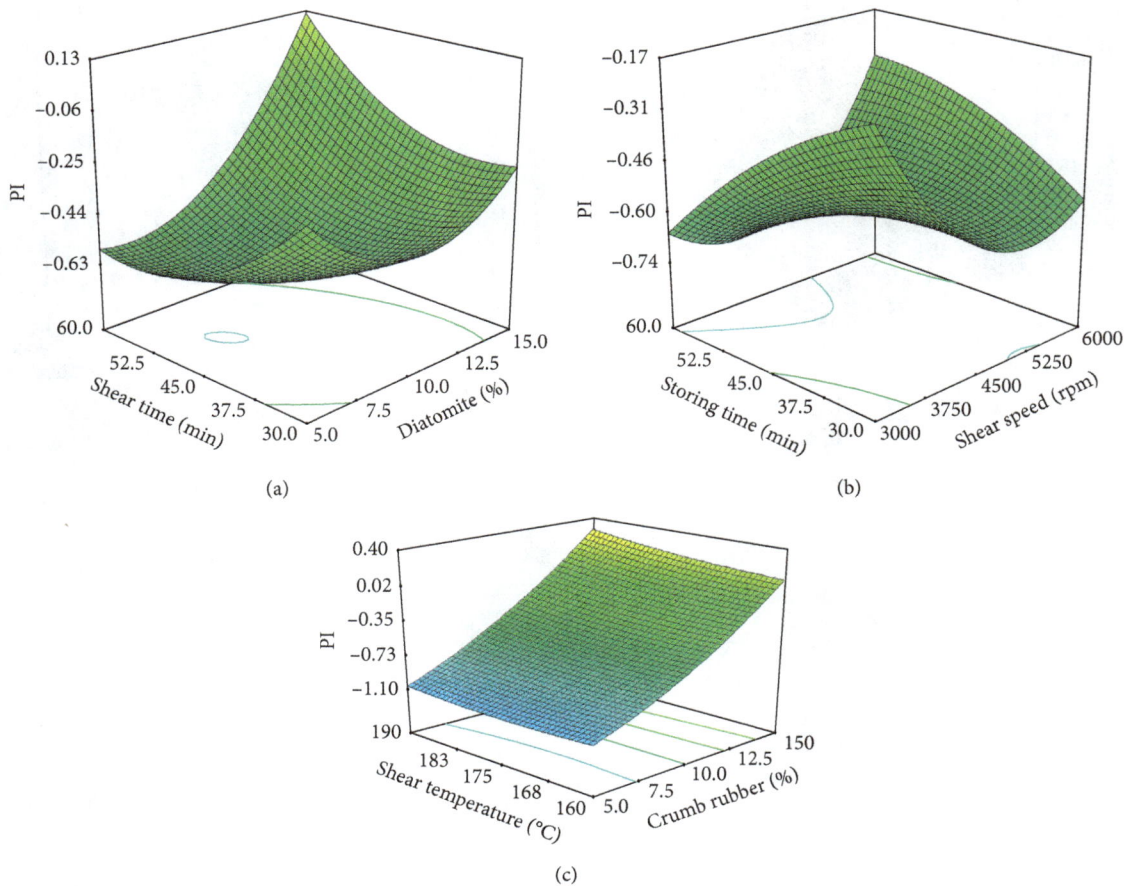

FIGURE 6: Response surface plots for the effect of preparation parameters on PI. (a) Shearing time * diatomite content, (b) storing time * shear speed, and (c) shear temperature * crumb rubber content.

TABLE 5: Optimum preparation parameters and corresponding predicted and experimental results.

Factors	X_1	X_2	X_3	X_4	X_5
Optimum value	13.8	6.2	55	5300	183
Responses	Y_1	Y_2	Y_3	Y_4	Y_5
Predicted results	54.15	55.17	307.4	2708	67.63
Experimental results	54.96	54.56	296.6	2814	68.64
Relative error (%)	1.5	1.1	3.5	3.9	1.5

The relationships between preparation parameters and PI were demonstrated in 3D response surface plots, which are shown in Figure 6. As can be seen from the figure, the PI increases with the increase of crumb rubber concentration. It remains relatively stable before 8–10% of diatomite concentration and increases after that. The relationships of PI with shear speed and shear time present the trend of decreasing firstly and then increasing. PI increases with the increase of shear temperature with little amplitude.

The results reveal that addition of crumb rubber and diatomite into asphalt can effectively improve its temperature stability. This is mainly because of the gel-like structure and anchorage structure. As for shear time, shear speed, shear temperature, and swelling time, their influences on temperature stability of binder are lower than on diatomite concentration and crumb rubber concentration.

3.8. *Multiresponse Optimization.* Preparation parameters present different effects on properties of crumb rubber and diatomite compound–modified asphalt binder. In order to obtain the modified asphalt binder with satisfactory properties, multiresponse optimization is conducted using the RSM. Taking the modified asphalt used in the cold region, for example, the corresponding optimization criteria are determined according to JTG F40-2004 [35] and *"Guide for Design and Construction of Asphalt Rubber and Mixtures"* [36] in China. The optimization was carried out by maximizing the softening point, ductility, elastic recovery, and minimizing penetration, while the viscosity is between 1000 and 3000 mPa·s. Viscosity should not be too high because of the workability of modified asphalt binder. PI is not considered because it can be calculated based on penetration and softening point results according to (2). The optimization process is determined and listed in Table 5.

In order to verify the equations as well as the optimum data obtained by the RSM, experimental tests were conducted using the same preparation parameters listed in Table 5. The experimental results are obtained and also shown in

Table 5. As shown in this table, the predicted and experimental results possess favorable consistency.

4. Conclusions

In this paper, crumb rubber and diatomite compound–modified asphalt binder was prepared. The effects of preparation parameters on properties of modified asphalt binder were investigated, and multiresponse optimization was conducted using the response surface method. The following conclusions can be obtained:

(1) Response surface method is a suitable method for preparation parameter analysis and optimization of crumb rubber and diatomite compound–modified asphalt binder, which can effectively reduce the number of groups.

(2) Crumb rubber concentration presents significant effects on all of the asphalt properties used in this study. Softening points, viscosity, elastic recovery, and PI increase, while penetration and ductility decrease with the increase of crumb rubber concentration. It reveals that the addition of crumb rubber can improve the high temperature susceptibility, viscosity, and elastic recovery ability of binder.

(3) Diatomite concentration presents significant effects on all of the asphalt properties except elastic recovery. Softening points, viscosity, and PI increase, while penetration and ductility decrease with the increase of diatomite concentration, which presents little influence on elastic recovery of binder.

(4) Shear temperature presented significant effects on penetration, softening point, viscosity, and ductility. Shear speed, shear time, and storing time have similar effects on binder properties because of their similar mechanism of action.

(5) Based on the model obtained from the RSM, optimized preparation parameters corresponding to specific criteria can be determined, which possessed favorable accuracy compared with experimental results.

Investigation of other properties such as bending beam rheometer (BBR) and dynamic shear rheometer (DSR) tests of crumb rubber and diatomite compound–modified asphalt binder needs to be conducted in the near future.

Conflicts of Interest

The authors declare that they have no conflicts of interest.

Acknowledgments

The authors express their appreciation for the financial support of the National Natural Science Foundation of China under Grant no. 51408258; the China Postdoctoral Science Foundation funded projects (nos. 2014M560237 and 2015T80305); the Fundamental Research Funds for the Central Universities (JCKYQKJC06); and Science and Technology Development Program of Jilin Province.

References

[1] V. W. Y. Tam and C. M. Tam, "A review on the viable technology for construction waste recycling," *Resources, Conservation and Recycling*, vol. 47, no. 3, pp. 209–221, 2006.

[2] M. N. Amin, M. I. Khan, and M. U. Saleem, "Performance evaluation of asphalt modified with municipal wastes for sustainable pavement construction," *Sustainability*, vol. 8, no. 12, p. 949, 2016.

[3] D. L. Presti, "Recycled tyre rubber modified bitumens for road asphalt mixtures: a literature review," *Construction and Building Materials*, vol. 49, pp. 863–881, 2013.

[4] M. C. Zanetti, S. Fiore, B. Ruffino, E. Santagata, D. Dalmazzo, and M. Lanotte, "Characterization of crumb rubber from end-of-life tyres for paving applications," *Waste Management*, vol. 45, pp. 161–170, 2015.

[5] A. Farina, M. C. Zanetti, E. Santagata, and G. A. Blengini, "Life cycle assessment applied to bituminous mixtures containing recycled materials: crumb rubber and reclaimed asphalt pavement," *Resources, Conservation and Recycling*, vol. 117, pp. 204–212, 2017.

[6] I. Bartolozzi, S. Mavridou, F. Rizzi, and M. Frey, "Life cycle thinking in sustainable supply chains: the case of rubberized asphalt pavement," *Environmental Engineering and Management Journal*, vol. 14, no. 5, pp. 1203–1215, 2015.

[7] A. D. La Rosa, G. Recca, D. Carbone et al., "Environmental benefits of using ground tyre rubber in new pneumatic formulations: a life cycle assessment approach," *Proceedings of the Institution of Mechanical Engineers, Part L: Journal of Materials: Design and Applications*, vol. 229, no. 4, pp. 309–317, 2013.

[8] B. Yu, L. Jiao, F. Ni, and J. Yang, "Evaluation of plastic–rubber asphalt: engineering property and environmental concern," *Construction and Building Materials*, vol. 71, pp. 416–424, 2014.

[9] F. Moreno, M. Sol, J. Martín, M. Pérez, and M. C. Rubio, "The effect of crumb rubber modifier on the resistance of asphalt mixes to plastic deformation," *Materials and Design*, vol. 47, pp. 274–280, 2013.

[10] S. K. Palit, K. S. Reddy, and B. B. Pandey, "Laboratory evaluation of crumb rubber modified asphalt mixes," *Journal of Materials in Civil Engineering*, vol. 16, no. 1, pp. 45–53, 2004.

[11] O. Xu, F. Xiao, S. Han, S. N. Amirkhanian, and Z. Wang, "High temperature rheological properties of crumb rubber modified asphalt binders with various modifiers," *Construction and Building Materials*, vol. 112, pp. 49–58, 2016.

[12] K. L. N. N. Puga and R. C. Williams, "Low temperature performance of laboratory produced asphalt rubber (AR) mixes containing polyoctenamer," *Construction and Building Materials*, vol. 112, pp. 1046–1053, 2016.

[13] T. Wang, F. Xiao, S. Amirkhanian, W. Huang, and M. Zheng, "A review on low temperature performances of rubberized asphalt materials," *Construction and Building Materials*, vol. 145, pp. 483–505, 2017.

[14] P. Cong, S. Chen, and H. Chen, "Effects of diatomite on the properties of asphalt binder," *Construction and Building Materials*, vol. 30, pp. 495–499, 2012.

[15] Y. Tan, Z. Lei, and X. Zhou, "Investigation of low-temperature properties of diatomite-modified asphalt mixtures," *Construction and Building Materials*, vol. 36, pp. 787–795, 2012.

[16] Y. Cheng, J. Tao, Y. Jiao et al., "Influence of the properties of filler on high and medium temperature performances of asphalt mastic," *Construction and Building Materials*, vol. 118, pp. 268–275, 2016.

[17] Q. Guo, L. Li, Y. Cheng, Y. Jiao, and C. Xu, "Laboratory evaluation on performance of diatomite and glass fiber compound modified asphalt mixture," *Materials and Design*, vol. 66, pp. 51–59, 2015.

[18] P. Cong, N. Liu, Y. Tian, and Y. Zhang, "Effects of long-term aging on the properties of asphalt binder containing diatoms," *Construction and Building Materials*, vol. 123, pp. 534–540, 2016.

[19] H. Liu, L. Fu, Y. Jiao, J. Tao, and X. Wang, "Short-term aging effect on properties of sustainable pavement asphalts modified by waste rubber and diatomite," *Sustainability*, vol. 9, no. 12, p. 996, 2017.

[20] C. Fang, X. Qiao, R. Yu et al., "Influence of modification process parameters on the properties of crumb rubber/EVA modified asphalt," *Journal of Applied Polymer Science*, vol. 133, no. 27, 2016.

[21] H. Yu, Z. Leng, Z. Zhou, K. Shih, F. Xiao, and Z. Gao, "Optimization of preparation procedure of liquid warm mix additive modified asphalt rubber," *Journal of Cleaner Production*, vol. 141, pp. 336–345, 2017.

[22] X. Shu and B. Huang, "Recycling of waste tire rubber in asphalt and Portland cement concrete: an overview," *Construction and Building Materials*, vol. 67, pp. 217–224, 2014.

[23] S. Kedarisetty, K. P. Biligiri, and J. B. Sousa, "Advanced rheological characterization of reacted and activated rubber (RAR) modified asphalt binders," *Construction and Building Materials*, vol. 122, pp. 12–22, 2016.

[24] Q. Wang, S. Li, X. Wu, S. Wang, and C. Ouyang, "Weather aging resistance of different rubber modified asphalts," *Construction and Building Materials*, vol. 106, pp. 443–448, 2016.

[25] J. Peralta, H. M. Silva, L. Hilliou, A. V. Machado, J. Pais, and R. C. Williams, "Mutual changes in bitumen and rubber related to the production of asphalt rubber binders," *Construction and Building Materials*, vol. 36, pp. 557–565, 2012.

[26] J. C. Munera and E. A. Ossa, "Polymer modified bitumen: optimization and selection," *Materials and Design*, vol. 62, pp. 91–97, 2014.

[27] L. H. Li, *Road Engineering Materials*, China Communications Press, Beijing, China, 5th edition, 2010.

[28] M. L. Huang, Y. H. Hung, and Z. S. Yang, "Validation of a method using Taguchi, response surface, neural network, and genetic algorithm," *Measurement*, vol. 94, pp. 284–294, 2016.

[29] I. Kaymaz and C. A. McMahon, "A response surface method based on weighted regression for structural reliability analysis," *Probabilistic Engineering Mechanics*, vol. 20, no. 1, pp. 11–17, 2005.

[30] O. Rezaifar, M. Hasanzadeh, and M. Gholhaki, "Concrete made with hybrid blends of crumb rubber and metakaolin: optimization using response surface method," *Construction and Building Materials*, vol. 123, pp. 59–68, 2016.

[31] H. İ. Odabaş and I. Koca, "Application of response surface methodology for optimizing the recovery of phenolic compounds from hazelnut skin using different extraction methods," *Industrial Crops and Products*, vol. 91, pp. 114–124, 2016.

[32] M. H. Esfe, H. Hajmohammad, R. Moradi, and A. A. A. Arani, "Multi-objective optimization of cost and thermal performance of double walled carbon nanotubes/water nanofluids

by NSGA-II using response surface method," *Applied Thermal Engineering*, vol. 112, pp. 1648–1657, 2017.

[33] Y. Song, J. Che, and Y. Zhang, "The interacting rule of diatomite and asphalt groups," *Petroleum Science and Technology*, vol. 29, no. 3, pp. 254–259, 2011.

[34] Y. Cheng, J. Tao, Y. Jiao, Q. Guo, and C. Li, "Influence of diatomite and mineral powder on thermal oxidative ageing properties of asphalt," *Advances in Materials Science and Engineering*, vol. 2015, Article ID 947834, 10 pages, 2015.

[35] JTG F40-2004, *Technical Specifications for Construction of Highway Asphalt Pavements*, Research Institute of Highway Ministry of Transport, Beijing, China, 2004, in Chinese.

[36] China Communications Press, *Guide for Design and Construction of Asphalt Rubber and Mixtures*, China Communications Press, Beijing, China, 2009, in Chinese.

17

Evaluation of Fracture Resistance of Asphalt Mixtures Using the Single-Edge Notched Beams

Biao Ding [ID],[1] **Xiaolong Zou** [ID],[2,3,4] **Zixin Peng**,[1] and **Xiang Liu**[5]

[1]*CCCC First Highway Consultants Co., Ltd., Xi'an, Shaanxi 710065, China*
[2]*School of Architecture and Civil Engineering, Xi'an University of Science and Technology, Xi'an, Shaanxi, China*
[3]*Guangxi Key Lab of Road Structure and Materials, Guangxi Transportation Research & Consulting Co., Ltd., Nanning, Guangxi, China*
[4]*Key Laboratory for Special Area Highway Engineering of Ministry of Education, Chang'an University, Xi'an, Shaanxi, China*
[5]*School of Highway, Chang'an University, Xi'an, Shaanxi, China*

Correspondence should be addressed to Xiaolong Zou; zouxiaolong_1234@163.com

Academic Editor: Hiroshi Noguchi

To determine and compare the fracture properties of different asphalt mixtures, single-edge notched beam (SENB) tests using three types of asphalt mixtures were applied in this study under the conditions of different notched depths and different temperatures. The effects of notched depths and temperatures on the fracture toughness and fracture energy were analyzed. The results indicate that the notch depth has no significant effects on the fracture toughness and the fracture energy, but the gradation has relatively obvious effects on the fracture energy, which the larger contents of course aggregate leads to increase the discreteness of the fracture energy of the specimen. The temperature has significant effects on the ultimate loads, fracture energy, and fracture toughness. The ultimate loads of the SENBs reach the peak value at 0°C, which could be resulted in that viscoelastic properties of asphalt mixture depend with temperatures. The fracture toughness at −20°C of continuously graded asphalt mixtures are higher than those of gap-graded asphalt mixtures. On the contrary, the fracture toughness of gap-graded asphalt mixtures is higher at temperatures from −10°C to 20°C. The fracture energy increases with temperatures, and the fracture energy of SMA-13 is significantly larger than those of AC-13 and AC-16.

1. Introduction

The research on the fracture characteristics of asphalt mixture is one of important topics on the properties of asphalt mixture. The main research methods for analysis and evaluation of the fracture characteristics include the numerical simulation method and fracture test method [1–3].

The numerical simulation is usually realized by the finite element method (FEM) and discrete element method (DEM). Two-dimensional (2D) micromechanical models using FEM and DEM have been developed to simulate microscale crack propagation of cemented particulate materials, which obtained well explanations of observed crack failures of the samples [1, 2, 4, 5]. The results of FEM simulation and the results of DEM simulation are usually compared to evaluate the similarities and differences. The findings show that the results of FEM simulation and DEM simulation have a fundamental similarity and, at the same time, have some basic differences [4]. Furthermore, three-dimensional (3D) model development has become a trend for the numerical simulation of fracture analysis.

The fracture test has three typical test methods, namely, the single-edge notched beam (SENB) test, the semicircular bending (SCB) test, and the disk-shaped compact tension (DC(T)) test, which are mainly applied to obtain the fracture characteristics of asphalt concrete [6–9]. These three methods have different specimen geometries, application occasions, and fracture models, so the results can hardly be compared directly [8, 10].

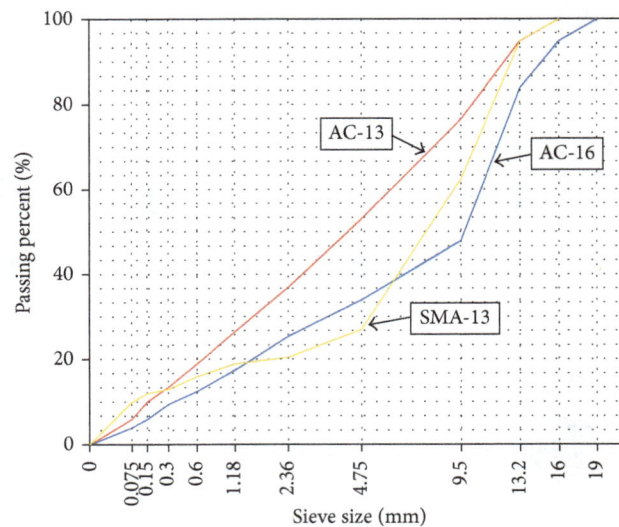

FIGURE 1: Experimental gradation curves for 0.45 power gradation graph.

TABLE 1: Properties of base asphalt.

Test item		Unit	Test result	Test method
Penetration (25°C, 5 s, 100 g)		0.1 mm	87.9	T0604-2011
Ductility (5°C, 5 cm/min)		cm	>150	T0605-2011
Softening point (R&B)		°C	48.8	T0606-2011
Residue after TFOT	Quality change	%	0.15	T0610-2011 T0609-2011
	Penetration ratio (25°C)	%	69.0	T0604-2011
	Ductility (15°C)	cm	34.8	T0605-2011

For asphalt mixtures, one kind of viscoelastic materials, the fracture characteristics are not only related to the initial crack depth but also to the temperature. Therefore, the aim of this study was to determine and compare the fracture properties of three typical surface layer asphalt mixtures, namely, two kinds of two continuously graded asphalt mixtures and a stone mastic asphalt (SMA), at different conditions of the initial crack depth and the temperature. Considering as the simple and widely used model, the SENB test was applied in this study and the SENB with different notched depths were tested at different temperatures.

2. Materials and Methods

2.1. Materials. Three asphalt mixtures used in this study included two continuously graded asphalt mixtures with nominal maximum aggregate size (NMAS) of 13.2 mm and 16 mm (AC-13 and AC-16) and a gap-graded stone mastic asphalt with NMAS of 13.2 mm (SMA-13). The experimental gradation curves of those three asphalt mixtures are shown in Figure 1. A base asphalt SK90 was used in AC-13 and AC-16, and a styrene-butadiene-styrene- (SBS-) modified asphalt binder produced by Shanxi Guolin Huatai Asphaltic Products Co., Ltd. was used in SMA-13. The properties of the base asphalt and the SBS-modified asphalt binder are shown in Tables 1 and 2, respectively.

The coarse aggregate and fine aggregate used in those three mixtures are basalt, and the properties are listed in Tables 3 and 4. The mineral filler used in this study is limestone powder, and the properties are listed in Table 5. Besides, lignin fibers were used as the stabilizer in SMA mixture. Table 6 lists the properties of fibers. Design asphalt contents were 5.0% for AC-13, 4.6% for AC-16, and 6.0% for SMA-13, which had 3.9%, 4.3%, and 4.1% void content, respectively. In addition, 0.3% of fiber was used in SMA-13.

2.2. Fabrication of Single-Edge Notched Beam. The slab specimens (300 mm × 300 mm × 50 mm) were fabricated using a rolling wheel compactor. The slab specimens were sawed into beams with the diameters of 250 mm (length) × 35 mm (height) × 30 mm (width).

A notch of designed depth, approximately 4 mm wide, with a square end, was sawed at the middle point of each beam, and single-edge notched beams (SENB) were obtained. A group of single-edge notched beams are shown in Figure 2.

2.3. Three-Point Bending Tests. Different notch depths and different test temperatures were considered in this study. The SENB with notch depth of 0 mm (without initial notch), 4 mm, 8 mm, 12 mm, and 16 mm was applied for three-point

TABLE 2: Properties of SBS-modified asphalt.

Test item		Unit	Test result	Test method
Penetration (25°C, 5 s, 100 g)		0.1 mm	66.5	T0604-2011
Ductility (5°C, 5 cm/min)		cm	38	T0605-2011
Softening point (R&B)		°C	77.4	T0606-2011
Kinematic viscosity (135°C)		Pa·s	2.0	T0625-2011
Flash point		°C	263	T0611-2011
Solubility		%	99.58	T0607-2011
Elastic recovery (25°C)		%	97	T0662-2000
Residue after TFOT	Quality change	%	0.09	T0610-2011 T0609-2011
	Penetration ratio (25°C)	%	73	T0604-2011
	Ductility (5°C)	cm	23	T0605-2011

TABLE 3: Properties of coarse aggregate.

Test item	Unit	Test result	Test method
Apparent relative density	/	2.953	T0304
Bulk density	/	2.886	T0304
Acicular content	%	3.95	T0312
Particle content of less than 0.075 mm	%	0.2	T0310
Soft stone content	%	1.0	T0320
Crushing value	%	12.9	T0316
Crushing value (200°C for 5 hours)	%	13.6	T0316
Firmness	%	2.0	T0314
Water absorption rate	%	0.78	T0314
Los Angeles wear value loss	%	13.7	T0317
Los Angeles wear value loss (200°C for 5 hours)	%	14.4	T0317
Polished value	PSV	44	T0321
Adhesion	Grade	5	T0305

TABLE 4: Properties of fine aggregate.

Test item	Unit	Test result	Test method
Apparent relative density	/	2.726	T0329
Bulk density	/	2.696	T0329
Sturdiness (>0.3 mm)	%	4	T0340
Particle content of less than 0.075 mm	%	9.9	T0340
Sand equivalent	%	82	T0334
Methylene blue value	g/kg	1.5	T0349
Angularity	s	38.6	T0345

TABLE 5: Properties of mineral filler.

Property	Unit	Result
Apparent specific gravity	—	2.693
Water content	%	0.34
Hydrophilic coefficient	—	0.58
Plasticity index	%	2.5

TABLE 6: Properties of lignin fibers.

Property	Unit	Result
Length	mm	<6
Ash content	%	19.64
pH	—	7.6
Hygroscopicity	%	6.57
Water content (by weight)	%	0.85

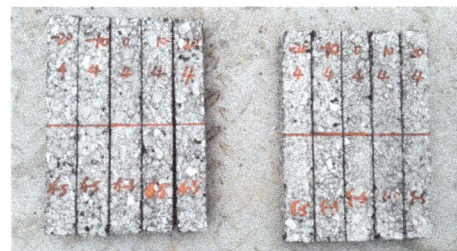

FIGURE 2: Single-edge notch beams.

bending tests at −10°C. The SENB with notch depth of 4 mm was applied for three-point bending tests at −20°C, −10°C, 0°C, 10°C, and 20°C.

A material test system (MTS-810) with an environmental chamber was used to perform the three-point bending tests. The configuration of the three-point bending test on a SENB is shown in Figure 3. The loading span is 200 mm. In a three-point bending test, the load using a constant displacement rate of 0.05 mm/min was directly applied at the point right above the notch on the upper surface. The midspan displacement δ was recorded at a sampling frequency of 10 Hz during the whole loading process until failure. The load-displacement curve can be drawn to obtain the peak loads.

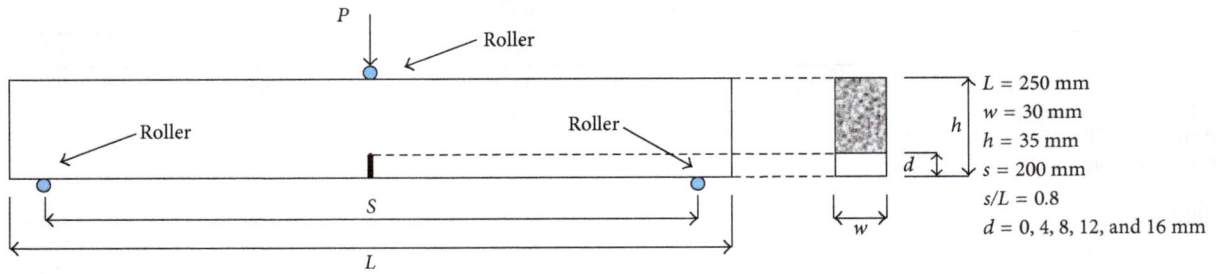

FIGURE 3: A schematic illustration of the three-point bending test on a single-edge notched beam.

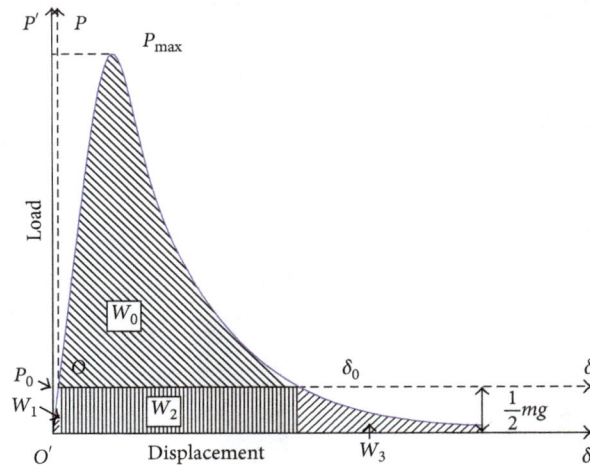

FIGURE 4: A load-displacement curve for a stable three-point bend test on a single-edge notched beam. W_0 defines the area under the load-displacement curve if there is no compensation for the energy supplied by the weight of the beam, m is the weight of the beam (between the supports), and $g = 9.81$ m/s^2 [17].

3. Theoretical Background

3.1. Fracture Toughness. The SENB specimen loaded in a three-point bending configuration and notched at the midpoint, as shown in Figure 3, is under the mode of tension, so the fracture toughness for a SENB specimen is given as follows [11, 12]:

$$K_I = \sigma_0 Y_I \sqrt{\pi a}, \tag{1}$$

where a is the notch depth, σ_0 is the applied stress, and Y_I is the normalized fracture toughness. The applied stress is given as follows:

$$\sigma_0 = \frac{3Ps}{2h^2 w}, \tag{2}$$

where P is the applied load, s is the loading span, and h and w are the specimen height and width, respectively, as seen in Figure 3. The normalized fracture toughness, Y_I, is given by the following analytical expression [3, 13–15]:

$$Y_I = \frac{1.99 - (a/W)(1 - (a/W))(2.15 - (3.93a/W) + (2.7a^2/W^2))}{\sqrt{\pi}(1 + (2a/W))(1 - (a/W))^{3/2}}. \tag{3}$$

3.2. Fracture Energy. Fracture energy, G_F, is defined as the area under the load-displacement curve divided by the ligament area, which could be expressed as follows [16]:

$$G_F = \frac{W}{A_{lig}}, \tag{4}$$

where W is the work of fracture for an entire crack propagation period and A_{lig} is the area of the ligament.

Figure 4 shows the work of fracture W for crack propagation, which can be expressed as (5), considering the effects of the self-weight of the beam [17, 18].

$$W = W_0 + W_1 + W_2 + W_3, \tag{5}$$

where W_0 is the work performed by the external force P for crack propagation and W_1, W_2, and W_3 are the additional works caused by the self-weight of the beam. Based on special

(a)   (b)

(c)

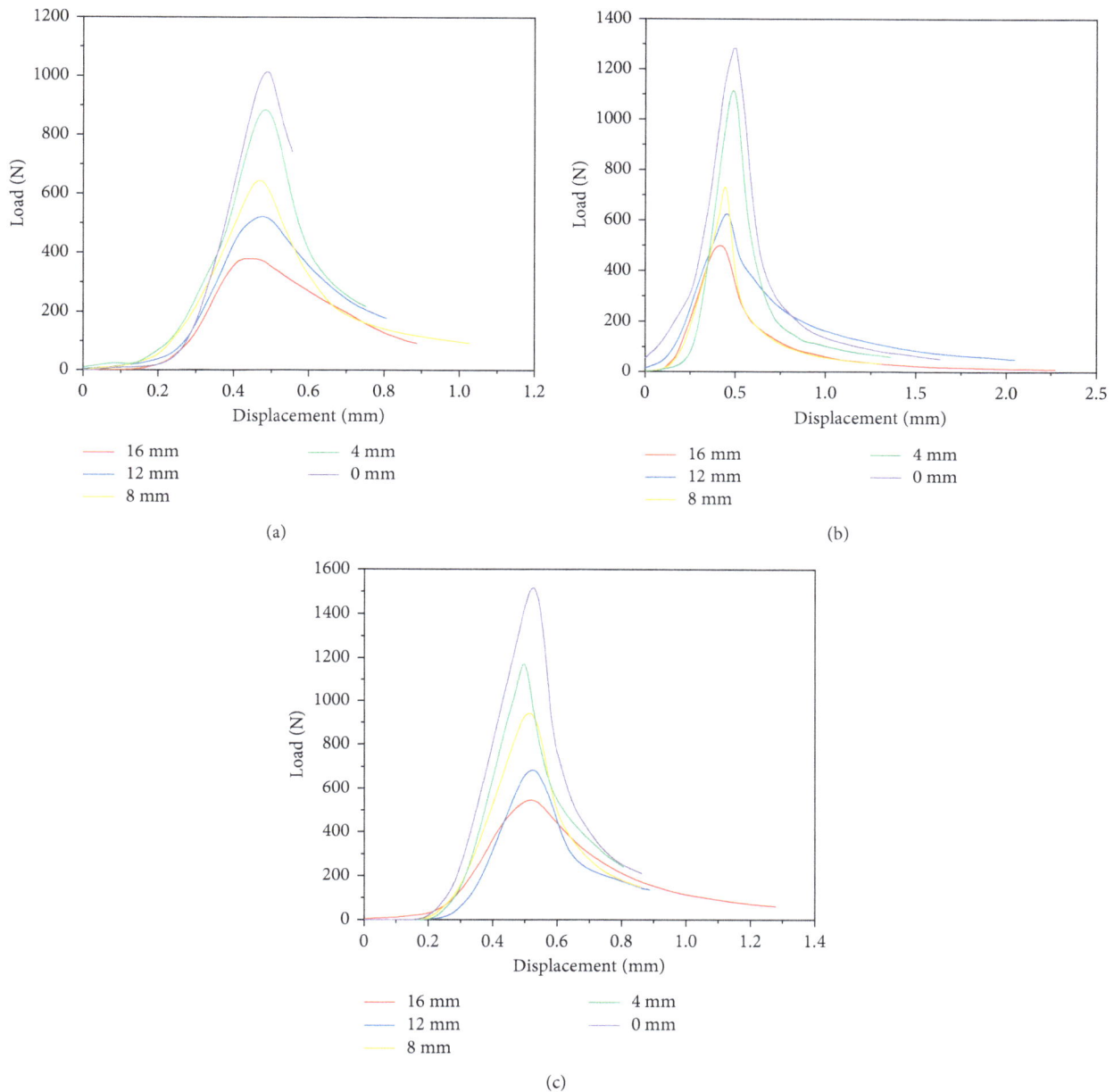

FIGURE 5: Load-displacement curves for different notch depths: (a) AC-13, (b) AC-16, and (c) SMA-13.

hypothetic situation, W_2 and W_3 can be calculated as follows [12, 17]:

$$W_2 = W_3 = \frac{1}{2}mg\delta_0. \qquad (6)$$

According to (4)–(6), fracture energy, G_F, can be expressed as (7) [16], so that the fracture energy of asphalt mixture could be calculated from a load-displacement curve recorded [19].

$$G_F = \frac{W}{A_{\text{lig}}} = \frac{\left[\int_0^{\delta_0} p(\delta)\,d\delta + mg\delta_0\right]}{A_{\text{lig}}}, \qquad (7)$$

where G_F = the fracture energy (N/m), $m = m_1 + m_2$ (kg), $m_1 = Ms/L$ (weight of the beam between the supports),

M = weight of the specimen, m_2 = weight of the part of the loading arrangement which is not attached to the machine but follows the beam until failure, δ_0 = the midspan displacement of the specimen at failure (m), δ = midspan displacement (m), $g = 9.81$ (m/s^2), and A_{lig} = area of the ligament. P, s, and L are shown in Figure 3.

4. Results and Discussions

4.1. Effects of Notch Depth

4.1.1. Displacement. The load-displacement curves were recorded by three-point bending tests. Figure 5 are shown the load-displacement curves of the three mixtures of SENB specimens with different notch depths. According to Figure 5,

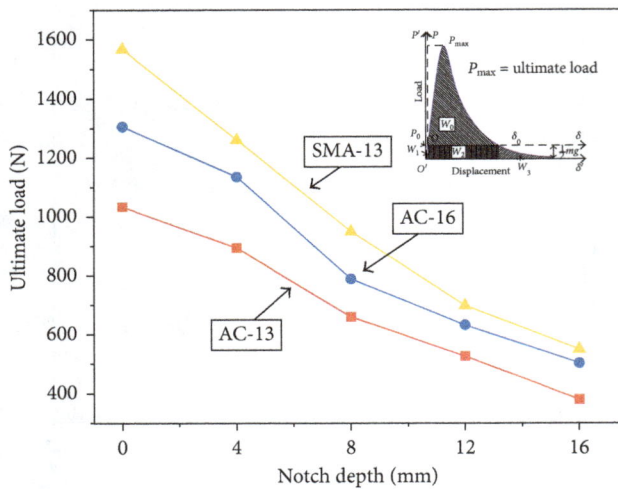

FIGURE 6: Ultimate loads for different notch depths.

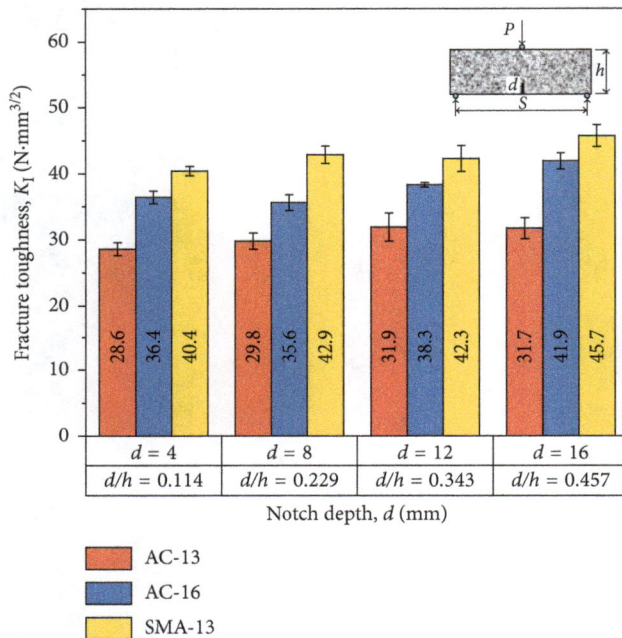

FIGURE 7: Fracture toughness for different notch depths.

the load-displacement curves of the mixtures present typical three stages. At the first stage, the load increases linearly with the displacement up to the peak. At the second stage, after the peak, the load decreases largely with the displacement and, at the same time, the fracture develops rapidly. At the third stage, the load decreases steadily until the specimen fracture failure.

4.1.2. Ultimate Load.

The ultimate loads of the three-point bending tests can be obtained through the load-displacement curves. Figure 6 shows the ultimate loads for the SENB specimens of the three mixtures with different notch depths at 10°C. From Figure 6, with the increase of notch depth, the ultimate loads of three kinds of asphalt mixture show a linear downward trend. In addition, under the same conditions of

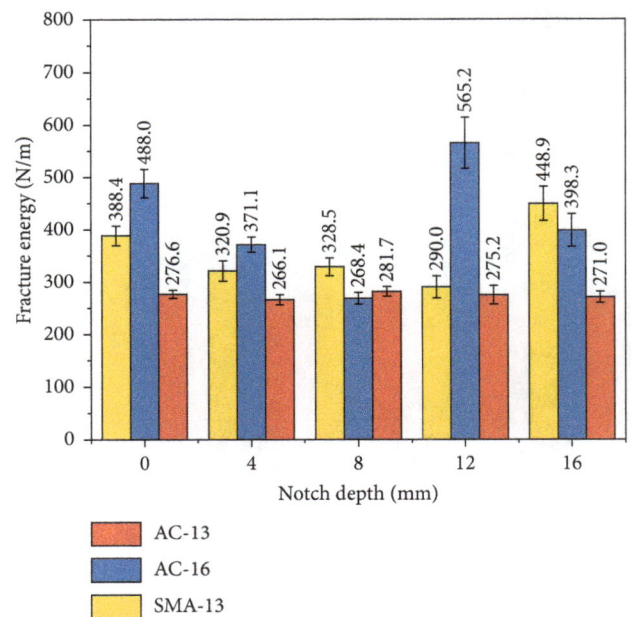

FIGURE 8: Fracture energy for different notch depths.

notch depth, the ultimate load of the AC-13 specimen is minimum, AC-16 is moderate, and SMA-13 is maximum among the three asphalt mixtures. It is analyzed that the higher ultimate load of the SMA-13 specimen is attributed to the material composition, which the utilization of the SBS-modified asphalt and the fiber can contribute to improving the tensile strength and the ultimate load.

4.1.3. Fracture Toughness.

Fracture toughness, K_I, for SENB specimens with different notch depths can be determined by (1)–(3). The calculations of fracture toughness for different notch depths are shown in Figure 7. With the increase of notch depth, for the three asphalt mixtures, the fracture toughness increases moderately. It is notable that the fracture toughness for SMA-13 is largest among the three asphalt mixtures at the same conditions of notch depth.

4.1.4. Fracture Energy.

From load-displacement curves, the fracture energy of mixtures was calculated by using (7). Fracture energy of the three-point bending test on SENB specimens with different notch depths is shown in Figure 8. From Figure 8, it can be seen that the notch depth has no significant effects on the fracture energy of AC-13 specimens. For AC-16 and SMA-13, there is no obvious regularity of the notch depth versus the fracture energy. The contents of fine aggregate (less than 2.36 mm) of AC-16 and SMA-13 are relatively small, and the contents of course aggregate are relatively large, which results in increasing the discreteness of the fracture energy of the specimen.

4.2. Effects of Temperature

4.2.1. Displacement.

The load-displacement curves were recorded by three-point bending tests. Figure 9 shows the

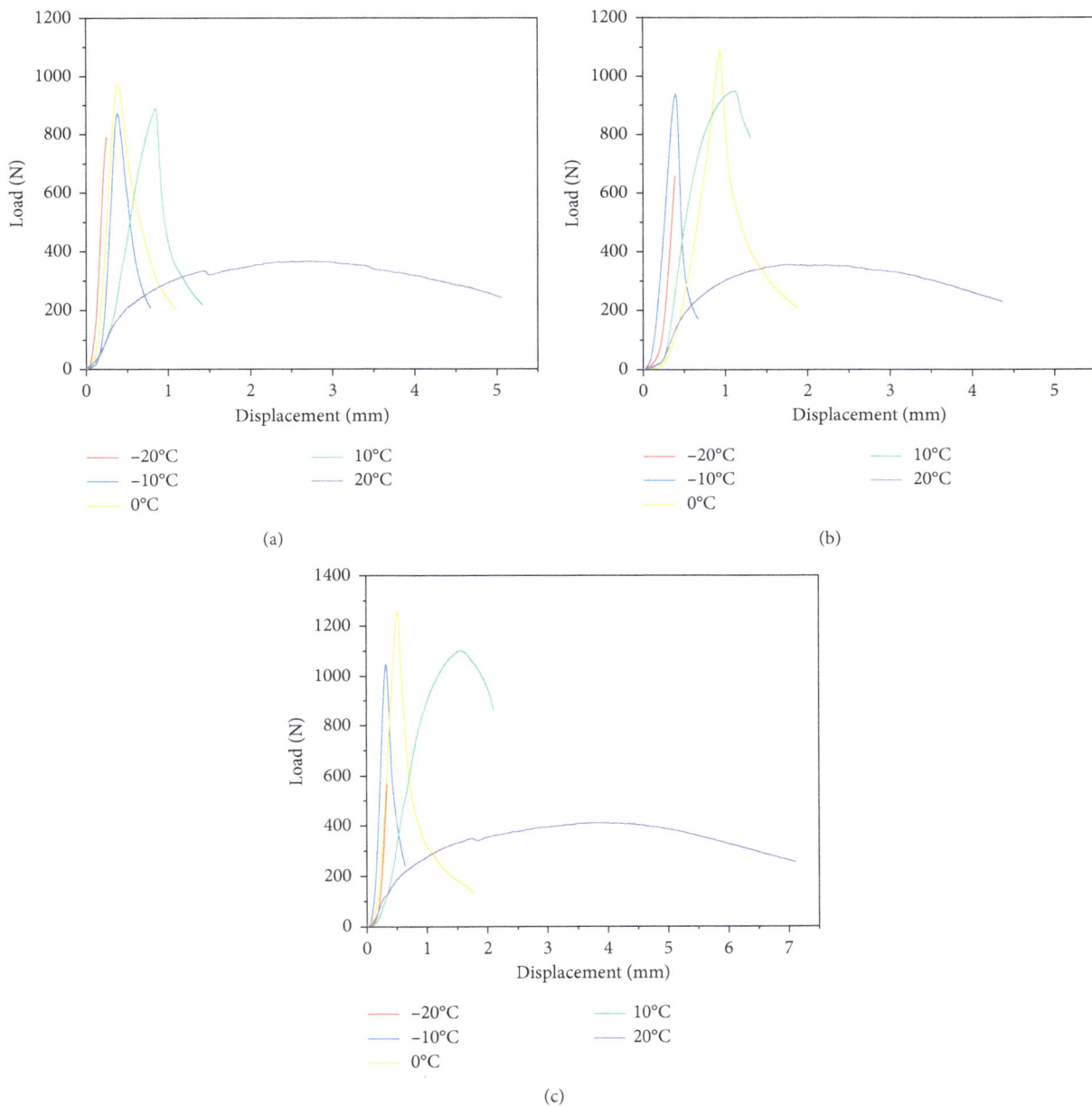

FIGURE 9: Load-displacement curves at different temperatures: (a) AC-13, (b) AC-16, and (c) SMA-13.

load-displacement curves at different temperatures. According to Figure 9, AC-13, AC-16, and SMA-13 have similar load-displacement curves at different temperatures. At −20°C, the loads rise rapidly, and there are no peak values until fracture failure which indicates that the SENB specimens are brittle failure in the three-point bending tests at −20°C. When the temperature increases (from −10°C to 10°C), the specimens present a certain toughness, the load of the beam increases first and then decreases with the displacement. In particular, when the temperature increases to 20°C, it is obvious that the load-displacement curves at 20°C are noticeably different from the curves at the lower temperatures. At 20°C, the linearly increase stage of the load reduces and the load rises steadily with the displacement and

the load decreases when the fracture grows to a certain extent and the specimen fracture failure occurs until the displacement reaches a relatively large value compared with that at lower temperatures.

4.2.2. Loading. Figure 10 shows the ultimate loads for the SENB specimens of the three mixtures at different temperatures. The ultimate loads increase at first and then decrease and reach the maximum at 0°C, which could be resulted in that asphalt mixture is a viscoelastic material that the mechanical properties depend with temperatures. The SENB specimens are brittle fracture at −20°C, and the ultimate load is small, and toughness of the SENB specimens

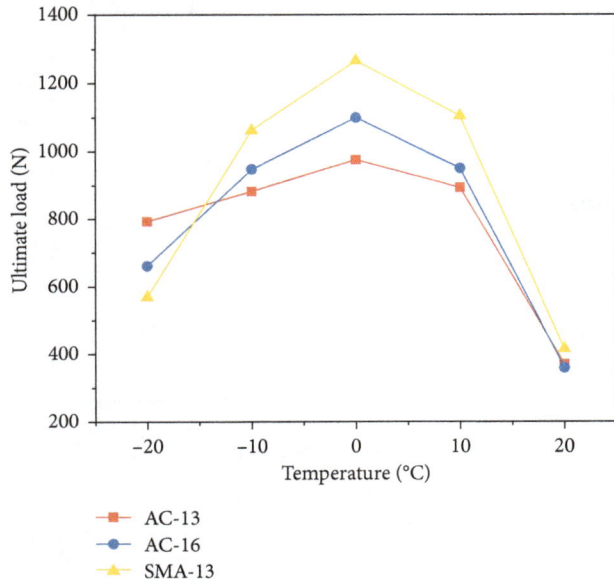

FIGURE 10: Ultimate loads at different temperatures.

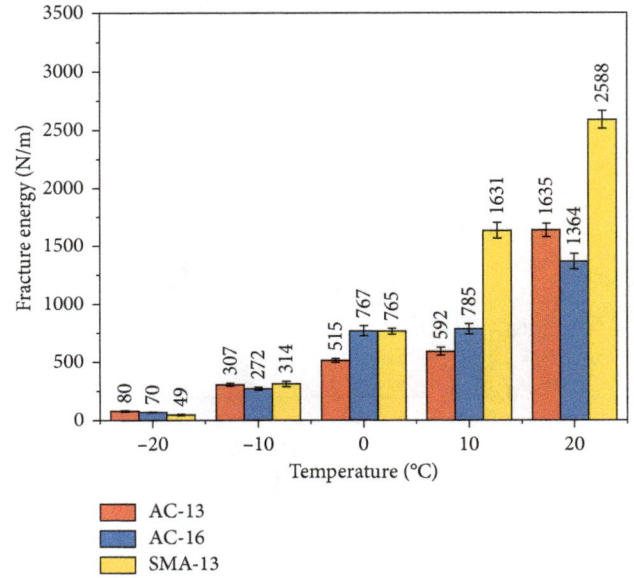

FIGURE 11: Fracture toughness at different temperatures.

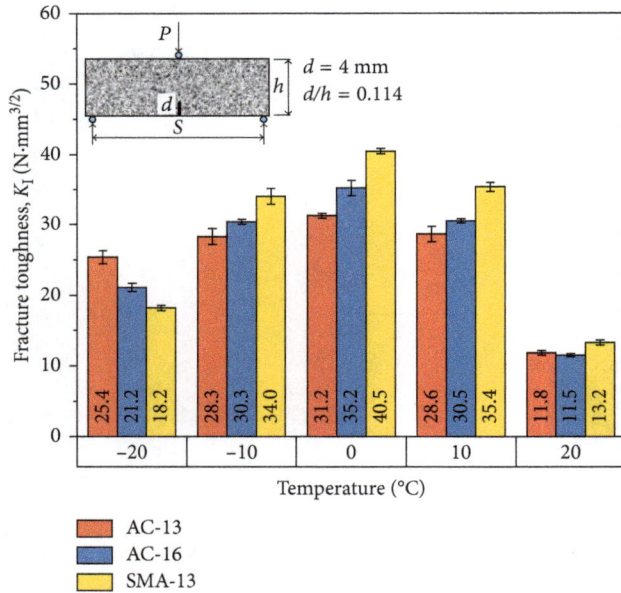

FIGURE 12: Fracture energy at different temperatures.

increases with temperatures which contributes to the ultimate load increase. The tensile strength of the SENB specimens decreases with temperatures after the critical temperature, and the ultimate load decreases.

4.2.3. Fracture Toughness. Figure 11 shows the fracture toughness at different temperatures for SENB specimens using the three asphalt mixtures. It can be found that the fracture toughness at −20°C of AC-13 and AC-16 is higher than that of SMA-13. However, it can be seen that SMA-13 has higher fracture toughness than AC-13 or AC-16 has at

temperatures from −10°C to 20°C, indicating its better resistance to fracture generally.

4.2.4. Fracture Energy. The fracture energy at different temperatures for SENB specimens using the three asphalt mixtures is shown in Figure 12. It can be seen from Figure 12 that the fracture energy increases with temperatures. At the lower temperatures, −20°C and −10°C, the asphalt mixtures present relatively notable elasticity and SENB specimens trend to brittle crack, consequently the three asphalt mixtures have similar fracture energy. At the higher temperatures, 0°C to 20°C, the asphalt mixtures present relatively notable viscosity especially at moderate temperatures, 10°C to 20°C, which the fracture energy greatly differs among the three asphalt mixtures. From Figure 12, the fracture energy of SMA-13 is significantly larger than those of AC-13 and AC-16.

5. Conclusions

This study adopted the SENB test to evaluate the fracture properties of three typical surface layer asphalt mixtures, AC-13, AC-16, and SMA-13, changing with variations of notched depths and temperatures. The following conclusions can be drawn:

(1) The notch depth has no significant effects on the fracture toughness and the fracture energy, but the gradation has relatively obvious effects on the fracture energy, which the larger contents of course aggregate results to increase the discreteness of the fracture energy of the specimen.

(2) The ultimate loads of the SENBs reach the maximum at 0°C, which could be resulted in that viscoelastic properties of asphalt mixture depend with temperatures.

(3) The fracture toughness at −20°C of continuously graded asphalt mixtures are higher than those of

gap-graded asphalt mixtures. On the contrary, the fracture toughness of gap-graded asphalt mixtures is higher at temperatures from −10°C to 20°C.

(4) Temperature has significant effects on the fracture energy, and the fracture energy increases with temperatures. The fracture energy of SMA-13 is significantly larger than those of AC-13 and AC-16.

Conflicts of Interest

The authors declare that they have no conflicts of interest.

Acknowledgments

This research was sponsored by the Natural Science Foundation of Shaanxi Province (2016JQ5115), the PhD Research Startup Foundation of Xi'an University of Science and Technology (2017QDJ024), the opening fund of Guangxi Key Lab of Road Structure and Materials (2017gxjgclkf-001), the opening fund of Key Laboratory for Special Area Highway Engineering of Ministry of Education (300102218512), and the Outstanding Youth Science Fund of Xi'an University of Science and Technology (2018YQ3-07). The results and opinions presented are those of the authors and do not necessarily reflect those of the sponsoring agencies.

References

[1] H. Kim, M. P. Wagoner, and W. G. Buttlar, "Simulation of fracture behavior in asphalt concrete using a heterogeneous cohesive zone discrete element model," *Journal of Materials in Civil Engineering*, vol. 20, no. 8, pp. 552–563, 2008.

[2] Q. Dai and K. Ng, "2D cohesive zone modeling of crack development in cementitious digital samples with microstructure characterization," *Construction and Building Materials*, vol. 54, pp. 584–595, 2014.

[3] I. Artamendi and H. A. Khalid, "A comparison between beam and semi-circular bending fracture tests for asphalt," *Road Materials and Pavement Design*, vol. 7, no. s1, pp. 163–180, 2011.

[4] M. H. Sadd and Q. Dai, "A comparison of micro-mechanical modeling of asphalt materials using finite elements and doublet mechanics," *Mechanics of Materials*, vol. 37, no. 6, pp. 641–662, 2005.

[5] W. Buttlar and Z. You, "Discrete element modeling of asphalt concrete: Microfabric approach," *Transportation Research Record: Journal of the Transportation Research Board*, vol. 1757, pp. 111–118, 2001.

[6] S. J. Sulaiman and A. F. Stock, "The use of fracture mechanics for the evaluation of asphalt mixes," in *Association of Asphalt Paving Technologists Technical Sessions, 1995*, pp. 500–531, Portland, OR, USA, 1995.

[7] J. M. M. Molenaar and A. A. A. Molenaar, "Fracture toughness of asphalt in the semi-circular bend test," in *Proceedings of the 2nd Eurasphalt & Eurobitume Congress*, pp. 509–517, Barcelona, Spain, September 2000.

[8] M. P. Wagnoner, W. G. Buttlar, and G. H. Paulino, "Disk-shaped compact tension test for asphalt concrete fracture," *Experimental Mechanics*, vol. 45, no. 3, pp. 270–277, 2005.

[9] G. S. Xeidakis, I. S. Samaras, D. A. Zacharopoulos, and G. E. Papakaliatakis, "Crack growth in a mixed-mode loading on marble beams under three point bending," *International Journal of Fracture*, vol. 79, no. 2, pp. 197–208, 1996.

[10] A. Braham, W. Buttlar, and F. Ni, "Laboratory mixed-mode cracking of asphalt concrete using the single-edge notch beam," *Road Materials and Pavement Design*, vol. 11, no. 4, pp. 947–968, 2011.

[11] J. E. Srawley, "Wide range fracture toughness expressions for ASTM E 399 standard fracture toughness specimens," *International Journal of Fracture*, vol. 12, no. 3, pp. 475-476, 1976.

[12] Y. Murakami and S. Aoki, *Stress Intensity Factors Handbook*, vol. 2, Pergamon, Oxford, UK, 1987.

[13] J. E. Srawley and W. F. Brown Jr., *Fracture Toughness Testing*, National Aeronautics and Space Administration, Lewis Research Center, Cleveland, OH, USA, 1965.

[14] J. E. Srawley and B. Gross, *Stress Intensity Factors for Crackline-Loaded Edge-Crack Specimens*, National Aeronautics and Space Administration, Washington, DC, USA, 1967.

[15] J. E. Srawley, M. H. Jones, and B. Gross, *Experimental Determination of the Dependence of Crack Extension Force on Crack Length for a Single-Edge-Notch Tension Specimen*, National Aeronautics and Space Administration, Washington, DC, USA, 1964.

[16] RILEM, "Determination of the fracture energy of mortar and concrete by means of three-point bend tests on notched beams," *Materials and Structures*, vol. 18, no. 4, pp. 285–290, 1985.

[17] P.-E. Petersson and L. Tekniska, *Crack Growth and Development of Fracture Zones in Plain Concrete and Similar Materials*, Lund Institute of Technology, Lund, Sweden, 1981.

[18] M. Hossain, S. Swartz, and E. Hoque, "Fracture and tensile characteristics of asphalt-rubber concrete," *Journal of Materials in Civil Engineering*, vol. 11, no. 4, pp. 287–294, 1999.

[19] G.-F. Peng, W.-W. Yang, J. Zhao, Y.-F. Liu, S.-H. Bian, and L.-H. Zhao, "Explosive spalling and residual mechanical properties of fiber-toughened high-performance concrete subjected to high temperatures," *Cement and Concrete Research*, vol. 36, no. 4, pp. 723–727, 2006.

18

Effect of Material Composition on Cohesion Characteristics of Styrene-Butadiene-Styrene-Modified Asphalt Using Surface Free Energy

Xing-jun Zhang,[1,2] **Hui-xia Feng,**[1] **Xiao-min Li,**[3] **Xiao-yu Ren,**[2,3] **Zhen-feng Lv,**[4] **and Bo Li**[3]

[1] *School of Petrochemical Engineering, Lanzhou University of Technology, Lanzhou 730050, China*
[2] *Gansu Province Highway Maintenance Engineering Research Center, Lanzhou 730070, China*
[3] *Key Laboratory of Road & Bridge and Underground Engineering of Gansu Province, Lanzhou Jiaotong University,*
 Lanzhou 730070, China
[4] *School of Highway, Chang'an University, Xi'an 710064, China*

Correspondence should be addressed to Bo Li; libolzjtu@hotmail.com

Academic Editor: Meor Othman Hamzah

Styrene-butadiene-styrene- (SBS-) modified asphalts were prepared by mixing different base asphalts, SBS modifier, extracting oil, and stabilizing agents. The contact angles between SBS-modified asphalt and distilled water, glycerol, and formamide were detected by the sessile drop method. Based on the surface energy theory, the surface free energy and cohesive power of SBS-modified asphalt were calculated. The influence of the raw materials composition, such as the virgin asphalt and SBS modifier types as well as the extracting oil and stabilizing agent contents, on the cohesive characteristics of SBS-modified asphalt was discussed. The results showed that virgin asphalt was compatible with SBS modifiers to improve cohesiveness. The cohesive power of branched SBS-modified asphalt was larger than that of linear SBS-modified asphalt. The cohesion of SBS-modified asphalt was improved as the SBS modifier and stabilizer contents increased but was reduced for excessive extraction oil contents. The cohesive characteristics of the SBS-modified asphalt were improved by the formation of stable three-dimensional network structures by cross-linking, winding, and grafting among different raw materials.

1. Introduction

Asphalt is a mixture of crude oil refining residue and various chemical components, widely used in the road pavement industry as an aggregate binder, because it offers good adhesion, viscoelasticity, and strength [1]. However, further applications are restricted by disadvantages including high-temperature rutting and low-temperature cracking [2]. Heavier vehicle loads, increased traffic volumes, and extreme weather conditions can cause pavement damage such as permanent deformation [3, 4]. In order to improve the quality of asphalt, various polymers can be incorporated by mechanical mixing or chemical reaction, thereby improving the mechanical properties, heat sensitivity, and aging resistance of the asphalt

[5, 6]. The most commonly used polymers are styrene-butadiene-styrene (SBS) block copolymers. SBS block copolymers are known to improve the low- and high-temperature performance of bitumen [7].

SBS-modified asphalt is a composite prepared by mixing neat asphalt, SBS modifier, extracting oil, and stabilizing agents [8]. For SBS block copolymers mixed with neat asphalt, the system gradually becomes a two-phase structure in which polymer phases formed by maltenes-swelling polymers are dispersed in the asphalt-rich phase [9]. Therefore, the compatibility between asphalt and SBS is considered critical, with a profound impact on the thermal mechanical properties, rheological properties, and morphology [10, 11]. Generally, asphalt comprises saturated hydrocarbons, aromatics, resins,

TABLE 1: Properties of neat asphalts.

Performance	Unit	Binder A	Binder B	Binder C
Penetration (25°C, 100 g, 5 s)	0.1 mm	85.5	84.3	87.6
Softening point (ring ball method)	°C	45.6	48.7	47
Ductility (5 cm/min, 15°C)	cm	>100	>100	>100
Residue after aging under rotary film oven (163°C, 85 min)				
Loss of quality	%	0.07	0.18	0.12
Penetration ratio	%	70	64	61
Ductility (5 cm/min, 10°C)	cm	9.0	6.5	8.0

and asphaltenes in solvents, the ratio of which is commonly referred to as the SARA fraction [1]. However, each component has a different solubility parameter and thus a different level of compatibility with the polystyrene (PS) and polybutadiene (PB) blocks in the SBS. Some efforts [12, 13] have investigated the compatibility between asphaltic components and SBS copolymers. As a rule of thumb, asphalt with high-aromatic contents or linear SBS can form compatible and stable SBS-modified asphalt [14, 15]. Moreover, incompatibility between asphalt and polymer can be avoided by adding aromatic oils or stabilizing agents to the mixture [12, 16]. While some authors have studied the effects of the raw materials on the performance of SBS-modified asphalt roads to guide the production and application of SBS-modified asphalts, the cohesion characteristics of SBS-modified asphalt are still unclear.

Moisture damage of asphalt pavement, related to the breakdown of the asphalt composite, can be caused by losses in asphalt-asphalt cohesion and/or aggregate-binder adhesion [17, 18]. Cohesion is the molecular attraction existing between two similar objects in close contact, such as the internal binding interactions in asphalt [19, 20]. The work of cohesion of the asphalt binder is strongly related to the fatigue cracking characteristics of asphalt mastics and mixtures [21]. The cohesive properties of the binder and mastic determine the fracture resistance of asphalt concrete [22]. The surface free energy of the asphalt binder can be used to characterize the work of cohesion [23, 24]. Cheng et al. found that aging processes can decrease the cohesion of asphalt via an investigation of the cohesion characteristics of asphalt binders based on surface free energy [25]. Tan and Guo tested the cohesion and adhesion of asphalt mastic using the surface free energy method, reporting that the work of cohesion of neat asphalt is greater than that of modified asphalt [26]. Most previous studies have focused on the surface free energy of neat and modified asphalt binders [27–29]. However, few have considered the influence of raw materials composition on the cohesion characteristics of SBS-modified asphalt.

The objective of this study was to investigate the influence of the raw materials composition on the cohesion characteristics of SBS-modified asphalt using the surface free energy. SBS-modified asphalts were prepared by mixing different base asphalts, SBS modifier, extracting oil, and stabilizing agent. The contact angles between SBS-modified asphalt and distilled water, glycerol, and formamide were detected by the sessile drop method. Based on surface energy theory, the

surface free energy and cohesive power of the SBS-modified asphalts were calculated. The influence of the raw materials composition, such as virgin asphalt, SBS modifier types, and contents of extracting oil and stabilizing agent on the cohesive characteristics were discussed for SBS-modified asphalt.

2. Raw Materials and Preparation Method of SBS-Modified Asphalt

2.1. Raw Materials

2.1.1. Neat Asphalt. Three kinds of asphalt used in the Gansu area, such as binder A, binder B, and binder C, were selected as virgin asphalts. They were provided by Gansu Luqiao Construction Group Maintenance Technology Co., Ltd. The main technical properties were tested according to Standard Test Methods of Bitumen and Bituminous Mixtures for Highway Engineering (JTG E20-2011) and results were shown in Table 1.

2.1.2. Modifier. SBS modifiers can be divided into linear and branched types according to their molecular structures. In this study, linear and branched SBS modifiers were used to prepare SBS-modified asphalt. The main technical indicators were tested according to thermoplastic elastomers styrene-butadiene block copolymer (SH/T 1610-2011) and results are shown in Table 2.

2.1.3. Furfural Extraction Oil. Furfural extraction oil was selected as a compatibility agent for preparing SBS-modified asphalt because it contains a high-aromatic fraction that can improve the proportion of matrix asphalt, promote compatibility between modifier and asphalt, and improve the compatibility of the SBS-modified asphalt. Furfural extraction oil can also improve the low-temperature plasticity and ductility of SBS-modified asphalt. In this study, the furfural extraction oil was provided by PetroChina's Lanzhou Refining Company. The main technical indicators of furfural extraction oil were tested according to ASTM D2007-91 MOD and results are shown in Table 3.

2.1.4. Stabilizer. Some stabilizer was added during the preparation of the SBS-modified asphalts in order to prevent internal phase separation and improve storage stability. In this study, a high-efficiency stabilizer developed by the Gansu

TABLE 2: Properties of SBS modifiers.

Index	Liner	Branched
Block ratio (S/B)	30/70	40/60
Volatile content %	0.50	0.50
300% Tensile stress, MPa	1.7	1.7
Tensile strength, MPa	12.0	12.0
Elongation at break %	700	600
Shore hardness, A	75 ± 7	82 ± 7
Melt flow rate (g/10 min)	0.50~5.00	0.00~1.00

TABLE 3: Properties of furfural extraction oil.

Appearance	Viscosity (40°C) (Pa·s)	Flash point, °C	Aromatic content, %
Brown-black	2500	200	55.0

TABLE 4: Properties of stabilizer.

Appearance	Apparent density (g/cm^3)	Melting point, °C	Residue at 80-mesh sieve, %	Water content, %
Gray-black	0.5~0.7	120	0.1	0.2

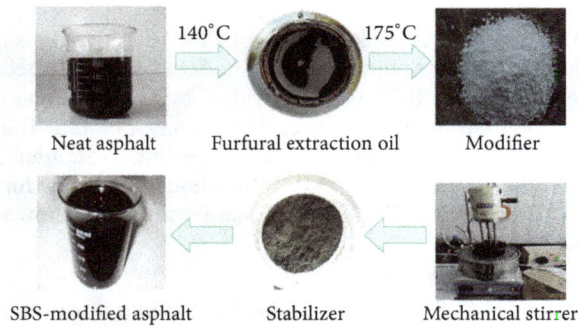

FIGURE 1: Preparation process of SBS-modified asphalts.

Provincial Engineering Research Center for Pavement Engineering is used. The main technical properties of the high-efficiency stabilizer were tested according to Synthetic Hydrotalcite Thermal Stabilizer (HG/T 4495-2013) and results are shown in Table 4.

2.2. Preparation of SBS-Modified Asphalts. In order to compare the influence of different raw materials and their contents on the adhesion properties of the SBS-modified asphalts, each SBS-modified asphalt sample was prepared using the same process. Figure 1 shows the preparation process of SBS-modified asphalts. First, furfural extraction oil was added to the flowing virgin asphalt and quickly heated to 175°C–180°C. Then, SBS modifier was added and the mixture was stirred for 5 min using a mechanical stirrer. The mixed asphalts were continuously sheared for 35 min at a speed of 5500 rpm using a shearing and dispersing emulsifier. The high-efficiency stabilizer was added in the last 10 min of the shearing process. The mixed asphalts were placed in an oven for 2 h at 170°C for growth.

3. Surface Free Energy Theory and Experiment

3.1. Surface Free Energy Theory. Surface free energy is defined as the work by a material in a vacuum necessary to produce a new interface per unit area. According to analyses by Fowkes and Good, the surface energy of a material mainly comprises a polar component and nonpolar dispersive component [30, 31]. Generally, the polar component of the surface free energy is the acid force and alkali force, while the dispersive component is composed of the Keesom orientation force, Debye induction force, and London dispersion force [32]. The surface free energies of liquid and solid materials can be expressed as follows [32]:

$$\gamma_l = \gamma_l^d + \gamma_l^p, \tag{1}$$

$$\gamma_s = \gamma_s^d + \gamma_s^p, \tag{2}$$

where γ_l is the surface free energy of a liquid material; γ_s is the surface free energy of a solid material; γ_l^d is the dispersive component of surface free energy for liquid materials; γ_l^p is the polar component of surface free energy for liquid materials; γ_s^d is the dispersive component of surface free energy for solid materials; and γ_s^p is the polar component of surface free energy for solid materials.

Fowkes indicated that the dispersive force between liquid and solid could be expressed as the geometric mean of the dispersive components of the liquid and solid surface free energies. Simultaneously, Owens and Wendt developed a similar method for the polar component [33]. Therefore, the surface free energy of the liquid-solid interface can be expressed as follows:

$$\gamma_{sl} = \gamma_s + \gamma_l - 2\sqrt{\gamma_s^d \gamma_l^d} - 2\sqrt{\gamma_s^p \gamma_l^p}. \tag{3}$$

TABLE 5: Surface free energy of three probe liquids at 25°C.

Test the liquid	γ_L	γ_L^d	γ_L^p	γ_L^+	γ_L^-
Distilled water	72.8	21.8	51.0	25.50	25.5
Glycerin	64.0	34.0	30.0	3.92	57.4
Formamide	58.0	38.0	19.0	2.28	39.6

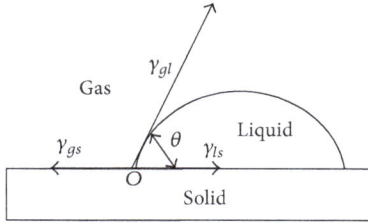

FIGURE 2: Young contact angle diagram.

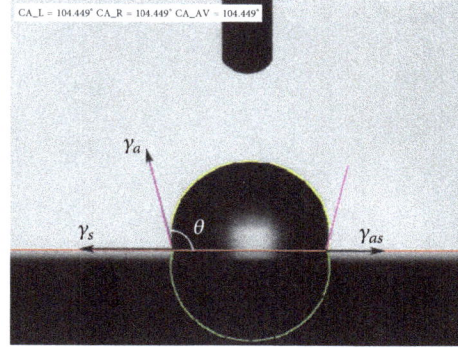

FIGURE 3: Schematic of the contact angle formed between a probe liquid and asphalt surface.

Figure 2 shows the Young contact angle diagram between solid and liquid. The polar and dispersive components of the surface free energy can be calculated using the Young equation, which relates the surface free energy and liquid-solid contact angle and is expressed as follows [33]:

$$\gamma_l \cos \theta = \gamma_s - \gamma_{sl}. \tag{4}$$

Fowkes proposed an equation to calculate the surface free energy though the dispersive component and contact angle. Oss calculated the surface free energy using the dispersion and polar components by developing (3) and (4) into (5) [33]:

$$1 + \cos \theta = 2\sqrt{\gamma_s^d}\left(\frac{\sqrt{\gamma_l^d}}{\gamma_l}\right) + 2\sqrt{\gamma_s^p}\left(\frac{\sqrt{\gamma_l^p}}{\gamma_l}\right). \tag{5}$$

The surface free energy of a solid can be calculated using (5), generally expressed as follows:

$$\frac{1 + \cos \theta}{2}\frac{\gamma_l}{\sqrt{\gamma_l^d}} = \sqrt{\gamma_s^d} + \sqrt{\gamma_s^p}\left(\frac{\sqrt{\gamma_l^p}}{\gamma_l^d}\right). \tag{6}$$

The dispersive component $\sqrt{\gamma_s^d}$ and polar component $\sqrt{\gamma_s^p}$ can be obtained through linear analysis by (6).

3.2. Cohesive Power of Asphalt. The work of cohesion is defined as the energy necessary to produce two new surfaces in a homogeneous material and is equal to twice the surface free energy. The equation for the work of cohesion is as follows [26]:

$$W_{\text{cohesion}} = 2\gamma_a, \tag{7}$$

where γ_a is the surface free energy of asphalt.

3.3. Contact Angle Measurement. In order to calculate the cohesion index of the asphalt, the surface free energy and its components must be obtained. According to the analysis of (5), the surface free energy and its components can be calculated using the contact angle between the solid asphalt and three liquids with known surface energy parameters [24, 31, 34]. In this study, the three liquids of distilled water, glycerin, and formamide were used as probe liquids in contact angle measurements. The surface free energy parameters of the three probe liquids are shown in Table 5.

The sessile drop method was used to conduct the surface free energy measurements in this study. This is an optical contact angle technique used to measure the contact angles between asphalt and the probe liquids [24, 34]. A schematic of the contact angle formed between the probe liquid and asphalt surface can be seen in Figure 2. The asphalt binder samples were prepared by heating at 163°C and then pouring into small plates, which had been previously placed on a heater to attain the constant temperature of 60°C. The plates with asphalt binder were heated by another heater to 163°C for approximately 5 min to create an even, thin film coating on the surface of the plate. Afterward, the samples were cooled to room temperature and kept in desiccators for 12 h at room temperature before testing. As shown in Figure 3, the instrument used to measure the contact angle is a drop shape analyzer (Binder A200KB) made in USA which is composed of an illumination device, a charge-coupled device (CCD) camera, three microsyringes with needles built into the machine, and image analysis software (Figure 4). The measurement was performed at room temperature. Each liquid drop was individually dropped at five different locations of the asphalt film and the contact angle was measured. The average contact angle from the five measurements per film was recorded.

TABLE 6: Contact angle and its variation coefficient for the SBS-modified asphalts.

	Distilled water		Glycerol		Formamide	
	Average value (°)	Coefficient of variation (%)	Average value (°)	Coefficient of variation (%)	Average value (°)	Coefficient of variation (%)
Binder A	102.6	0.43	92.6	0.36	86.1	0.32
Binder B	100.6	0.48	93.8	0.28	86.9	0.63
Binder C	103.8	0.25	97.8	0.48	92.0	0.47
Binder A + linear SBS	94.0	0.29	89.6	0.39	82.0	0.18
Binder B + linear SBS	96.6	0.44	89.2	0.35	83.0	0.39
Binder C + linear SBS	95.4	0.50	90.6	0.50	84.1	0.29
Binder A + branched SBS	92.4	0.53	87.2	0.17	79.6	0.47
Binder B + branched SBS	95.6	0.35	88.0	0.57	80.9	0.48
Binder C + branched SBS	94.7	0.44	89.6	0.50	82.5	0.57
Binder A + 3.5% SBS	96.2	0.31	89.8	0.38	83.8	0.46
Binder A + 4.0% SBS	94.9	0.35	89.4	0.64	82.9	0.52
Binder A + 4.5% SBS	94.0	0.29	89.6	0.39	82.0	0.18
Binder A + 5.0% SBS	93.6	0.31	86.8	0.39	78.9	0.38
Binder A + 0% extraction oil	96.8	0.25	90.4	0.40	84.3	0.31
Binder A + 2.5% extraction oil	93.3	0.37	88.0	0.52	81.5	0.43
Binder A + 3.5% extraction oil	94.0	0.29	89.6	0.39	82.0	0.18
Binder A + 4.5% extraction oil	93.4	0.33	87.4	0.32	81.0	0.39
Binder A + 0% stabilizer	97.2	0.35	91.0	0.28	85.2	0.17
Binder A + 0.1% stabilizer	95.3	0.45	88.6	0.52	82.5	0.30
Binder A + 0.2% stabilizer	94.0	0.29	89.6	0.39	82.0	0.18
Binder A + 0.3% stabilizer	92.8	0.40	86.1	0.25	78.5	0.52

4. Results and Discussion

4.1. Contact Angle and Its Variation Coefficient. The contact angles and variation coefficients for the SBS-modified asphalts prepared with different raw materials and the three probe liquids were measured by the sessile drop method. The variation coefficient is the ratio of the standard deviation to the average value of the contact angle between the SBS-modified asphalts and each of the three probe liquids, used to investigate whether the contact angle test has good repeatability. The results are shown in Table 6. From statistical analysis, it is found that the coefficient of variation of the

contact angle test results is in the range of 0.18%–0.64%. These results indicate that the contact angle test results of the SBS-modified asphalts prepared with different raw materials and three probe liquids have good reproducibility.

4.2. Surface Energy of Modified Asphalt. The surface free energy parameters of the three test liquids and the contact angles between them and each of the SBS-modified asphalts were substituted into (5). The component values of the surface free energy of each SBS-modified asphalt were calculated by solving (6). According to (2), the surface free energies of the individual SBS-modified asphalts were obtained. The

TABLE 7: Surface free energy of SBS-modified asphalts.

Asphalt type	Surface energy and its components		
	Surface free energy	Dispersion component	Polar component
Binder A	18.47	17.17	1.30
Binder B	18.84	18.06	0.77
Binder C	15.10	14.00	1.10
Binder A + linear SBS	22.14	21.96	0.18
Binder B + linear SBS	19.60	17.40	2.20
Binder C + linear SBS	19.68	18.39	1.29
Binder A + branched SBS	23.38	22.63	0.74
Binder B + branched SBS	22.12	20.74	1.38
Binder C + branched SBS	21.10	20.14	0.96
Binder A + 3.5% SBS	18.93	16.58	2.35
Binder A + 4.0% SBS	20.15	18.49	1.66
Binder A + 4.5% SBS	22.14	21.96	0.18
Binder A + 5.0% SBS	24.12	23.27	0.85
Binder A + 0% extraction oil	18.84	16.75	2.09
Binder A + 2.5% extraction oil	20.89	19.02	1.88
Binder A + 3.5% extraction oil	22.14	21.96	0.18
Binder A + 4.5% extraction oil	20.91	18.66	2.25
Binder A + 0% stabilizer	18.12	15.87	2.25
Binder A + 0.1% stabilizer	19.72	17.30	2.42
Binder A + 0.2% stabilizer	22.14	21.96	0.18
Binder A + 0.3% stabilizer	23.89	22.68	1.21

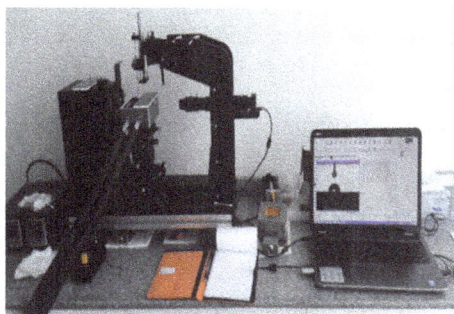

FIGURE 4: Drop shape analyzer.

surface free energies of the different SBS-modified asphalts range from 15.10 mJ/m^2 to 23.38 mJ/m^2, which is close to the range of surface free energy reported in the literature [35]. As shown in Table 7, the dispersion component is the major part of the surface free energy of the asphalt, compared to the polar component. Wei et al. also reported a similar conclusion and attributed it to the main component of the asphalt being nonpolar hydrocarbons [28]. In addition, with the increase of the content of SBS modifier, the total surface energy and dispersion component of the SBS-modified asphalt are gradually increased, while the polarity components of the SBS-modified asphalts generally decline at the different degrees. The surface free energy of the SBS-modified asphalt first increases and then decreases with increasing extraction oil content. For the extraction oil content of 3.5%, the surface free

energy of SBS-modified asphalt is maximized. The surface free energy of SBS-modified asphalt increases as the amount of stabilization agent increases.

The reliability and effectiveness of the surface free energy results were evaluated using a method developed by Kwok and Neumann [34]. They conclude that the values of $\gamma_l \cos\theta$ and γ_l should show a linear relationship for a given solid with a variety of liquids. If the resulting curve is nonlinear, the results must be remeasured. Using the surface free energy result of asphalts with different SBS modifier contents as an example, the illustration of $\gamma_l \cos\theta$ and γ_l for asphalt is given in Figure 5. It is observed that each asphalt sample shows a good linear fit between $\gamma_l \cos\theta$ and γ_l, with the coefficient of determination (R^2) values varying from 0.9612 to 0.9995. According to this method, $\gamma_l \cos\theta$ and γ_l of other groups of SBS-modified asphalts and test liquids are regressed; all show linear correlation coefficients >0.95. This indicates that the contact angles between the SBS-modified asphalts prepared with different raw materials and the three probe liquids by the sessile drop method are accurate and that the surface free energy results of the SBS-modified asphalts can be used.

4.3. The Influence of Neat Asphalt Type on Cohesive Work of SBS-Modified Asphalt. Figure 6 presents the results of the cohesive work of different neat asphalts and SBS-modified asphalts. In general, the cohesive work of binder B asphalt is the highest among the three neat asphalts, while the cohesive work of binder C asphalt is the lowest. For asphalts with linear modifiers, the cohesive work of the binder A + SBS-modified

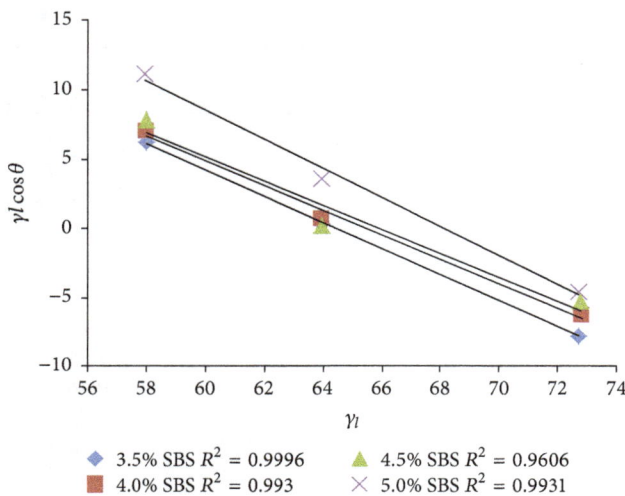

FIGURE 5: Plot of γ_l versus $\gamma_l \cos \theta$.

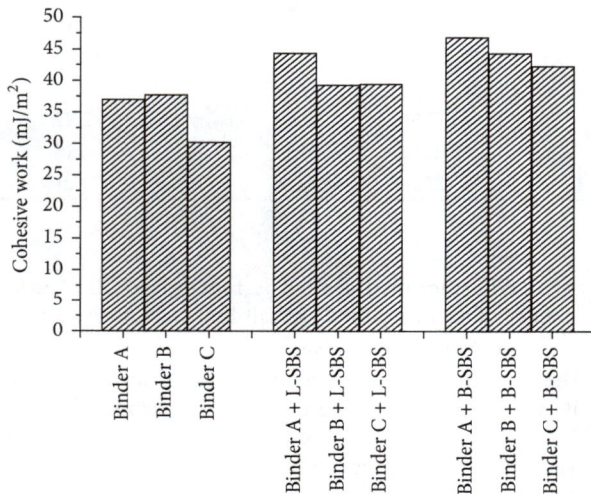

FIGURE 6: The cohesive work of SBS-modified asphalt prepared from different neat asphalts.

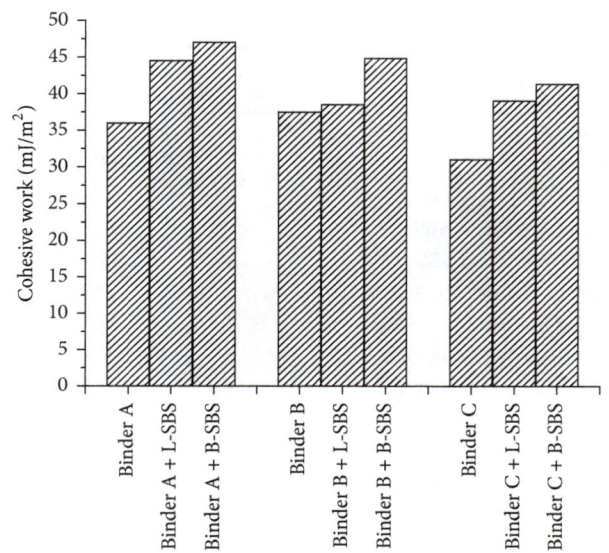

FIGURE 7: The cohesive work of SBS-modified asphalts prepared by different SBS types.

asphalt is the highest, while that of binder B + SBS-modified asphalt is equal to that of binder C + SBS-modified asphalt. The cohesion of binder A, binder B, and binder C asphalts is increased by 19.86%, 4.04%, and 30.34%, respectively, when the same amounts of linear SBS modifiers are added. For the three types of branched SBS-modified asphalt, the cohesive power ranks in the order of binder A + branched SBS-modified asphalt, binder B + branched SBS-modified asphalt, and binder C + SBS-modified asphalt. From the above results, it can be seen that the same SBS modifier has different modification effects on different neat asphalts, demonstrating the compatibility problems between SBS modifiers and neat asphalts from the perspective of cohesive properties.

4.4. Effect of SBS Type on Cohesive Work of SBS-Modified Asphalts. In order to compare and analyze the influence of SBS type on the surface energy of SBS-modified asphalt, the cohesive work values of the nine asphalt samples obtained in the previous section are arranged according to different SBS modifiers with the same kinds of asphalt (Figure 7). The results are shown in Figure 6. The cohesive work of the asphalts prepared with the linear SBS modifier is higher than that of the neat asphalt but lower than that of the asphalt modified by the branched SBS modifier. The results show that the SBS modifier type has a significant effect on the cohesive properties of asphalt. Specifically, the branched SBS-modified asphalt has the strongest resistance to self-cracking, the linear SBS-modified asphalt has intermediate resistance, and the neat asphalt has the least resistance to internal cracking during use. This is because the SBS modifiers comprise styrene and butadiene block copolymers, which form stable three-dimensional network structures by interlocking between styrene and butadiene blocks with the help of the highly aromatic extraction oil and stabilizer. The network structures enhance the cohesive properties of asphalt. Simultaneously, the SBS modifier absorbs the light components in the neat asphalt, which increases the asphalt polarity. Eventually, the cohesion of the asphalt increases with the addition of SBS modifier [36]. In addition, the branched SBS modifier has better modifying effects on asphalt because it has a higher molecular weight and a more compact structure, which promote the best resistance to cracking.

4.5. Effect of SBS Contents on Cohesive Work of SBS-Modified Asphalts. Figure 8 shows the relationship between the SBS content and cohesive work of SBS-modified asphalt. The cohesive work of SBS-modified asphalt gradually increases with the increase of SBS modifier. Asphalt modified with 5% SBS shows the highest resistance to cracking within the range of SBS contents, which has good resistance to water damage. Increased contents of SBS modifiers in a certain range strengthen the cross-linking and winding between the

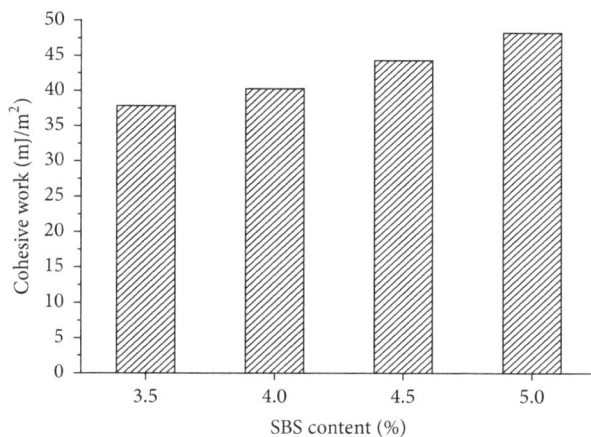

FIGURE 8: The cohesive work versus SBS content of SBS-modified asphalts.

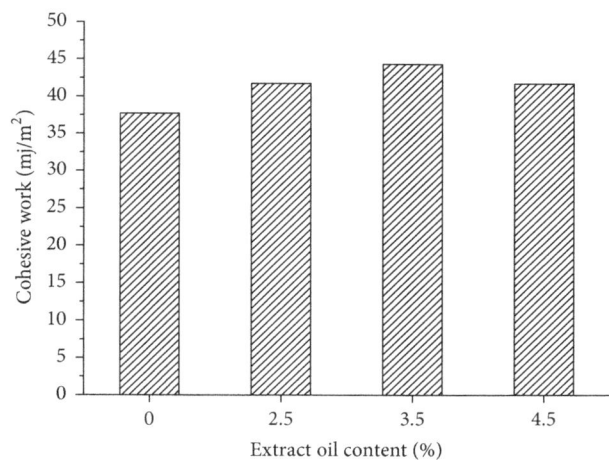

FIGURE 9: The cohesive work versus furfural extraction oil contents of SBS-modified asphalts.

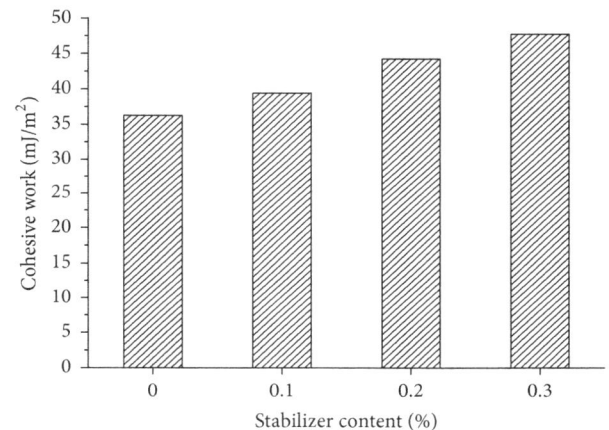

FIGURE 10: The cohesive work versus stabilizer content of SBS-modified asphalt.

modifier and pitch, which reinforces the three-dimensional network structure. Simultaneously, increasing the level of SBS modifier absorbs the aromatic light components of neat asphalt, which further extends the effect of SBS modification and enhances the internal intermolecular forces of the SBS-modified asphalts. Any amount of SBS improves the cohesive performance of the modified asphalt. In addition, the growth rate of the asphalt cohesive work is 6.48%, 9.85%, and 8.93% for SBS content increases of 3.5% to 4%, 4% to 4.5, and 4.5% to 5%, respectively. This is because the modifier fully absorbs the aromatic components and promotes dispersibility when the amount of SBS is small, but the network structure in this range is weak, which causes cohesiveness to increase at a low rate [37].

4.6. Effect of Extraction Oil Content on Cohesive Work of SBS-Modified Asphalt.

Figure 9 illustrates the changes in the cohesive work of SBS-modified asphalts with increasing contents of extraction oil. It can be seen that the cohesive work of SBS-modified asphalt increases and then decreases with increasing extraction oil contents. With 2.5% extraction oil, the cohesive work of the SBS-modified asphalt is increased by 10.87% relative to oil-free asphalt, while the cohesive work increases by 5.97% for the addition of 3.5% extraction oil compared to that with 2.5%. That is because the optimal amount of furfural extraction oil promotes the swelling and dispersing of the SBS modifier, thereby assisting microstructural formation with uniform particle size, obtaining better three-dimensional network structures by interparticle interactions, and increasing the cohesive work of SBS-modified asphalt [38]. It should be noted that the cohesive work of the SBS-modified asphalt decreases for oil contents exceeding 3.5%, dropping by 5.57% with 4.5% furfural oil. This is because of the excessive dilution of light components when the content of the highly aromatic extraction oil exceeds that required for SBS swelling and dispersing. Negative correlation relationships exist between the surface energy and aromatic content that decrease the cohesive power of SBS-modified asphalt. It can be concluded

from the above results that the SBS-modified asphalt with the best cohesive work can be obtained by adding the proper amount of extraction oil as a compatibility agent.

4.7. Effect of Stabilizer Content on Cohesive Work of SBS-Modified Asphalt.

Figure 10 illustrates the changes in the cohesive work of SBS-modified asphalt with increasing stabilizer contents. The cohesive work of the SBS-modified asphalt clearly increases with increasing contents of stabilizer. The increments of cohesive work of the SBS-modified asphalts are 8.82%, 12.28%, and 7.9% with 0.1%, 0.2%, and 0.3% stabilizer, respectively. This is because the microstructure of the SBS-modified asphalt from the addition of stabilizer causes reactions between the SBS and neat asphalt. The distribution of SBS modifier in asphalt is changed from bead-like structures to a fine network structure, and interfacial adsorption layers form between the polymer phase and the neat asphalt phase in the modified asphalt [39]. Thus, the cohesive properties of the SBS-modified asphalt are effectively improved.

5. Conclusions

(1) When evaluating surface free energy of the SBS-modified asphalt, the sessile drop method is an appropriate method because the contact angle measurements show good repeatability for the SBS-modified asphalts prepared with different raw materials and three probe liquids.

(2) The cohesive work of SBS-modified asphalt can be increased by selection the most suitable bitumen and SBS modifier to improve the compatibility. And the cohesive properties of asphalt are maximal with branched SBS, intermediate with linear SBS, and minimal in neat asphalt.

(3) The addition content of admixtures, including SBS modifier, stabilizer, and extraction oil can affect cohesive work of SBS-modified asphalt differently. The cohesive work of SBS-modified asphalt increases with the increase of SBS modifier and stabilizer contents, while it increases and then decreases with increases of extraction oil content.

(4) The writers envision that continued development of other intuitive detection method to cohesion characteristics of SBS-modified asphalt in the future will contribute toward implementation of the moisture susceptibility.

Conflicts of Interest

The authors declare that there are no conflicts of interest regarding the publication of this paper.

Acknowledgments

The research work reported in this paper was supported by the National Natural Science Foundation of China (51408287 and 51668038), Rolls Supported by Program for Changjiang Scholars and Innovative Research Team in University (IRT_15R29), the Distinguished Young Scholars Fund of Gansu Province (1606RJDA318), the Natural Science Foundation of Gansu Province (1506RJZA064), Excellent Program of Lanzhou Jiaotong University (201606), and Foundation of A Hundred Youth Talents Training Program of Lanzhou Jiaotong University.

References

[1] B. Li, J. Yang, Z. Chen, and H. Li, "Microstructure morphologies of asphalt binders using atomic force microscopy," *Journal Wuhan University of Technology, Materials Science Edition*, vol. 31, no. 6, pp. 1261–1266, 2016.

[2] G. D. Airey, "Styrene butadiene styrene polymer modification of road bitumens," *Journal of Materials Science*, vol. 39, no. 3, pp. 951–959, 2004.

[3] X. Zhao, S. Wang, Q. Wang, and H. Yao, "Rheological and structural evolution of SBS modified asphalts under natural weathering," *Fuel*, vol. 184, pp. 242–247, 2016.

[4] J. T. Kim, J. Baek, H. Lee, and Y. Ji, "Fatigue performance evaluation of SBS modified mastic asphalt mixtures," *Construction & Building Materials*, vol. 48, no. 11, pp. 908–916, 2013.

[5] F. Gahvari, "Effects of thermoplastic block copolymers on rheology of asphalt," *Journal of Materials in Civil Engineering*, vol. 9, no. 3, pp. 111–116, 1997.

[6] G. Polacco, S. Filippi, F. Merusi, and G. Stastna, "A review of the fundamentals of polymer-modified asphalts: asphalt/polymer interactions and principles of compatibility," *Advances in Colloid and Interface Science*, vol. 224, pp. 72–112, 2015.

[7] A. Modarres, "Investigating the toughness and fatigue behavior of conventional and SBS modified asphalt mixes," *Construction and Building Materials*, vol. 47, pp. 218–222, 2013.

[8] D. Niu, S. Han, K. Chen, and O. Xu, "Study on influences of key process parameters on SBS modified asphalt," *Changan Daxue Xuebao*, vol. 34, no. 3, p. 16, 2014.

[9] W. Yin, F. Ye, and H. Lu, "Establishment and experimental verification of stability evaluation model for SBS modified asphalt: Based on quantitative analysis of microstructure," *Construction and Building Materials*, vol. 131, pp. 291–302, 2017.

[10] F. Dong, W. Zhao, Y. Zhang et al., "Influence of SBS and asphalt on SBS dispersion and the performance of modified asphalt," *Construction and Building Materials*, vol. 62, pp. 1–7, 2014.

[11] D. O. Larsen, J. L. Alessandrini, A. Bosch, and M. S. Cortizo, "Micro-structural and rheological characteristics of SBS-asphalt blends during their manufacturing," *Construction and Building Materials*, vol. 23, no. 8, pp. 2769–2774, 2009.

[12] H. Fu, L. Xie, D. Dou, L. Li, M. Yu, and S. Yao, "Storage stability and compatibility of asphalt binder modified by SBS graft copolymer," *Construction and Building Materials*, vol. 21, no. 7, pp. 1528–1533, 2007.

[13] T. Wang, T. Yi, and Z. Yuzhen, "The compatibility of SBS-modified asphalt," *Petroleum Science and Technology*, vol. 28, no. 7, pp. 764–772, 2010.

[14] P. Yang, D. Liu, F. Yan et al., "Application of the compatibility theory and the solubility parameter theory in SBS modification asphalt," *Petroleum Science and Technology*, vol. 20, no. 3-4, pp. 367–376, 2002.

[15] M. Liang, P. Liang, W. Fan et al., "Thermo-rheological behavior and compatibility of modified asphalt with various styrene-butadiene structures in SBS copolymers," *Materials and Design*, vol. 88, pp. 177–185, 2015.

[16] L. Wang and C. Chang, "Rheological evaluation of polymer modified asphalt binders," *Journal Wuhan University of Technology, Materials Science Edition*, vol. 30, no. 4, pp. 695–702, 2015.

[17] R. A. Tarefder and A. M. Zaman, "Nanoscale evaluation of moisture damage in polymer modified asphalts," *Journal of Materials in Civil Engineering*, vol. 22, no. 7, pp. 714–725, 2010.

[18] M. Arabani and G. H. Hamedi, "Using the surface free energy method to evaluate the effects of liquid antistrip additives on moisture sensitivity in hot mix asphalt," *International Journal of Pavement Engineering*, vol. 15, no. 1, pp. 66–78, 2014.

[19] G. Xu and H. Wang, "Study of cohesion and adhesion properties of asphalt concrete with molecular dynamics simulation," *Computational Materials Science*, vol. 112, pp. 161–169, 2016.

[20] B. Li, Z. Zhang, and X. Liu, "Adhesion in SBS modified asphalt containing warm mix additive and aggregate system based on surface free theory," *Cailiao Daobao*, vol. 31, no. 2, pp. 115–120, 2017.

[21] E. Masad, C. Zollinger, R. Bulut et al., "Characterization of HMA moisture damage using surface energy and fracture properties," in *Proceedings of the Association of Asphalt Paving Technologists -Proceedings of the Technical Sessions 2006 Annual Meeting*, pp. 713–754, March 2006.

[22] Y. Veytskin, C. Bobko, C. Castorena, and Y. R. Kim, "Nanoindentation investigation of asphalt binder and mastic cohesion," *Construction and Building Materials*, vol. 100, no. 4, pp. 163–171, 2015.

[23] A. Bhasin, D. N. Little, K. L. Vasconcelos, and E. Masad, "Surface free energy to identify moisture sensitivity of materials for asphalt mixes," *Transportation Research Record*, vol. 2001, no. 1, pp. 37–45, 2007.

[24] J. Wei, Y. Zhang, and J. Youtcheff, "Determination of the surface free energy of asphalt binders by sessile drop method," in *Acta Petrolei Sinica (Petroleum Processing Section)*, vol. 25, pp. 207–215, 2 edition, 2009.

[25] D. Cheng, D. N. Little, R. L. Lytton et al., "Use of surface free energy properties of the asphalt-aggregate system to predict moisture damage potential," in *Proceedings of the Asphalt Paving Technology 2002*, pp. 59–88, March 2002.

[26] Y. Tan and M. Guo, "Using surface free energy method to study the cohesion and adhesion of asphalt mastic," *Construction and Building Materials*, vol. 47, no. 5, pp. 254–260, 2013.

[27] M. R. Kakar, M. O. Hamzah, M. N. Akhtar, and D. Woodward, "Surface free energy and moisture susceptibility evaluation of asphalt binders modified with surfactant-based chemical additive," *Journal of Cleaner Production*, vol. 112, pp. 2342–2353, 2016.

[28] J. Wei, Y. Zhang, and J. S. Youtcheff, "Effect of polyphosphoric acid on the surface free energy of asphalt binders," *Acta Petrolei Sinica (Petroleum Processing Section)*, vol. 27, no. 2, pp. 280–285, 2011.

[29] F. Ma, J.-X. Hao, Z. Fu, L.-L. Wang, and L.-B. Wang, "Surface free energy analysis of asphalt modified with natural asphalt," *Journal of Traffic and Transportation Engineering*, vol. 15, no. 1, pp. 18–24, 2015.

[30] F. Fowkes, "Attractive forces at interfaces," *Industrial & Engineering Chemistry*, vol. 56, no. 12, pp. 40–52, 1964.

[31] R. J. Good, "Contact Angle, Wetting, and Adhesion: A Critical Review," *Journal of Adhesion Science and Technology*, vol. 6, no. 12, pp. 1269–1302, 1992.

[32] C. J. van Oss, R. J. Good, and M. K. Chaudhury, "Additive and nonadditive surface tension components and the interpretation of contact angles," *Langmuir*, vol. 4, no. 4, pp. 884–891, 1988.

[33] D. K. Owens and R. C. Wendt, "Estimation of the surface free energy of polymers," *Journal of Applied Polymer Science*, vol. 13, no. 8, pp. 1741–1747, 1969.

[34] D. Y. Kwok and A. W. Neumann, "Contact angle measurement and contact angle interpretation," *Advances in Colloid and Interface Science*, vol. 81, no. 3, pp. 167–249, 1999.

[35] A. W. Hefer, A. Bhasin, and D. N. Little, "Bitumen surface energy characterization using a contact angle approach," *Journal of Materials in Civil Engineering*, vol. 18, no. 6, pp. 759–767, 2006.

[36] S. Li, Q. Lin, and S. Dong, "Progress in mechanism of SBS-modified asphalts," *Chinese Polymer Bulletin*, vol. 34, no. 5, pp. 14–19, 2008.

[37] J.-A. Yuan, D. Ji, and Z.-G. Zhu, "Effect of different dosage SBS on properties of modified asphalt," *Journal of Chang'an University (Natural Science Edition)*, vol. 25, no. 3, pp. 19–48, 2005.

[38] X. Li and J. Cheng, "Influence of furfural extract oil on property of SBS modified asphalt," *Petroleum Asphalt*, vol. 25, no. 3, pp. 65–68, 2011.

[39] P. Xiong and P. Hao, "Measures and mechanism analysis of improving the storage stability of styrene-butadiene-styrene polymer modified asphalt," *Journal of Tongji University (Natural Science)*, vol. 34, no. 5, pp. 613–618, 2006.

19

Aging of Rejuvenated Asphalt Binders

Mojtaba Mohammadafzali,[1] Hesham Ali,[1] James A. Musselman,[2] Gregory A. Sholar,[3] and Wayne A. Rilko[3]

[1]*Department of Civil and Environmental Engineering, Florida International University, Miami, FL, USA*
[2]*Oldcastle Materials Group, FL, USA*
[3]*State Material Office, Florida Department of Transportation, Gainesville, FL, USA*

Correspondence should be addressed to Mojtaba Mohammadafzali; mmoha020@fiu.edu

Academic Editor: Hainian Wang

An important concern that limits the RAP content in asphalt mixtures is the fact that the aged binder that is present in the RAP can cause premature cracking. Rejuvenators are frequently added to high RAP mixtures to enhance the properties of the binder. There is no existing method to predict the longevity of a rejuvenated asphalt. This study investigated the aging of rejuvenated binders and compared their durability with that of virgin asphalt. Various samples with different types and proportions of RAP, virgin binder, and rejuvenator were aged by RTFO and three cycles of PAV. DSR and BBR tests were conducted to examine the high-temperature and low-temperature rheological properties of binders. Results indicated that the type and dosage of the rejuvenator have a great influence on the aging rate and durability of the binder. Some rejuvenators make the binder age slower, while others accelerate aging. These observations confirm the importance of evaluating the long-term aging of recycled binders. For this purpose, critical PAV time was proposed as a measure of binder's longevity.

1. Introduction

To maintain over 2.4 million miles of asphalt surfaced roads in the US, over one hundred million tons of asphalt pavement is milled and resurfaced every year. The milled asphalt material, which is reclaimed asphalt pavement (RAP), is entirely recyclable [1]. Despite an abundance of this valuable resource, some portions of RAP are still wasted in landfills, and yet another part of it is used in nonasphalt applications, such as an embankment, subbase, base, and shoulder. Therefore, the majority of the RAP is not recycled to its highest potential.

Asphalt mixtures with more than 25% RAP are often identified as high RAP mixtures [2, 3]. One of the most important obstacles for using high RAP asphalt mixtures is a lack of confidence in their performance [4]. A high RAP mixture can perform differently from a conventional mixture, and currently, there are no adequate methods to predict and assess all aspects of these differences. Aging of the asphalt binder is among most critical parameters that make recycled asphalt mixtures different from new material. When

pavement ages, the asphalt binder becomes hard, brittle, and prone to cracking [5–7]. The reason for this phenomenon is that lighter components of the asphalt or maltenes are partially lost due to evaporation and oxidation. Rejuvenating is the process of restoring the original properties of an aged asphalt binder by adding a rejuvenator or recycling agent (RA). Rejuvenation is often perceived as simply softening the hard asphalt. Therefore, in most cases, specifications call for a target range of penetration, viscosity, or performance grade (PG) to verify the effectiveness of rejuvenation [8, 9]. However, there are other concerns that differentiate a rejuvenated binder from a virgin binder that cannot be addressed by these requirements. The durability of the binder is one of these issues.

Today, decades after recycling of asphalt pavement became a common practice in the late 1970s, rejuvenated mixtures are aging again, and their aging behavior is not necessarily similar to that of virgin mixtures. Therefore, it is important to study the aging of recycled asphalt and compare it to that of virgin binders. To achieve a high RAP mixture with durability similar to that of new pavement material, the rejuvenated asphalt binder should not age faster than a virgin asphalt.

Terrel and Fritchen compared the durability of recycled and virgin mixtures in 1978 [10]. They simulated long-term moisture damage using a vacuum-submerged conditioning procedure, followed by several freeze-thaw thermal cycles. Results showed that the performance of recycled asphalt concrete samples was similar to new samples. In another study, virgin and recycled mixtures were subjected to long-term oven aging [11]. Dynamic modulus tests showed that samples that contained RAP aged slower. It should be noted that no rejuvenator was added to the mixtures in this study. The emergence of the Superpave performance grade system and its tests and aging procedures provided new tools for evaluating the changes in asphalt properties as it ages. A study that evaluated the performance of recycled asphalt binders using the performance grade system concluded that generally recycled asphalt binders perform similar to or better than virgin binders [12]. Aging was simulated by oven heating of mixtures. Results from this study confirmed the ability of rejuvenators to effectively lower the PG of samples containing up to 45% RAP. It was concluded from a series of shear modulus master curves that using recycling agents in mixtures containing RAP improves fatigue cracking resistance without adversely affecting rutting resistance. Also, it has been showed that the use of an excessively aged binder in recycled mixtures can also lead to segregation problems due to a reduction in binder's adhesion [13].

While the majority of the previous researches on the durability of recycled asphalt show a better longevity for recycled material, there are instances that conclude vice versa. According to a study that used Rolling Thin Film Oven (RTFO) and Pressure Aging Vessel (PAV) for simulating the aging and Fourier Transformed Infrared Spectroscopy for evaluating the level of aging, rejuvenated binder generally aged faster than virgin binder [14]. There is a need to study the parameters that affect the longevity of rejuvenated binder and cause them age faster or slower in comparison with virgin asphalt. In a previous research by the authors, the aging of 100% recycled asphalt binders was evaluated [15, 16]. In that study, the PAV was used to simulate long-term aging. The results showed that aging of recycled asphalt could be either faster or slower than that of virgin asphalt, depending on the rejuvenator used. The difference in the service lives of binders rejuvenated by different products was estimated to be up to ten years. Also, it was shown that the standard PAV aging time was not adequate to evaluate aging behavior. Therefore, aging for a longer period of time is required to ensure the durability of recycled asphalt.

There is a well-recognized need to increase the use of RAP in the pavement. However, this cannot be achieved if asphalt rejuvenation continues to be perceived as simply softening the binder. This paper presents an effort to overcome this shortcoming by addressing the need for an effective rejuvenation. The objective of this research was to evaluate the durability of asphalt binders containing RAP. Rheological properties of the binder were used to assess the level of aging. The long-term aging of binders with 20 and 40 percent RAP was compared to that of virgin asphalt binders. The effect of different rejuvenators on the durability was also investigated.

2. Materials and Methods

2.1. Experimental Approach. The experimental approach was designed based on Superpave PG tests and aging procedures. The aging was simulated by the PAV, which exposes the asphalt to heat and high pressure. The standard PAV aging time is 20 hours. There is no definite correlation between PAV aging and actual field aging time, but a study performed in Florida for this purpose estimated that the aging caused by a 20-hour PAV cycle is equivalent to 8 years of service [17]. This estimate was used to provide an approximate correlation between PAV time and field aging. Furthermore, it was aimed to look at the aging beyond the first 8 years. Therefore, samples were subjected to three PAV cycles to increase the PAV aging time to 60 hours and simulate almost 24 years of in-service aging.

One of the objectives of this research was to introduce a quantitative description for binder durability. For this purpose, the PAV time that increases the high-temperature PG of each sample from 70°C to 95°C was considered as a measure of aging that makes the binder too hard to perform well. This value is referred to as *critical PAV time* in this paper. Based on the Florida Department of Transportation's testing of over 21 RAP stockpiles, a high-temperature PG of 95°C is the typical grade of a RAP binder in Florida. Each sample was subjected to four levels of aging, which included RTFO aging and three cycles of PAV aging. After each level, the samples underwent Dynamic Shear Rheometer (DSR) testing. The primary parameter used for characterizing the stiffness of the binder and thereafter the level of aging was the continuous high-temperature performance grade of the binder. This parameter is briefly referred to as *high PG* in this paper. Also, the low-temperature properties of the binders were assessed by Bending Beam Rheometer (BBR) tests after standard (20 hours) and extended (60 hours) PAV aging.

2.2. Material. Two types of nonmodified RAP, two virgin binders, and two rejuvenators were used. Samples were prepared with 20% and 40% content of RAP binder.

2.2.1. RAP Binders. A medium-aged RAP and a hard RAP, recently milled in Florida, were used. These are referred to as *RAP 1* and *RAP 2*, respectively. The RAP binder was recovered using a centrifuge extractor and a rotary evaporator in accordance with ASTM D2172 and ASTM D5404. Trichloroethylene was used as the solvent for the binder recovery process.

2.2.2. Virgin Binders. Two types of virgin binders were used. These are referred to as *VB1* and *VB2*. Although both binders had an incremental grade of PG 67-22, the continuous grade of VB1 was slightly higher than VB2. Table 1 presents results from the high-temperature DSR tests on RAP and virgin binders and the resulting high PG values. The RTFO mass loss was 0.61% for VB1 and 0.43% for VB2.

TABLE 1: DSR test results for RAP and virgin binders.

Binder	Test temperature (°C)	δ (°)	$G^*/\sin\delta$ (kPa)	High PG (°C)
RAP 1 (medium)	82	82	3.25	91.95
	88	84	1.60	
RAP 2 (hard)	82	76	14.60	104.24
	88	79	7.08	
VB1 (not aged)	67	86	1.48	70.19
	76	86	0.49	
VB2 (not aged)	67	88	1.29	69.18
	76	88	0.45	
VB1 (RTFO aged)	67	83	3.68	70.97
	73	85	1.69	
VB2 (RTFO aged)	67	85	2.77	68.86
	73	87	1.32	

2.2.3. Recycling Agents. Two recycling agents (rejuvenators) were used. These are commercial products that are commonly used in Florida, referred to as RA1 and RA2.

RA1 is a dark yellow heavy paraffinic oil with high aromatic content that gives it good softening power. The rejuvenator contains no asphaltene. This helps restore the maltene to asphaltene ratios that are reduced by aging. The flash point of this oil was 420°F (216°C) as determined by the Cleveland Open Cup Test (ASTM D92). The material has good high-temperature stability and does not emit much smoke at mixing temperatures. However, its high aromatic content allows it to evaporate quickly during the mixing procedures. The RTFO mass loss was determined to be as high as 1.92% for RA1.

RA2 is a semisolid black substance with an asphalt odor. This product is manufactured by rerefining used oils through vacuum distillation. Using a rerefined product as the rejuvenator is a step toward enhancing the use of recycled material. This rejuvenator has a high flash point of 522°F (272°C) and does not release much smoke at high temperatures. It also evaporates much less than RA1 at mixing temperatures, and its RTFO mass loss is only 0.21%.

2.3. Sample Preparation. Sixteen samples were prepared by varying the RAP content and the type of RAP, virgin binder, and recycling agent. The two virgin binders were used as controls. The samples were prepared by mixing a soft binder with the RAP binder. The soft binder is a mixture of a virgin binder and a recycling agent. This sequence is consistent with the practice that is often followed by the industry. Similar initial high PG values were required to facilitate the comparison between samples. Thus, the target grade for the samples was set similar to the high PG of virgin binders ±1°C. To determine the proportion of the components that would produce samples with those target grades, the following three steps were followed.

2.3.1. Step 1: Determination of Soft Binder Grade. The first step was to determine the grade of the soft binder in such a way that, after blending with the RAP binder, a sample with the target grade is achieved. For this purpose, a linear interpolation was used to estimate the grade of the soft binder, along with the RAP content, the high PG of the RAP, and the target high PG. This is in accordance with the method recommended in ASTM D4887. Figure 1 shows this interpolation for each combination of virgin and RAP binder.

2.3.2. Step 2: Establishing Softening Curves. Softening curves were established for each combination of virgin binder and recycling agent, as shown in Figure 2. The dotted lines represent the linear trend lines. The softening power of the RA2 was considerably lower than the RA1. Therefore, large doses were needed to soften the binder to the desired grade. This fact makes RA2 an inappropriate choice when the high RAP content is considered, especially when the RAP is highly aged.

2.3.3. Step 3: Calculating the Proportion of Material. The results from the two preceding steps made it possible to calculate the proportion of the RAP binder, virgin binder, and recycling agent for each sample. Table 2 presents the factorial design of samples and the composition of each. Figure 3 is a flowchart that shows the steps for the sample preparation process.

3. Results and Discussions

3.1. RTFO Aging. The RTFO simulates the aging that the binder undergoes during construction. This aging is primarily due to the evaporation of lighter components of the asphalt binder when it is heated. Table 3 shows the results of the DSR tests and the high PG of the samples before and after RTFO aging and the resulting high PG values. The high PG values for nonaged and RTFO-aged samples were determined differently based on the corresponding criteria (formulas (1) and (2)).

$$\text{Original (nonaged) sample: } \frac{G^*}{\sin\delta} \geq 1.0 \, \text{kPa} \qquad (1)$$

$$\text{RTFO-aged sample: } \frac{G^*}{\sin\delta} \geq 2.2 \, \text{kPa}. \qquad (2)$$

The results showed that the degree of aging caused by the RTFO depended on the type of asphalt and recycling agent. VB1 lost more weight in the RTFO (0.61% compared to 0.43% for VB2) and experienced more aging. Its RTFO grade (based on formula (2)) was 0.77°C higher than its nonaged grade (based on formula (1)). On the other hand, the RTFO grade was 0.79°C less than the nonaged grade for VB2. Generally, RA1 increased the RTFO aging, while RA2 did not influence it significantly. There is a meaningful correlation between the percentage of RA1 and the extent of RTFO aging (see the correlation in Figure 4). This was expectable, because RA1 has a high aromatic content and RTFO mass loss. Faster RTFO aging is not necessarily a negative quality. In fact, PG

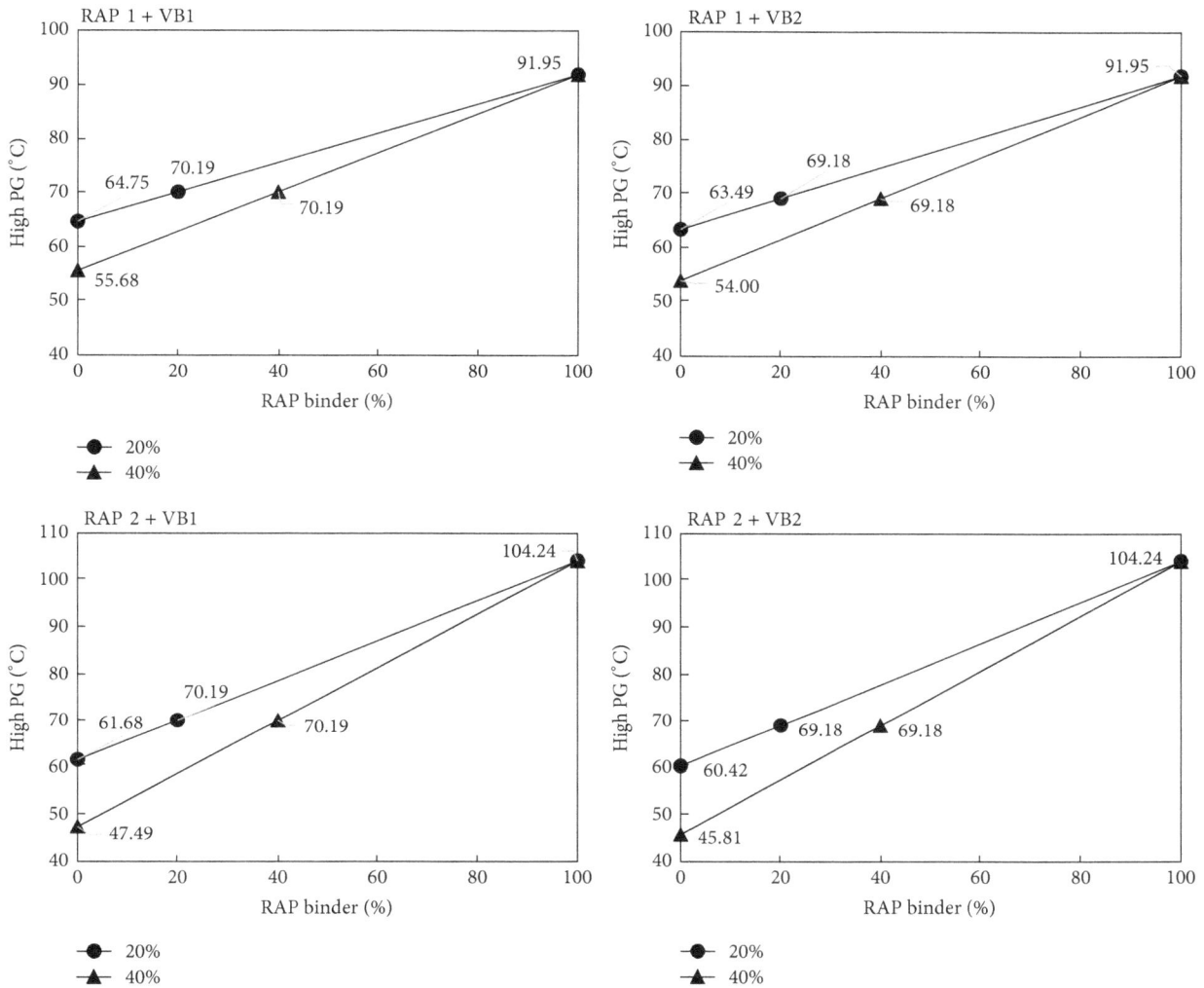

FIGURE 1: Soft binder grade determination.

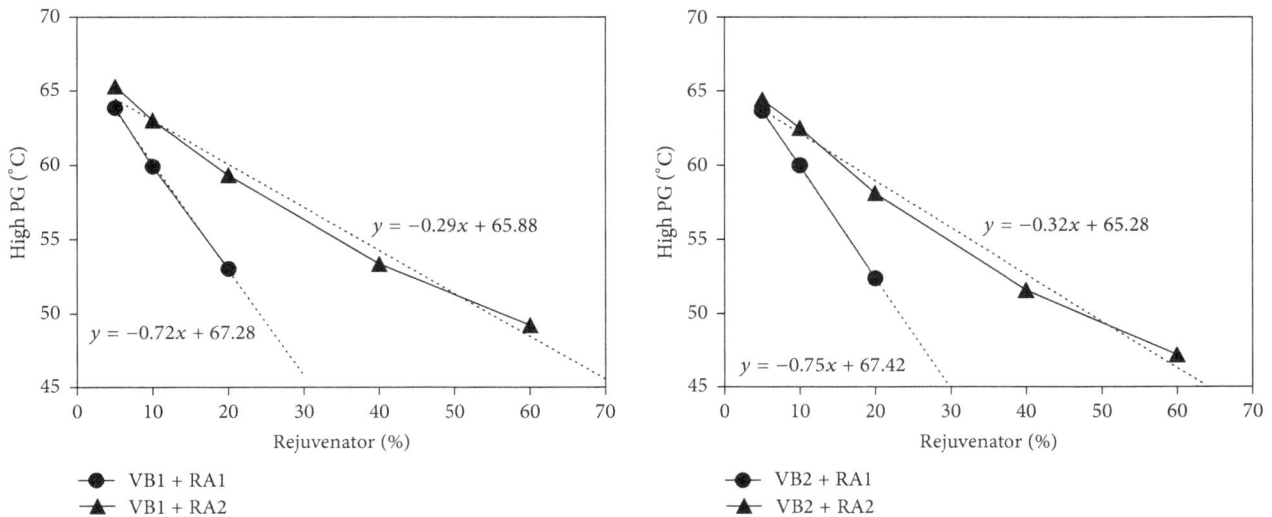

FIGURE 2: Softening curves.

TABLE 2: Factorial design and composition of samples.

Sample number	Sample composition						Target high PG (°C)	Soft binder high PG (°C)	RA/VB (%)
	VB	RA	RAP	VB%	RA%	%RAP			
S01			RAP1	76.6	3.4	20	70.19 ± 1	64.75	4.3
S02		RA1		50.3	9.7	40	70.19 ± 1	55.68	16.1
S03			RAP2	73.8	6.2	20	70.19 ± 1	61.68	7.7
S04	VB1			43.2	16.8	40	70.19 ± 1	47.49	28.0
S05			RAP1	75.1	4.9	20	70.19 ± 1	64.75	6.2
S06		RA2		40.7	19.3	40	70.19 ± 1	55.68	32.2
S07			RAP2	69.1	10.9	20	70.19 ± 1	61.68	13.6
S08				19.8	40.2	40	70.19 ± 1	47.49	67.0
S09			RAP1	75.9	4.1	20	69.18 ± 1	63.49	5.2
S10		RA1		49.3	10.7	40	69.18 ± 1	54.00	17.9
S11			RAP2	72.5	7.5	20	69.18 ± 1	60.42	9.4
S12	VB2			42.8	17.2	40	69.18 ± 1	45.81	28.6
S13			RAP1	74.1	5.9	20	69.18 ± 1	63.49	7.4
S14		RA2		40.5	19.5	40	69.18 ± 1	54.00	32.6
S15			RAP2	68.2	11.8	20	69.18 ± 1	61.42	14.7
S16				20.2	39.8	40	69.18 ± 1	45.81	66.3

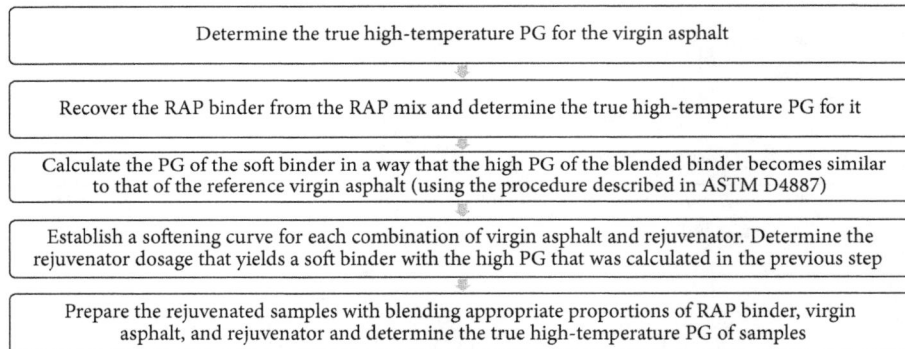

Determine the true high-temperature PG for the virgin asphalt

Recover the RAP binder from the RAP mix and determine the true high-temperature PG for it

Calculate the PG of the soft binder in a way that the high PG of the blended binder becomes similar to that of the reference virgin asphalt (using the procedure described in ASTM D4887)

Establish a softening curve for each combination of virgin asphalt and rejuvenator. Determine the rejuvenator dosage that yields a soft binder with the high PG that was calculated in the previous step

Prepare the rejuvenated samples with blending appropriate proportions of RAP binder, virgin asphalt, and rejuvenator and determine the true high-temperature PG of samples

FIGURE 3: The flowchart for the sample preparation process.

specifications call for a minimum stiffness for the pavement to have the adequate strength after construction. However, if a rejuvenator causes faster aging, this should be known and considered during the mix design phase.

The phase angle is relatively small for samples with a large dosage of RA2. For instance, samples 12 and 16 have almost similar magnitudes of $G^*/\sin\delta$, but the phase angle is 8° smaller for sample 16. Therefore, RA2 decreases the viscous portion of the complex modulus.

3.2. *PAV Aging.* Three 20-hour cycles of the PAV aging with a temperature of 100°C and a pressure of 2.1 MPa were applied. Table 4 displays the results from the DSR tests on samples after each PAV cycle. These samples were already RTFO-aged. Therefore, the criterion for the RTFO samples (formula (2)) was used to determine their high PG. DSR testing on samples 8 and 16 after 60 hours of aging did not result in valid data. Large complex modulus values were measured

during the first few iterations, but the measurements dropped rapidly and finally converged to very low values. In some cases, the target strain of 10% was not achieved with the maximum stress that the DSR could apply. These samples also exhibited unusual physical behavior. Although the samples are expected to be extremely hard after 60 hours of aging, they could be easily cut off by a spatula in a brittle manner at room temperature. This is an indication of weak cohesion and shear and tensile strengths of the binder. These samples had very low values of δ even after the first PAV cycle. This infers that they have a less viscous behavior when compared to conventional asphalt binders.

Table 5 summarizes the results of the PAV aging experiment and shows the increase in the high PG that takes place in each stage. Critical PAV time values are also presented. The critical PAV time was calculated for samples as a measure of durability. This parameter is defined as the PAV aging time it takes to increase the high PG from 70°C to 95°C. PAV times

TABLE 3: High PG of samples based on nonaged and RTFO-aged criteria.

Sample	Total RA%	No aging				RTFO				Difference (RTFO-no aging) (°C)
		Temp. (°C)	δ (°)	$G^*/\sin\delta$ (kPa)	High PG (°C)	Temp. (°C)	δ (°)	$G^*/\sin\delta$ (kPa)	High PG (°C)	
VB1	0	67	86	1.48	70.19	67	83	3.68	70.97	0.77
		76	86	0.49		73	85	1.69		
S01	3.42	67	86	1.35	69.47	67	82	3.19	70.17	0.70
		73	87	0.65		73	84	1.58		
S02	9.68	67	85	1.46	69.97	67	80	4.08	71.88	1.91
		73	87	0.68		73	83	1.91		
S03	6.19	67	85	1.62	70.96	67	82	4.25	72.10	1.14
		73	87	0.78		73	84	1.96		
S04	16.82	67	84	1.53	70.72	67	79	4.88	73.58	2.86
		73	86	0.77		73	82	2.36		
S05	4.93	67	85	1.46	70.55	67	80	3.79	71.32	0.77
		73	86	0.77		73	82	1.78		
S06	19.33	67	83	1.35	69.74	67	67	3.31	70.65	0.90
		73	85	0.70		73	73	1.69		
S07	10.48	67	84	1.61	70.87	67	79	3.80	71.92	1.04
		73	86	0.77		73	82	1.95		
S08	40.18	67	74	1.42	70.59	67	69	3.34	70.65	0.06
		73	75	0.79		73	72	1.68		
VB2	0	67	88	1.29	69.18	67	85	2.62	68.39	−0.79
		76	88	0.45		73	87	1.23		
S09	4.14	67	86	1.42	69.96	67	84	2.98	69.37	−0.59
		73	87	0.70		73	85	1.38		
S10	10.73	67	86	1.26	69.04	67	83	2.97	69.42	0.38
		73	87	0.63		73	84	1.41		
S11	7.53	67	86	1.34	69.36	67	84	2.92	69.33	−0.02
		73	88	0.64		73	86	1.41		
S12	17.18	67	85	1.43	70.07	67	81	3.42	70.63	0.56
		73	87	0.71		73	84	1.65		
S13	5.88	67	84	1.15	68.29	67	82	2.52	68.09	−0.20
		73	85	0.60		73	84	1.19		
S14	19.54	67	82	1.36	69.55	67	79	2.70	68.61	−0.94
		73	84	0.66		73	81	1.26		
S15	11.75	67	85	1.38	69.85	67	82	2.85	69.16	−0.69
		73	86	0.70		73	84	1.39		
S16	39.79	67	77	1.44	70.00	67	74	2.74	68.90	−1.10
		73	80	0.69		73	77	1.37		

corresponding to high PGs of 70°C and 95°C were obtained by interpolation or extrapolation of the results.

As a general trend, RA1 caused slower aging of the binders, while RA2 accelerated the aging. A meaningful

correlation existed between increasing the dosage of RA1 and the critical PAV time. Also, there was a reverse correlation between the dosage of RA2 and the critical PAV time (see Figure 5). The type of virgin binder also had a significant

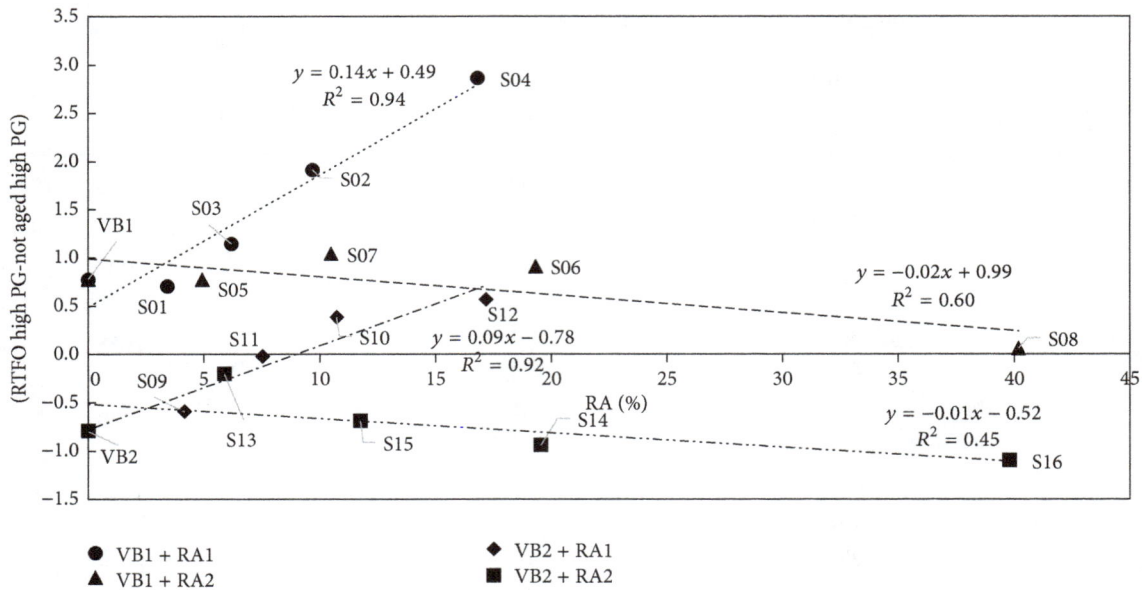

FIGURE 4: Change in the high PG by RTFO (RTFO PG, not aged PG).

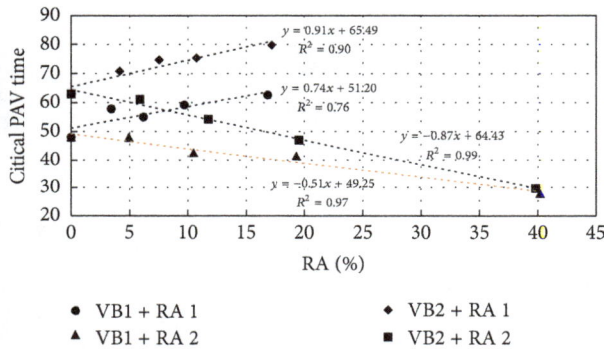

FIGURE 5: Variations of critical PAV time with rejuvenator content.

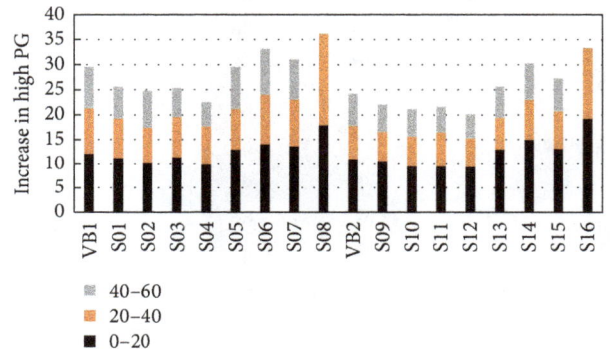

FIGURE 6: Rise of high PG in each stage of PAV aging.

influence on the rate of aging and the critical PAV time. VB1 aged faster than VB2, and its aging was more influenced by rejuvenators.

The rate of aging decreased with the increase in PAV time. The first cycles increased the high PG by an average of 12°C. This increase was, respectively, 9°C and 6°C for the second and third cycles. Figure 6 shows the increase in high PG for each sample in each stage of aging.

A 20-hour cycle of PAV simulates almost 8 years of field aging [17]. Therefore, to estimate pavement service life (the service time before excessive binder aging), every hour of PAV aging time was assumed to correspond to 0.4 years of field aging. Based on this assumption, the field longevity of binders was estimated, as illustrated in Figure 7. The right vertical axis in this figure indicates service life.

Samples 8 and 16, which contained large quantities of RA2, aged extremely fast. Their aging after 40 hours was more

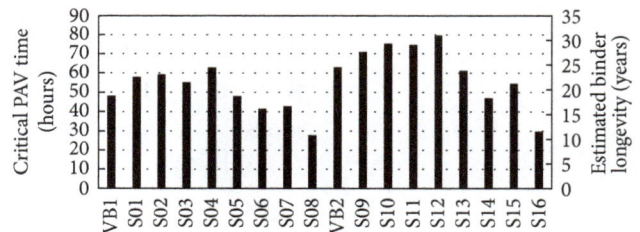

FIGURE 7: Critical PAV time (left axis) and estimated longevity (right axis) of samples.

than that of any other sample after 60 hours. Also, relatively small phase angles were obtained.

3.3. BBR Tests. The BBR test evaluates an asphalt binder's low-temperature cracking resistance. The stiffness is obtained by applying a point load on a small asphalt beam and

TABLE 4: DSR tests results for PAV-aged samples.

Sample	20-hour PAV				40-hour PAV				60-hour PAV			
	Temp. (°C)	δ (°)	$G^*/\sin\delta$ (kPa)	High PG (°C)	Temp. (°C)	δ (°)	$G^*/\sin\delta$ (kPa)	High PG (°C)	Temp. (°C)	δ (°)	$G^*/\sin\delta$ (kPa)	High PG (°C)
VB1	82	78	2.51	83.05	82	74	8.47	92.36	82	66	22.79	100.53
	88	81	1.18		88	78	3.88		88	70	10.69	
S01	76	78	4.31	81.39	82	76	5.35	89.51	82	69	13.79	95.78
	82	81	2.04		88	79	2.63		88	73	6.2	
S02	82	81	2.22	82.07	82	75	5.54	89.36	82	69	11.28	96.63
	88	83	1.03		88	78	2.61		88	73	5.77	
S03	82	80	2.60	83.43	82	73	7.00	91.78	82	69	14.82	97.46
	88	83	1.29		88	77	3.44		88	73	7.07	
S04	82	79	2.80	83.50	82	73	6.64	91.36	82	69	13.31	96.08
	88	82	1.36		88	77	3.27		88	73	6.18	
S05	82	79	2.84	84.28	82	71	8.64	92.59	82	64	23.20	100.87
	88	81	1.45		88	75	3.98		88	68	10.97	
S06	82	75	3.05	84.77	82	66	10.66	94.77	82	59	30.65	103.73
	88	78	1.50		88	70	5.08		88	62	14.81	
S07	82	77	3.40	85.58	82	68	11.25	95.07	82	62	26.36	102.99
	88	80	1.64		88	72	5.32		88	66	12.96	
S08	82	74	4.27	88.57	82	68	9.50	106.82		Invalid data		
	88	74	2.33		88	68	6.67					
VB2	76	82	3.65	79.86	82	81	3.82	86.76	82	74	9.67	93.12
	82	85	1.66		88	83	1.91		88	78	4.35	
S09	76	81	3.72	80.28	82	80	3.83	86.47	82	76	7.36	91.80
	82	84	1.78		88	82	1.82		88	80	3.51	
S10	76	81	3.20	78.89	82	80	3.19	85.20	82	75	6.68	90.57
	82	83	1.47		88	83	1.59		88	78	3.07	
S11	76	80	3.79	78.82	82	80	3.59	85.93	82	75	7.03	90.96
	82	83	1.19		88	81	1.70		88	79	3.23	
S12	76	79	3.56	79.97	82	78	3.55	86.03	82	75	6.98	90.71
	82	82	1.72		88	81	1.74		88	78	3.15	
S13	76	76	4.15	81.08	82	75	4.35	87.58	82	71	9.00	93.74
	82	80	1.96		88	79	2.09		88	75	4.38	
S14	82	74	2.71	83.70	82	67	7.57	91.86	82	61	17.34	98.95
	88	77	1.30		88	72	3.57		88	66	8.35	
S15	82	80	2.27	82.28	82	71	5.83	89.97	82	69	12.91	96.41
	88	82	1.15		88	76	2.80		88	72	6.18	
S16	82	66	4.20	87.30	82	62	8.58	101.49		Invalid data		
	88	69	2.02		88	63	5.64					

measuring the deflection at 8, 15, 30, 60, 120, and 240 seconds. The output of the BBR consists of two parameters:

(i) *Creep stiffness (S),* which is a measure of thermal stresses in the asphalt due to contraction.

(ii) The *m-value,* which is the slope of the creep stiffness master curve and indicates the ability of the asphalt to relieve stresses through plastic deformation.

The BBR test was performed on all samples after the standard (20 hours) and the ultimate (60 hours) PAV aging. The

results are presented in Table 6. The PG system specifies the following requirement at 60 seconds and at the temperature 10°C higher than the low-temperature specification. This is based on the time-temperature superposition principle that allows shortening the loading time by increasing the temperature. For a PG 67-22 binder, these requirements should be met at −12°C for a 20-hour PAV-aged residue.

$$S \leq 300\,\text{MPa} \tag{3}$$

$$m\text{-value} \geq 0.300. \tag{4}$$

TABLE 5: The increase in high PG of samples after each level of aging and resulting critical PAV time.

Sample	High PG (°C)					Increase in high PG (°C)			Critical PAV time (hours)
	Not aged	RTFO aged	20-hour PAV	40-hour PAV	60-hour PAV	0–20 hours	20–40 hours	40–60 hours	
VB1	70.19	70.97	83.05	92.36	100.53	12.08	9.31	8.17	48.07
S01	69.47	70.17	81.39	89.51	95.78	11.22	8.11	6.27	57.83
S02	69.97	71.88	82.07	89.36	96.63	10.19	7.29	7.27	59.21
S03	70.96	72.10	83.43	91.78	97.46	11.33	8.35	5.69	55.04
S04	70.72	73.58	83.50	91.36	96.08	9.92	7.86	4.72	62.65
S05	70.55	71.32	84.28	92.59	100.87	12.96	8.31	8.28	47.86
S06	69.74	70.65	84.77	94.77	103.73	14.12	10.00	8.96	41.42
S07	70.87	71.92	85.58	95.07	102.99	13.67	9.49	7.91	42.62
S08	70.59	70.65	88.57	106.32	—	17.92	18.25	—	27.77
VB2	69.18	68.39	79.86	86.76	93.12	11.47	6.91	6.36	62.97
S09	69.96	69.37	80.28	86.47	91.80	10.91	6.19	5.33	70.84
S10	69.04	69.42	78.89	85.20	90.57	9.47	6.32	5.37	75.29
S11	69.36	69.33	78.82	85.93	90.96	9.48	7.11	5.03	74.66
S12	70.07	70.63	79.97	86.03	90.71	9.34	6.06	5.72	79.68
S13	68.29	68.09	81.08	87.58	93.74	12.99	6.50	6.16	61.16
S14	69.55	68.61	83.70	91.86	98.95	15.09	8.16	7.09	47.01
S15	69.85	69.16	82.28	89.97	96.41	13.11	7.69	6.44	54.32
S16	70.00	68.90	87.30	101.49	—	18.40	14.19	—	29.66

For all samples, the m-value criterion was more critical and dominated the determination of the low temperate PG. Virgin binders did not meet the m-value requirement for PG 67-22, but they were very close to it (0.299 for VB1 and 0.291 for VB2). VB1 had a better low-temperature performance, compared to VB2. It had a smaller creep stiffness and a higher m-value, despite VB1's higher stiffness at high-temperature and greater high-temperature PG.

The samples with RA1 passed the criteria for PG 67-22. Samples with RA2, on the other hand, did not meet these criteria and yielded lower m-values. Generally, the addition of RA1 did not significantly change the creep stiffness. The RA2, however, caused a fast drop in the stiffness. The higher the dosage of RA2, the smaller the values of creep stiffness (Figure 8). A smaller amount of low-temperature creep stiffness shows that less thermal stresses are expected. However, the very small stiffness found in the samples that contain a large dosage of RA2 is an indication of the detrimental behavior of RA2 when applied at a large dosage.

The RA1 increased the m-value, and the RA2 decreased it. Therefore, samples with RA1 had a lower low-temperature PG. This effect was more significant when the RA content was higher (Figure 8). A higher m-value shows a binder with a more viscous behavior and a greater ability to relieve stresses. The less viscous behavior of samples containing RA2 is in line with the observations from DSR tests where these samples had lower phase angles.

Similar to the DSR experiment, samples 8 and 16 did not output valid data. The samples were very soft, and they broke under the BBR load at −6°C. Also, their results at −12° and −18°C yielded critical temperature values that were out of acceptable ranges.

Applying the extended PAV aging (60 hours) increased the creep stiffness and decreased the m-value. The change in the m-value was more significant. While the creep stiffness critical temperature increased by 3°C on average, the average rise in the m-value critical temperature was 9°C. Since the 60-hour aged samples were excessively hard, it was difficult to pour the BBR mold with these samples. Therefore, they were heated to 175°C for 10 minutes to achieve the required fluidness.

Unlike DSR tests, BBR tests on samples with extended PAV aging did not add important information about the effects of using RAP and rejuvenation. Therefore, performing BBR tests on samples with standard aging is adequate for durability evaluation.

4. Conclusion

This research evaluated the durability of recycled asphalt binders. For this purpose, two virgin binders and 16 samples containing RAP binder and rejuvenator were aged at four stages: one RTFO and three PAV cycles. The samples were different in the type of RAP binder, virgin binder, and recycling agent. The RAP contents of 20% and 40% were

TABLE 6: BBR test for 20-hour PAV-aged samples.

Sample	RA%	Temp. (°C)	20-hour PAV			60-hour PAV			Low temp. PG increase (°C) 20 to 60 hours
			S	m-value	Low temp. PG (°C)	S	m-value	Low temp. PG (°C)	
VB1	0	−6	161	0.362	−21.90	147	0.282	−14.48	7.43
		−12	205	0.299		235	0.211		
		−18	391	0.256		448	0.178		
1	3.42	−6	109	0.370	−23.37	159	0.292	−14.67	8.70
		−12	191	0.313		281	0.256		
		−18	367	0.267		465	0.230		
2	9.68	−6	78.7	0.377	−24.00	135	0.298	−15.74	8.26
		−12	173	0.319		248	0.251		
		−18	332	0.262		480	0.182		
3	6.19	−6	116	0.358	−23.10	152	0.297	−15.54	7.56
		−12	203	0.309		273	0.258		
		−18	388	0.253		468	0.211		
4	16.82	−6	90.4	0.364	−23.68	149	0.301	−16.17	7.51
		−12	194	0.314		269	0.266		
		−18	404	0.251		461	0.232		
5	4.93	−6	74.4	0.345	−20.50	126	0.273	−12.40	8.10
		−12	153	0.285		193	0.228		
		−18	282	0.255		343	0.190		
6	19.33	−6	39.6	0.340	−19.29	94.9	0.270	−11.00	8.29
		−12	91	0.267		144	0.234		
		−18	137	0.248		212	0.202		
7	10.48	−6	54.2	0.321	−18.57	108	0.268	−11.32	7.25
		−12	114	0.272		161	0.227		
		−18	193	0.246		282	0.194		
8	40.18	−6	Invalid	Invalid	Invalid	Invalid	Invalid	Invalid	—
		−12	34.1	0.241		39.4	0.245		
		−18	55.1	0.231		61.3	0.230		
VB2	0	−6	149	0.352	−21.11	184	0.276	−13.64	7.48
		−12	301	0.291		341	0.215		
		−18	454	0.262		487	0.183		
9	4.14	−6	112	0.360	−22.62	182	0.294	−14.80	7.82
		−12	215	0.303		319	0.264		
		−18	393	0.274		455	0.232		
10	10.73	−6	83.6	0.365	−24.84	144	0.305	−16.86	7.98
		−12	180	0.327		267	0.270		
		−18	365	0.270		445	0.239		
11	7.53	−6	98.4	0.351	−23.96	173	0.298	−15.48	8.48
		−12	181	0.317		307	0.275		
		−18	387	0.265		472	0.248		
12	17.18	−6	85.4	0.373	−25.63	137	0.312	−17.89	7.73
		−12	175	0.329		261	0.274		
		−18	370	0.281		470	0.246		

TABLE 6: Continued.

Sample	RA%	Temp. (°C)	20-hour PAV			60-hour PAV		Low temp. PG (°C)	Low temp. PG increase (°C) 20 to 60 hours
			S	m-value	Low temp. PG (°C)	S	m-value		
13	5.88	−6	76.3	0.346	−20.68	129	0.259	−10.98	9.70
		−12	169	0.287		264	0.210		
		−18	311	0.268		358	0.191		
14	19.54	−6	36.6	0.345	−21.19	77.9	0.256	−4.00	17.19
		−12	74.8	0.293		126	0.234		
		−18	135	0.260		185	0.214		
15	11.75	−6	51.2	0.329	−19.16	81.2	0.261	−9.31	9.85
		−12	115	0.274		193	0.226		
		−18	189	0.235		277	0.197		
16	39.79	−6	Invalid	NA	Invalid	45.8	0.238	Invalid	Invalid
		−12	37.5	0.271		71.9	0.219		
		−18	47.5	0.266		107	0.200		

used. The samples were prepared so that their initial high PG was similar to that of the virgin binder they contained. After each stage of aging, DSR tests were performed, and the high-temperature PG was determined. The following is a summary of findings obtained from the described experiment:

(1) The rejuvenators have a significant effect on the aging rate of the binder. A recycled binder can age either faster or slower than a virgin binder, depending on the rejuvenator that is used. In this experiment, RA1 caused slower aging, and RA2 caused faster aging. The higher the percentage of recycling agent, the greater its effect on the aging of the binder.

(2) The type and amount of rejuvenator can considerably affect the longevity of the binder. The use of a fast-aging rejuvenator can reduce the life of the binder to less than one-half. Conversely, a slow aging rejuvenator can increase the life of the binder by up to 30%.

(3) The extent of construction aging, which was simulated by the RTFO, is affected by the rejuvenator. A recycled binder containing a rejuvenator with a higher aromatic content is expected to undergo more aging due to construction heating.

(4) The effectiveness of RA2 for rejuvenating high RAP mixtures is questionable. This rejuvenator has relatively low softening power. Therefore, a large quantity of it is required to soften a binder with a high RAP content. In addition, binders containing a high volume of this recycling agent age very quickly and have a short life span before they become extensively aged again.

(5) RA1 has desirable properties for recycling high RAP mixtures. It has a high softening power, and a relatively small quantity is enough to rejuvenate a highly aged binder. Also, it has an advantageous aging behavior. It makes the binder age faster during construction and gives extra strength to the pavement immediately after construction when the strength is most needed. Afterwards, it decelerates aging and gives the binder a longer life span before excessive aging.

(6) The low-temperature behavior of recycled binders is significantly affected by the type and dosage of the rejuvenator. In this experiment, samples with RA1 had higher m-values, indicating their greater ability to relieve stresses. As a result, despite the virgin binder not passing the low-temperature criteria for PG 67-22, all RA1 samples did pass. Samples with the RA2, on the other hand, had smaller m-values. This correlates with the smaller viscous portion of the complex modulus, which was observed for RA2 samples in high-temperature DSR tests. It can be concluded from both DSR and BBR tests that RA2 causes a reduction in the viscous behavior of the binder.

(7) It is necessary to differentiate between rejuvenators that reduce the longevity of the binder and those that increase it. To achieve this, a quantitative description of durability is needed. Critical PAV time can serve as a measure of the longevity of the binder. Using this parameter to set a durability criterion for rejuvenation effectiveness is recommended.

Conflicts of Interest

The authors declare that there are no conflicts of interest regarding the publication of this paper.

Acknowledgments

This research was sponsored by the Florida Department of Transportation (FDOT). The authors express appreciation to the department for funding this effort. Special thanks are

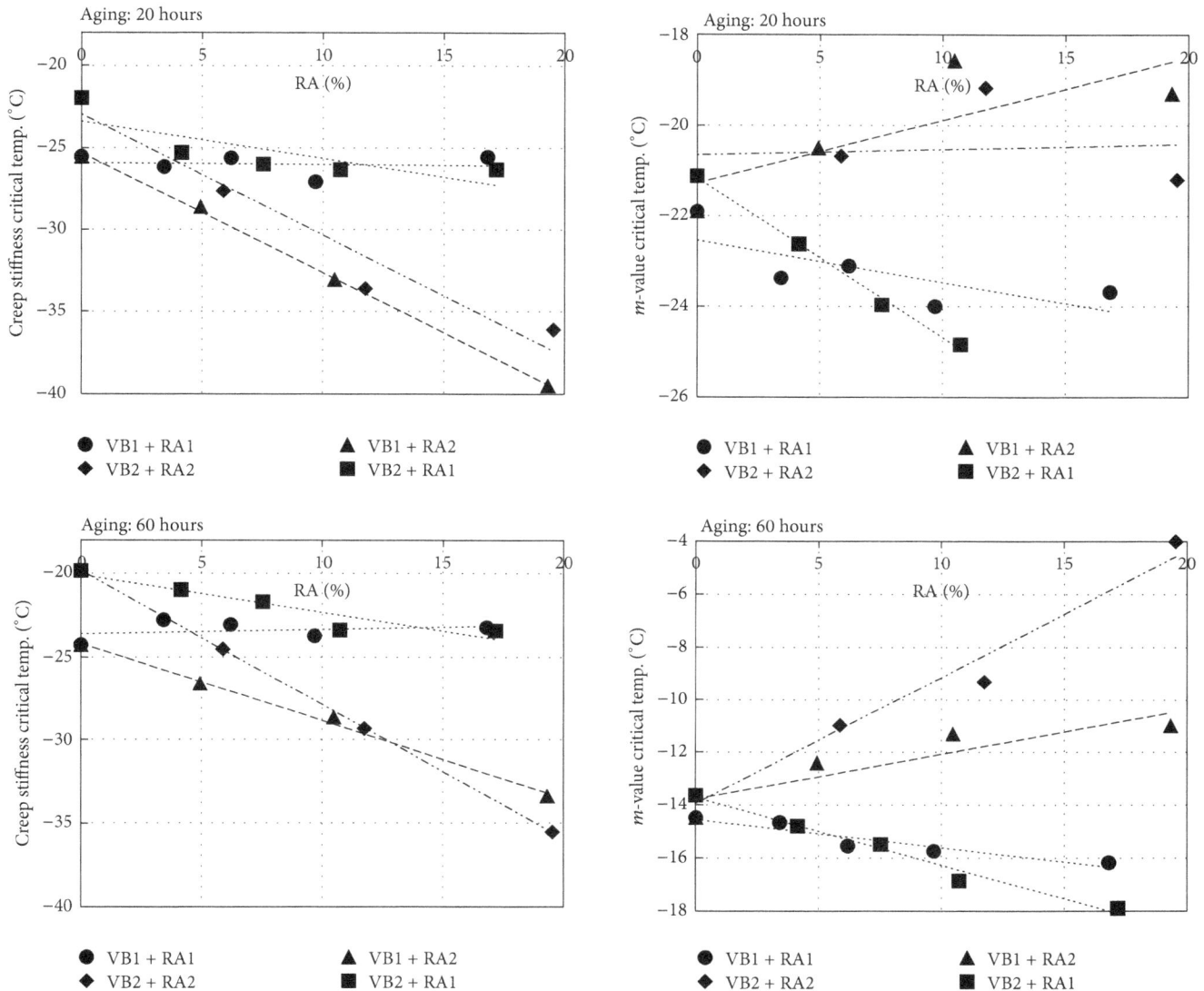

FIGURE 8: Variations of creep stiffness and *m*-value critical temperatures with the rejuvenator content.

due to the State Material Office, Bituminous Section, for its support and invaluable technical contribution.

References

[1] R. B. Mallick, K. A. O'Sullivan, M. Tao, and R. Frank, "Why not use rejuvenator for 100% RAP recycling?" in *Proceedings of the Ransportation Research Board 89Th Annual Meeting*, Washington DC, USA, 2010.

[2] J. R. Willis and M. Marasteanu, "Improved mix design, evaluation, and materials management practices for hot mix asphalt with high reclaimed asphalt pavement content," Tech. Rep. NCHRP Report 752, Transportation Research Board of the National Academies, Washington DC, Wash, USA, 2013.

[3] H. Haghshenas, H. Nabizadeh, Y. Kim, and K. Santosh, "Research on high-rap asphalt mixtures with rejuvenators and WMA additives," Tech. Rep., Department of Roads Research Reports, Lincoln, Neb, USA, 2016.

[4] A. Copeland, Reclaimed Asphalt Pavement in Asphalt Mixtures: State of Practice," McLean, VA, 2011.

[5] E. Rahmani, M. K. Darabi, D. N. Little, and E. A. Masad, "Constitutive modeling of coupled aging-viscoelastic response of asphalt concrete," *Construction and Building Materials*, vol. 131, pp. 1–15, 2017.

[6] M. Z. Alavi, E. Y. Hajj, and P. E. Sebaaly, "Significance of oxidative aging on the thermal cracking predictions in asphalt concrete pavements," in *8th RILEM International Conference on Mechanisms of Cracking and Debonding in Pavements*, vol. 13, 2016.

[7] M. Hasaninia and F. Haddadi, "The characteristics of hot mixed asphalt modified by nanosilica," *Petroleum Science and Technology*, vol. 35, no. 4, pp. 351–359, 2017.

[8] C. Purdy, Y. Mehta, A. Nolan, and A. Ali, "Methodology to determine optimum rejuvenator dosage for 50 percent high-rap mixture," in *Transportation Research Board 96th Annual Meeting*.

[9] M. Zaumanis, R. B. Mallick, and R. Frank, "Determining optimum rejuvenator dose for asphalt recycling based on Superpave performance grade specifications," *Construction and Building Materials*, vol. 69, pp. 155–166, 2014.

[10] R. L. Terrel and D. R. Fritchen, "Laboratory performance of recycled asphalt concrete," in *Recycling of Bituminous Pavements*, pp. 104–122, 1978.

[11] D. Singh, M. Zaman, and S. Commuri, "A laboratory investigation into the effect of long-term oven aging on RAP mixes using dynamic modulus test," *International Journal of Pavement Research and Technology*, vol. 5, no. 3, pp. 142–152, 2012.

[12] A. W. Ali, Y. A. Mehta, A. Nolan, C. Purdy, and T. Bennert, "Investigation of the impacts of aging and RAP percentages on effectiveness of asphalt binder rejuvenators," *Construction and Building Materials*, vol. 110, pp. 211–217, 2016.

[13] M. Baqersad, A. Hamedi, M. Mohammadafzali, and H. Ali, "Asphalt mixture segregation detection: digital image processing approach," *Advances in Materials Science and Engineering*, vol. 2017, 6 pages, 2017.

[14] A. Ongel and M. Hugener, "Impact of rejuvenators on aging properties of bitumen," *Construction and Building Materials*, vol. 94, pp. 467–474, 2015.

[15] M. Mohammadafzali, H. Ali, J. A. Musselman, G. A. Sholar, S. Kim, and T. Nash, "Long-term aging of recycled asphalt binders: a laboratory evaluation based on performance grade tests," in *Airfield and Highway Pavements 2015*, Miami, Fla, USA.

[16] H. Ali and M. Mohammadafzali, "Long-term aging of recycled binders," Project number BDV29 Two 977-01, Florida Department of Transportation, 2015.

[17] U. Bahia Hussain and A. David, "The Pressure aging vessel (pav): a test to simulate rheological changes due to field aging," in *ASTM Special Technical Publication 1241*, pp. 67–88, 1995.

20

Asphalt Mixture Segregation Detection: Digital Image Processing Approach

Mohamadtaqi Baqersad, Amirmasoud Hamedi,
Mojtaba Mohammadafzali, and Hesham Ali

Department of Civil and Environmental Engineering, Florida International University, 10555 West Flagler Street, Miami, FL 33174, USA

Correspondence should be addressed to Mohamadtaqi Baqersad; mbaqe001@fiu.edu

Academic Editor: Luigi Nicolais

Segregation determination in the asphalt pavement is an issue causing many disputes between agencies and contractors. The visual inspection method has commonly been used to determine pavement texture and in-place core density test used for verification. Furthermore, laser-based devices, such as the Florida Texture Meter (FTM) and the Circular Track Meter (CTM), have recently been developed to evaluate the asphalt mixture texture. In this study, an innovative digital image processing approach is used to determine pavement segregation. In this procedure, the standard deviation of the grayscale image frequency histogram is used to determine segregated regions. Linear Discriminate Analysis (LDA) is then implemented on the obtained standard deviations from image processing to classify pavements into the segregated and nonsegregated areas. The visual inspection method is utilized to verify this method. The results have demonstrated that this new method is a robust tool to determine segregated areas in newly paved FC9.5 pavement types.

1. Introduction

Characteristic of asphalt material impacts on asphalt quality and driving safety [1]. Segregation of aggregates alters material properties and accelerates deterioration's rate [2, 3]. Pavement segregation is one of the main concerns affecting the performance of asphalt pavements. Segregation is defined as separation of aggregates gradation, so that coarse and fine aggregates are separated in the asphalt mixture [4]. Also, existence of enough bituminous and stickiness between asphalt materials may reduce deterioration of asphalt concrete [5]. Detection of segregated areas in pavements has always been a disputable issue between agencies and contractors. Segregated areas are primarily detected by the visual inspection and verified by core density [4]. Segregation can also be determined by measuring the surface texture of the pavement. Florida Texture Meter (FTM) and Circular Track Meter (CTM) are laser-based devices that can be used for this purpose [6, 7].

Image processing is another tool that is used to qualify pavement texture [8]. Scanners and cameras are two types of digitalizing devices to produce a 3D image model [9]. Digitalizing can be used to measure different size and scale objects [10]. Blais (2004) developed a scanning machine, which captures 3D data from the surface, and then categorized the road texture using two algorithms [10]. The first algorithm calculates the estimated texture depth (ETD), and the second calculates the texture profile level (TPL). The measurements from nine pavements were collected to validate these algorithms. The results show that there is a good agreement between the traditional Sand Patch method results and the 3D scanning prototype equipment [11].

The feasibility of using image processing to determine the gradation without separation of bitumen and aggregates was studied by Bruno et al. [12]. The image, at first, subdivided into different distinct areas where each distinct area has homogeneity with the area around it. This procedure helps to separate voids, aggregates, and bitumen areas. Frequency histogram of each distinct area has been used to detect the gradation of the asphalt mixture. Dividing the image into distinct regions can also be used to detect pavement distresses [13]. In this case, the transform method to designate lines and pixels is used to define the distresses in each distinct area. The Discriminant Analysis, *K*-Nearest Neighbor, and Discrete Choice Method have also been used to classify each area.

More recently, the concrete pavement texture has been evaluated through digital image processing [14]. Visual inspection has been used to verify this method. This validation approved that the flatbed scanner can be used to measure the pavement texture accurately. The scanner was used to take digital images of the surface texture. Simplicity to control the light source is the advantage of scanner application because the source of light, during the photography, has an effect on the color frequency. The colored digital image that has been produced by scanning the pavement surface was later converted to a grayscale image to generate a grayscale image frequency histogram. The pavement texture, then, has been classified based on the standard deviation of the frequency histogram. Bug-hole was argued as the major parameters that affect the frequency histogram standard deviation. In other words, pavement surface with more bug-holes would show higher standard deviation in comparison with surface textures with fewer bug-holes.

Furthermore, Chen et al. performed an investigation on the determination of pavement texture and pavement Mean Profile Depth (MPD) using image processing [15]. Pavement MPD was determined using the scanned pavement cross-sections. Stationary Laser Profilometer (SLP) was then used to verify asphalt texture detected with discrete Fourier transform method. This technique was shown to be useful to design pavement when texture noise and friction are important.

Although image processing techniques have frequently been used to measure pavement texture in previous studies, little or no work has used image processing to determine segregation. The method introduced in this research is a non-destructive tool and potentially can be used as a smartphone application to detect asphalt pavement segregation in place.

1.1. Digitalization. Pictures consist of divided elements named pixel. Each pixel has some information which shows color at that particular point. In other words, the color in each part of the image is represented by a pixel. Pixels are stored in bits which can be 0 or 1. The intensity of pixels in an image shows the quality of that image. The more pixels in the image, the higher quality and resolution it has [16].

Photos can be either colored or black and white. While colored images have a color level of pixels, black and white pictures have a gray level. The gray level can vary from 0 to 255 resulting in 256 color spectra. The basic colors to produce a color image are red, green, and blue (RGB) which form 16.7 million spectra [17]. It is possible to add some other colors to the basic colors of a colored image. For example, CMYK color image which contains cyan, magenta, yellow, and black is another kind of color image which has 4.3 billion spectra colors. Extracting this numerical information from an image is named image digitization [18].

Using a black and white image to extract numerical information from the image is easier than the colored image because colored images have three layers (one layer for each basic color) and millions and billions of spectra; therefore, categorizing information extraction from a color image is difficult. On the other hand, using black and white color image is preferable because it has 256 spectra colors. The image frequency histogram can be applied to extract numerical information from any picture.

2. Segregation Detection Methodology

In this study, the frequency histogram was used to conduct digital image processing. Frequency histogram of an image is a graph which uses vertical columns to show the number of repetitions of each spectrum in an image [18]. When the asphalt mixture frequency histogram is used, it is expected that nonsegregated area's pictures have uniform colors. In other words, it is supposed that most of the colors in nonsegregated asphalt image have gray colors. In contrast, in segregated pavement surface, the separation of bitumen and aggregates or fine and coarse aggregates may occur. This separation can cause darker or lighter color intensity in segregated areas of the image which can lead to a higher standard deviation of segregated pavement images in comparison with nonsegregated pavements images. In the following, this hypothesis will be investigated by using a MATLAB program which defines the frequency histogram standard deviation. Then, these results will be classified through Linear Discriminate Analysis (LDA).

LDA is a useful tool to classify and predict the group's membership based on the linear distinction between measured factors. This theory first was developed by R. A. Fisher in 1936. The robustness of this theory is defining the probability of an expected portion of data belonging to each group. However, this method demonstrates the threshold and membership of each group almost the same as the Analysis of Variance (ANOVA) [19].

In ANOVA means of groups are verified to be the same; so ANOVA performs an eigendecomposition and finds its eigenvalues. In contrast, LDA performs eigendecomposition and looks at the eigenvectors. These eigenvectors define directions in the variable space and are called discriminant axes. Therefore, the advantage of implementing LDA is in predicting the membership of each group in generating a big gap between groups [20].

The concept of LDA is searching for a linear combination of variables that best separates two classes. In order to define each group member, it first defines a linear equation between variables, which determines a Discriminate Factor (DF). Then, the data is classified based on the calculated DF and the cutoff score. The cutoff score defines the threshold between two groups of data.

The LDA classification method has been used to predict each group's membership and classify the segregated and nonsegregated pavements. In this case, the standard deviation of each picture was determined and classified. Two scenarios were considered for the LDA: (1) each site location standard deviation result was classified and the threshold line was defined separately and (2) the standard deviation results from both locations were analyzed together and the threshold line was designated. These two scenarios were considered

FIGURE 1: Image capturing table.

to identify if there is the same threshold for separating the segregated asphalt and nonsegregated asphalt for all the test sites. To perform LDA analysis according to the considered scenarios, the SPSS software was used.

3. Experiments

Two sites in Florida that were recently paved with FC9.5 asphalt mixture were chosen for this study because FC9.5 is a commonly paved asphalt mixture type in Florida [21, 22]. FC9.5 is a Superpave mixture with the nominal maximum aggregate size of 9.5 millimeters [23]. In this research, newly paved pavements were selected to eliminate the effect of pavement aging and contamination on results. Experts from the Florida Department of Transportation (FDOT) designated the segregated and nonsegregated areas on the pavement by the visual inspection method. Also, the pictures were evaluated by the other three FDOT experts. The average of these experts inspection was considered for the spot locations where the inspectors' ideas did not match each other. Fifteen nonsegregated and five segregated points were selected at each test site. Since the type and resolution of the camera may affect the results, all the pictures were taken by the same camera and the same operator (same quality of the pictures) from the same elevation. The setup shown in Figure 1 was used to capture the pictures. In the center of the table, there is a hole where the camera was placed to take the picture.

To take pictures, the pavement surface was inspected, and clean and dry surfaces were selected. Field surfaces were free of loose particles and the surfaces were cleaned of the residue by brush. Then, the segregated and nonsegregated areas were chosen and marked within the same asphalt lane. The table was placed exactly on the marked surface areas. The pavement surface was flat and the size of the table was approximately 45 ∗ 45 centimeter, which was small enough that the pavement slope had no effect on the pictures. Then, the camera was placed on the table and the picture was taken in such a way that it was oriented with the direction of the pavement.

As mentioned before, the light source is an important factor and may affect frequency histogram. In this study, all the pictures were taken at the same time, so that the light was same for all photos and also the pictures were free of shadows. To control this problem, the standard deviation of frequency histogram is used to detect segregation. In fact, the ambient light may affect the whole picture. The ambient light source can impact on the mean of frequency histogram, but it cannot affect the standard deviation of frequency histogram [24]. Thus, the standard deviation of frequency histogram is used to detect segregation areas in asphalt pavement.

4. Results and Discussion

A MATLAB code was used to find image frequency histogram. The program imports the colored picture, converts it to a grayscale image, and obtains the frequency histogram. Afterward, the frequency histogram standard deviation is calculated as a symbol representing the pavement surface texture to be used to detect pavement segregation. Figure 2 represents the frequency histogram of the asphalt grayscale image which is the conversion of color asphalt image. All the pictures follow this trend. Figure 3 shows the standard deviation of segregated and nonsegregated frequency histogram in both site locations, respectively.

As shown in Figure 3, nonsegregated areas have a lower standard deviation in comparison with segregated areas. This trend makes it possible to define a threshold between pictures indicating segregated and nonsegregated areas. This difference between segregated and nonsegregated frequency histogram standard deviation can be caused by the closer texture of nonsegregated pavements in comparison with the segregated pavement. In other words, as shown in Figure 4, when segregation occurs, on one hand, coarse aggregates may separate from fine aggregates [25]; therefore, aggregates in segregated locations may not have enough interlock to each other which can lead to an increase of voids between the aggregates in pavement texture [26]. This rise in voids in the pavement surface can cause brighter or darker colors, which can, in turn, result in higher standard deviation of segregated areas.

The standard deviation of both site locations is represented together in Figure 5. The threshold line between segregated and nonsegregated areas is not clearly obvious as shown in Figure 3. Due to this fact, the LDA classification method has been used to classify the segregated and nonsegregated pavements.

The LDA analysis was conducted using SPSS software. The results and prediction of each group membership are represented in Tables 1 and 2, respectively. Table 3 presents the membership prediction of both projects based on standard deviation of two projects using LDA method.

As shown in Tables 1 and 2, the threshold line which separates segregated and nonsegregated pavements is different. But, the overall estimation of segregated and nonsegregated group's membership in each site location is high enough. In other words, by comparison of Tables 1 and 2 with Table 3, the overall estimation decreased in Table 3, because each group has a specific threshold line to separate segregated and nonsegregated pavements which are different from the other. Therefore, the threshold which can separate segregated and

(a)

(b)

(c)

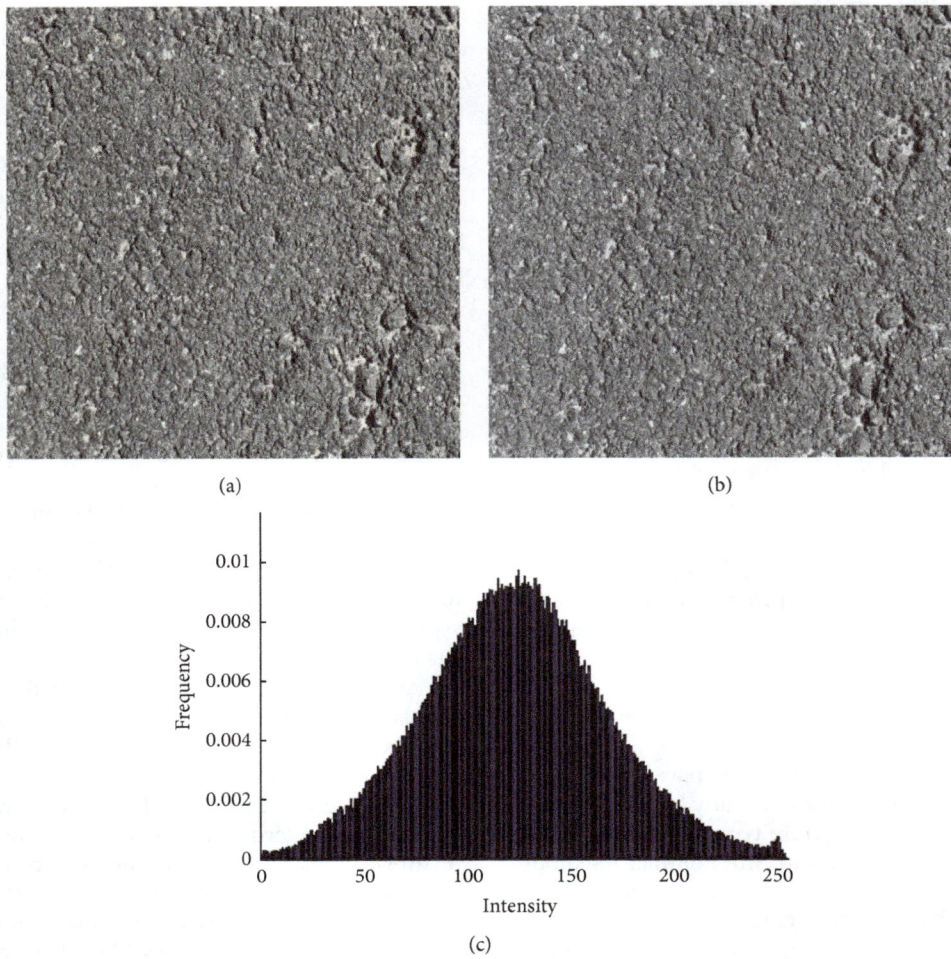

FIGURE 2: (a) Color asphalt Image, (b) grayscale asphalt image, and (c) frequency histogram of grayscale image.

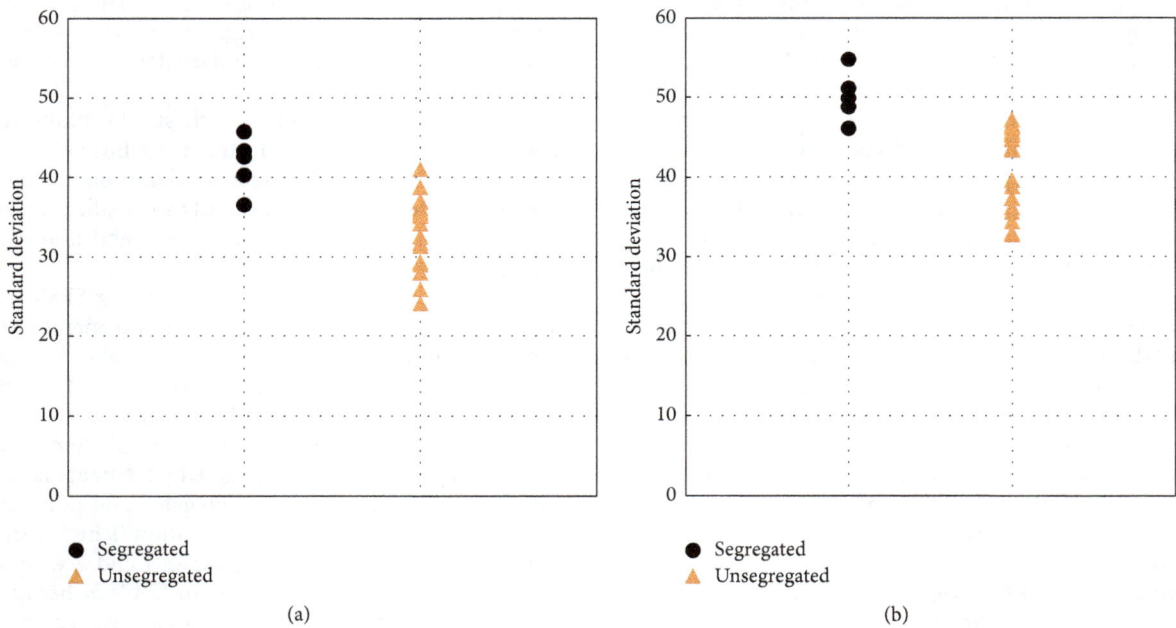

(a)

(b)

FIGURE 3: Standard deviation of frequency histogram in (a) site location 1 and (b) site location 2.

FIGURE 4: (a) Nonsegregated asphalt surface image and (b) segregated asphalt surface image.

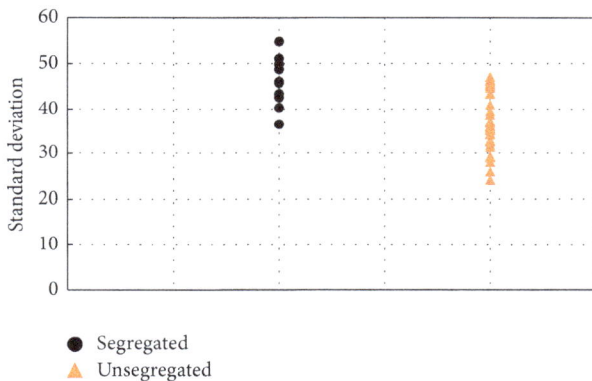

FIGURE 5: Frequency histogram standard deviation of both sites together.

TABLE 1: Classification results[a,b] for site location 1.

Standard deviation	Predict group membership		Total
	Segregated	Nonsegregated	
Segregated%	80	20	100
Nonsegregated%	7.7	93.3	100

[a]90.0% of original grouped cases correctly classified.
[b]Threshold line is in standard deviation of 40.

TABLE 2: Classification results[a,b] for site location 2.

Standard deviation	Predict group membership		Total
	Segregated	Nonsegregated	
Segregated%	100	0	100
Nonsegregated%	26.7	73.3	100

[a]80.0% of original grouped cases correctly classified.
[b]Threshold line is in standard deviation of 46.

TABLE 3: Classification results[a,b] for both sites together.

Standard deviation	Predict group membership		Total
	Segregated	Nonsegregated	
Segregated%	80	20	100
Nonsegregated%	20.7	73.3	100

[a]75.0% of original grouped cases correctly classified.
[b]Threshold line is in standard deviation of 41.

nonsegregated pavements is site-dependent. For each site, first the threshold line and membership prediction analysis should be conducted, and the results then can be used to determine segregated and nonsegregated pavement areas. In this research, for these two projects, it is obvious that the overall accuracy to determine the pavement segregation in each site location is beyond 80%.

5. Conclusion

A new image processing-based method to detect segregation in asphalt pavements was introduced. In the proposed method, the frequency histogram standard deviation of the grayscale image was used to identify segregated asphalt mixture pavement areas. The LDA was then performed to classify segregated and nonsegregated image processing results. The visual inspection has been applied by FDOT experts to verify this new method's result.

This new method, which is nondestructive, can provide an alternative and decent solution to overcome disputes between contractors and agencies on the determination of segregated areas. Results from this work showed that images from segregated pavement areas have a higher standard deviation in comparison with those from nonsegregated areas. This outcome is due to the voids increase in segregated asphalt surface which can lead to brighter colors in pavement images. It is indicated that, in nonsegregated area's image, the color distribution is uniform and spectra include gray level colors while segregated area's image has brighter or darker colors. To apply these results to categorize segregated areas, first, the threshold line between segregated and nonsegregated pavements should be identified through LDA. The results then can be used to define segregated areas. For two studied projects in this research, LDA results show over 80% accuracy for detecting segregated areas.

Disclosure

The work represented herein was the result of a team effort. An earlier version of this work was presented as a poster at the 96th Annual Meeting of the Transportation Research Board 2017.

Conflicts of Interest

The authors declare that they have no conflicts of interest.

Acknowledgments

The authors would like to acknowledge FDOT's State Materials Office, District 2 Materials Office, and District 4 Materials Office for their assistance with the data collection effort and technical advice.

References

[1] N. Nabiun and M. M. Khabiri, "Mechanical and moisture susceptibility properties of HMA containing ferrite for their use in magnetic asphalt," *Construction and Building Materials*, vol. 113, pp. 691–697, 2016.

[2] J. Sadeghi and M. Fesharaki, "Importance of nonlinearity of track support system in modeling of railway track dynamics," *International Journal of Structural Stability and Dynamics*, vol. 13, no. 1, Article ID 1350008, 2013.

[3] M. Baqersad, A. E. Haghighat, M. Rowshanzamir, and H. M. Bak, "Comparison of coupled and uncoupled consolidation equations using finite element method in plane-strain condition," *Civil Engineering Journal*, vol. 2, no. 8, pp. 375–388, 2016.

[4] M. Stroup-Gardiner and E. R. Brown, "Segregation in asphalt mixture pavements," NCHRP Report 441, Transportation Research Board, National Research Council, Washington, DC, USA, 2000.

[5] M. Moravej, B. Hajra, P. Irwin, I. Zisis, and A. G. Chowdhury, "An experimental investigation on the effects of building height on velocity coefficients and local wind pressure on building Roofs," in *Proceedings of the AAWE Workshop*, Miami, Fla, USA, 2016.

[6] H. S. Lee, P. Upshaw, C. Holzschuher, B. Choubane, and T. Ruelke, "Detection of asphalt concrete segregation using laser texturemeters," in *Proceedings of the Transportation Research Board 93rd Annual Meeting*, no. 14-1988, 2014.

[7] D. I. Hanson and B. D. Prowell, "Evaluation of circular texture meter for measuring surface texture of pavements," NCAT Report 04-05, National Center for Asphalt Technology, 2004.

[8] R. Elunai, V. Chandran, and P. Mabukwa, "Digital image processing techniques for pavement macro-texture analysis," in *Proceedings of the 24th ARRB Conference: Building on 50 Years of Road Transport Research*, no. 0572-1, pp. 1–5, ARRB Group Ltd, 2010.

[9] G. Sansoni, M. Trebeschi, and F. Docchio, "State-of-the-art and applications of 3D imaging sensors in industry, cultural heritage, medicine, and criminal investigation," *Sensors*, vol. 9, no. 1, pp. 568–601, 2009.

[10] F. Blais, "Review of 20 years of range sensor development," *Journal of Electronic Imaging*, vol. 13, no. 1, pp. 231–243, 2004.

[11] J. L. Vilaça, J. C. Fonseca, A. C. M. Pinho, and E. Freitas, "3D surface profile equipment for the characterization of the pavement texture—TexScan," *Mechatronics*, vol. 20, no. 6, pp. 674–685, 2010.

[12] L. Bruno, G. Parla, and C. Celauro, "Image analysis for detecting aggregate gradation in asphalt mixture from planar images," *Construction and Building Materials*, vol. 28, no. 1, pp. 21–30, 2012.

[13] H. N. Koutsopoulos, V. I. Kapotis, and A. B. Downey, "Improved methods for classification of pavement distress images," *Transportation Research Part C: Emerging Technologies*, vol. 2, no. 1, pp. 19–33, 1994.

[14] A. L. de Oliveira and L. R. Prudêncio Jr., "Evaluation of the superficial texture of concrete pavers using digital image processing," *Journal of Construction Engineering and Management*, vol. 141, no. 10, Article ID 04015034, 2015.

[15] D. Chen, N. Roohi Sefidmazgi, and H. Bahia, "Exploring the feasibility of evaluating asphalt pavement surface macro-texture using image-based texture analysis method," *Road Materials and Pavement Design*, vol. 16, no. 2, pp. 405–420, 2015.

[16] J. Sachs, *Digital Image Basics*, Digital Light & Color, 1996–1999.

[17] K. R. Castleman, *Digital Imaging Processing*, Prentice Hall, 1996.

[18] B. Jähne, *Digital Image Processing*, Springer, 2005.

[19] M. Welling, *Fisher Linear Discriminant Analysis*, vol. 3, Department of Computer Science, University of Toronto, 2005.

[20] R. Johnson and D. Wichern, *Applied Multivariate Statistical Methods*, Prentice Hall, Englewood Cliffs, NJ, USA, 3rd edition, 1992.

[21] H. Ali and M. Mohammadafzali, *Asphalt Surface Treatment Practice in Southeastern United States*, Louisiana Transportation Research Center, 2014.

[22] A. Massahi, H. Ali, F. Koohifar, and M. Mohammadafzali, "Analysis of pavement raveling using smartphone," in *Proceedings of the Transportation Research Board 95th Annual Meeting*, no. 16-6155, 2016.

[23] Florida Department of Transportation, *Standard Specifications for Road and Bridge Construction*, Florida Department of Transportation, 1973.

[24] M. A. Sutton, J. J. Orteu, and H. Schreier, *Image Correlation for Shape, Motion and Deformation Measurements: Basic Concepts, Theory and Applications*, Springer, 2009.

[25] M. Esmaeili and N. Rezaei, "In situ impact testing of a light-rail ballasted track with tyre-derived aggregate subballast layer," *International Journal of Pavement Engineering*, vol. 17, no. 2, pp. 176–188, 2016.

[26] R. C. Williams, G. Duncan, and T. White, "Sources, measurements, and effects of segregated hot mix asphalt pavement," Joint Transportation Research Program 193, 1996.

21

Study of Antiultraviolet Asphalt Modifiers and Their Antiageing Effects

Jinxuan Hu,[1] Shaopeng Wu,[1] Quantao Liu,[1] María Inmaculada García Hernández,[1] Wenbo Zeng,[1] and Wenhua Xie[2]

[1]*State Key Laboratory of Silicate Materials for Architectures, Wuhan University of Technology, Wuhan, China*
[2]*Wuhan Youfeng Moulding Co. Ltd., Wuhan, China*

Correspondence should be addressed to Shaopeng Wu; wusp@whut.edu.cn

Academic Editor: Frederic Dumur

Ultraviolet (UV) radiation causes serious ageing problems on pavement surface. In recent years, different UV blocking materials have been used as modifiers to prevent asphalt ageing during the service life of the pavement. In this study, three different materials have been used as modifiers in base asphalt to test their UV blocking effects: layered double hydroxides (LDHs), organomontmorillonite (OMMT), and carbon black (CB). UV ageing was applied to simulate the ageing process and softening point, penetration, ductility, DSR (Dynamic Shear Rheometer) test, and Fourier Transform Infrared Spectroscopy (FTIR) test were conducted to evaluate the anti-UV ageing effects of the three UV blocking modifiers. Physical property tests show that base asphalt was influenced more seriously by UV radiation compared to the modified asphalt. DSR test results indicate that the complex modulus of asphalt before UV ageing is increased because of modifiers, while the complex modulus of base asphalt after UV ageing is higher than that of the modified asphalt, which shows that the UV blocking modifiers promote the antiageing effects of asphalt. FTIR test reveals that the increment of carbonyl groups and sulfoxide groups of modified asphalt is less than that in base asphalt. Tests indicate the best UV blocking effect results for samples with LDHs and the worst UV blocking effect results for samples with CB.

1. Introduction

Asphalt has been used for road construction for more than a century [1–3]. Compared with cement pavement, asphalt pavement has remarkable advantages in comfortableness and smoothness [4, 5]. However, the UV radiation leads to a shorter lifespan of the pavement. A set of complex physicochemical processes happen because of the exposition of the asphalt pavement to UV radiation and result in a harder and more brittle asphalt [6, 7]; as a result, low temperature cracking and fatigue cracks are more likely to occur on the pavement [8, 9]. Under these circumstances, it is an urgent task to develop the method to prevent the ageing of asphalt.

Thermooxidative degradation of asphalt has been investigated deeply, but the effects of UV radiation on asphalt binder ageing have been given little attention in previous researches [10]. Although some researchers [11, 12] showed that UV radiation only affects the upper layers of asphalt pavement, the effect of UV radiation on asphalt cannot be ignored. Studies [13] have shown that UV radiation after RTFOT could age a thin film of asphalt to the same ageing level as the one aged by PAV in a few hours.

Researchers adopted modifiers such as UV absorbents [14, 15], LDHs [16–18], and nanomaterials [19] to improve the UV ageing resistance of asphalt. CB is an organic protective material with high absorbance to UV radiation, as a UV absorber has the potential to improve the UV ageing resistance and enhances the low temperature properties [20]. Cong et al. [10] used CB to reflect and absorb UV radiation in asphalt. Results show that low temperature properties of CB modified asphalt were better than unmodified asphalt after UV ageing.

Montmorillonite (MMT) nanocomposite is a phyllosilicate nanomaterial which has been used for the modification of polymers [21, 22]. Nanosize layers of MMT could be dispersed into the polymer matrix due to the fact that polymer

TABLE 1: Physical properties of BA.

Physical properties	BA
Softening point (°C)	45.4
Penetration (25°C, 0.1 mm)	90.9
Ductility (10°C, 1 cm/min)	170
Viscosity (60°C, pa·s)	205
Viscosity (135°C, pa·s)	0.439

chains can intercalate into the interlayer of MMT [23, 24]. Previous researches investigated the effects of OMMT on the physical properties of asphalt and found a positive effect [25, 26]. Zhang et al. [27, 28] modified SBS asphalt with OMMT and found that OMMT could improve the UV ageing resistance of asphalt.

In recent years the use of LDHs as antiageing agent to improve the asphalt ageing resistance has attracted much attention [17]. LDHs are a anionic compound with sheet-like structure [29]. Previous researches indicate that LDHs could improve the UV ageing resistance of asphalt because their shielding physical structure and their chemical UV light absorption properties [18, 30]. UV blocking materials have been used by some researchers to improve the UV ageing resistance of asphalt. Liu et al. [31, 32] used LDHs as a UV blocking and absorbing material in asphalt. In their research, the rheological properties and fatigue properties of LDHs modified asphalt and its mixture were improved greatly with LDHs. However, there is no research about the antiageing effects of these UV blocking materials. So in this study, LDHs, OMMT, and CB were used as asphalt modifiers to prepare anti-UV ageing asphalt. Softening point, penetration, ductility, DSR test, and FTIR tests were conducted to evaluate anti-UV ageing effects of the three UV blocking materials.

2. Experiment Materials

2.1. Materials. Three UV blocking materials (LDHs, OMMT, and CB) were used as modifiers in this research to prepare anti-UV ageing asphalt, because they were frequently used as antiageing materials in the previous research. Base asphalt (BA) with 80/100 pen grade was the matrix. The physical properties of the asphalt used are listed in Table 1.

The molecule structure of modifiers is illustrated in Figures 1–3. As shown in Figure 1, LDHs are a kind of anion layered material which consists of interlayer galleries filled with electrically balanced anions, along with hydration water molecules and basal layers with positively charged metal hydroxide. The chemical formula of LDHs could be described as $[M^{2+}_{1-X}M^{3+}_{X}(OH)_2]^{X+}\text{-}(A^{n-}_{X/n})\cdot mH_2O$, where M^{2+} was divalent metal cation and M^{3+} was trivalent. The subscript X was the molar ratio equal to $M^{3+}/(M^{2+} + M^{3+})$, which ranges from 0.2 to 0.33. A^{n-} represents the exchangeable n-valent anion. A part of M^{2+} in the layers is replaced by M^{3+} because of isomorphous substitution, which let the layers have a positive electric charge. And A^{n-} between the layers maintains the system charge balance. The special structural composition of LDHs results in regulation of composition in the layers and anion between the layers [18, 33].

LDH structure

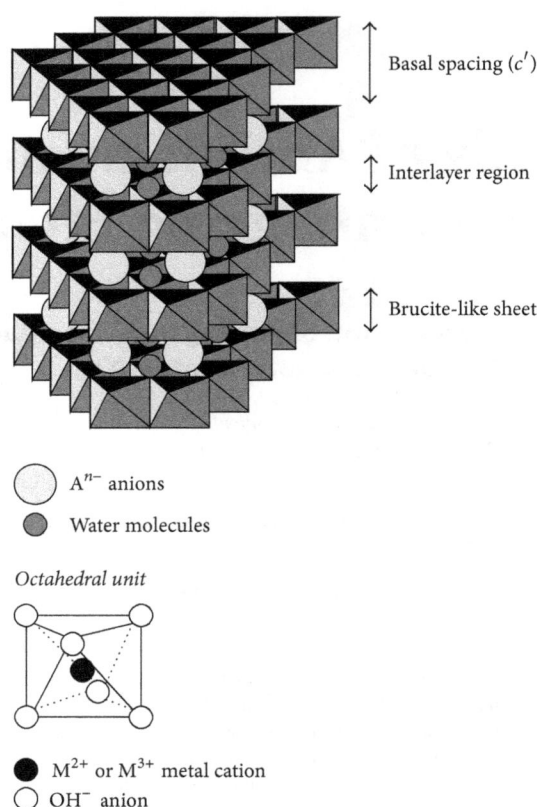

FIGURE 1: The molecule structure of LDHs [33].

MMt is a layered silicate mineral, whose chemical formula is $(Na,Ca)_{0.33}(Al,Mg)_2(Si_4O_{10})(OH)_2$. It consists of a 2:1 type layer structure, one octahedral sandwiched by two tetrahedrons, as shown in Figure 2. The thickness of a single MMt layer is around 1 nm and the cross-sectional area is 100 nm². Between the layers, there are some hydrated cations, such as Na^+. In order to make the MMt more compatible with polymer, it is usually modified by some surfactants, such as quaternary ammonium salt. After being modified, the interlayer hydrated cations will be replaced by organic cations. In this way, organophilic will replace the normally hydrophilic silicate surface [34, 35], and the Mt becomes OMMT.

CB could be generated when gaseous or liquid hydrocarbon conducts incomplete combustion or hot crack under insufficient oxygen. The component of CB consists of a great deal of carbon and a little bit of oxygen, hydrogen, sulfur, water, and other impurities, as shown in Figure 3. The basic building blocks of CB particles are the individual graphitic layers. In a single CB particle, small crystallites are randomly oriented, which is different from graphite. The fundamental units are aggregates comprised of particles but the particles appear to be continuous on the lattice scale within the same aggregate [36, 37].

2.2. Preparation of Modified Asphalt. Melting asphalt was added to a mixing container where the bitumen was modified

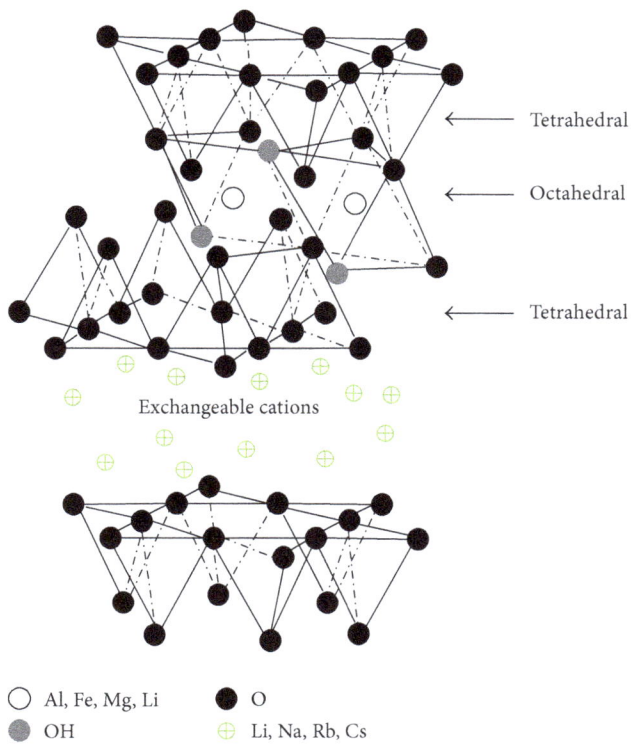

FIGURE 2: The molecule structure of MMT [34].

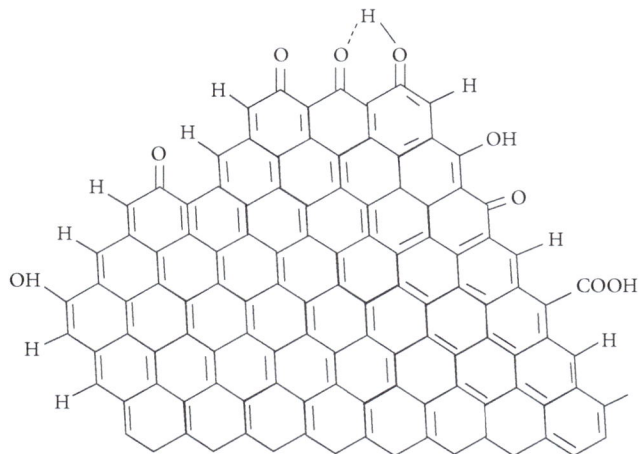

FIGURE 3: The molecule structure of CB [37].

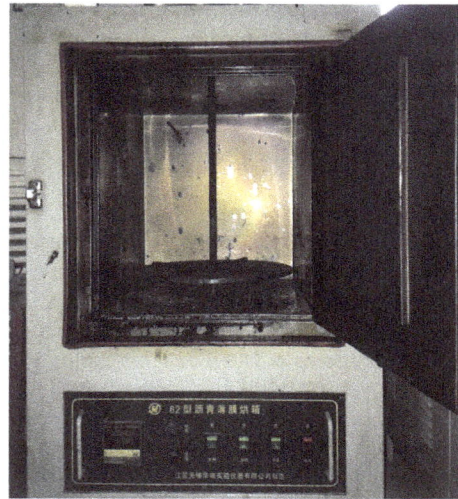

FIGURE 4: TFOT oven.

FIGURE 5: UV chamber.

2.3. Ageing Procedure. A UV chamber was used to simulate the UV ageing of asphalt. In the top of the chamber, there are four UV lamps on four corners in order to ensure homogeneous UV intensity. The samples previously experienced TFOT (ASTM D 1754) and were placed on a rotary table in the UV chamber. The main wavelength of UV lamp was 365 nm. And the average density of UV radiation on asphalt surface is about 500 $\mu w/cm^2$. This procedure was performed for the samples for 10 days as the UV ageing time. The figures of TFOT oven and UV chamber are shown in Figures 4 and 5. Asphalt was poured into Φ (140 ± 0.5) mm iron pan and the thickness of samples was 1.3 mm.

3. Test Methods

3.1. Physical Properties Test. The penetration, softening point, and ductility of the samples were tested in accordance with the standards ASTM D36-76, ASTM D5-13, and ASTM D113-99, respectively.

3.2. Dynamic Shear Rheometer (DSR) Test. Rheological properties of asphalt can be measured by DSR test. In this paper, strain sweep test, high temperature sweep test, and

with 3 wt% of UV blocking material. An oil-bath heating container was applied to heat the asphalt at about 150 ± 5°C. When temperature reached the target temperature, high speed shear mixer was used to disperse the modifier. Finally, 4000 r/min rotation speed for 60 min [31] was applied to ensure a homogeneous modified asphalt. BA was also treated with the same process to eliminate the preparation effects, which might lead to experimental error. LMA, OMA, and CMA are the acronyms for LDHs modified asphalt, OMMT modified asphalt, and CB modified asphalt, respectively.

low temperature sweep test were performed. DSR tests were adopted under the strain-controlled mode, whose constant load frequency was 10 rad/s. In the strain sweep test, the strain level was conducted applying first 0.005% strain amplitude to a specimen, which was continuously increased until nonlinearity appeared in the response. The test temperature was 60°C.

During the high temperature sweep test, temperature increased from 30°C to 80°C with 2°C/min increment. The diameter of the plate was 25 mm and the gap between parallel plates was 1 mm. During the low temperature sweep test, temperature increased from −10°C to 30°C with 2°C/min increment. The diameter of the plate was 8 mm and the gap between parallel plates was 2 mm. The strain level applied during the high temperature sweep tests and low temperature tests was 0.5% and 0.05% strain amplitude, respectively. Complex modulus (G^*) and phase angle (δ) could be obtained from the DSR test, which are important evaluation indices for asphalt rheological properties before and after UV ageing.

3.3. Fourier Transform Infrared Spectroscopy. The ageing index of asphalt can be conveniently measured by FTIR tests. Asphalt binders were dissolved into carbon disulfide to prepare 5 wt% solutions. Then a drop of solution was added to a KBr cell. After carbon disulfide evaporated, the thin asphalt film was prepared as FTIR sample. The scan range is from 3000 cm^{-1} to 600 cm^{-1} with a resolution of 4 cm^{-1}.

The carbonyl function C=O (1700 cm^{-1}) and sulfoxide S=O (1032 cm^{-1}) are calculated by the peak areas of themselves. The carbonyl function index and sulfoxide index can be calculated by the following equations [38, 39]:

$$I_{C=O}$$
$$= \frac{\text{Area of carbonyle band centered around 1700 cm}^{-1}}{\sum \text{Area of spectral bands between 2000} \wedge 600\,\text{cm}^{-1}}$$

$$I_{S=O}$$
$$= \frac{\text{Area of ethylene band centered around 1030 cm}^{-1}}{\sum \text{Area of spectral bands between 2000} \wedge 600\,\text{cm}^{-1}}.$$

(1)

4. Results and Discussion

4.1. Softening Point. Figure 6 shows that all the three anti-UV ageing materials increased the softening point of the asphalt binder and CB modified asphalt shows the biggest increment of the softening point temperature compared to the BA sample. The softening point temperatures of LDHs and OMMT modified asphalt are 45.3°C and 45.6°C, respectively. The reason why CB modified asphalt shows higher increment is that the higher specific surface area of CB would absorb light components (aromatic and saturation); this leads to a worse deforming ability and a higher softening point value. After the UV ageing process, all the samples present an increment in the softening point and the BA showed the highest increment compared to the three modified asphalt samples, which means that the anti-UV ageing materials could alleviate the UV ageing of asphalt. And the softening

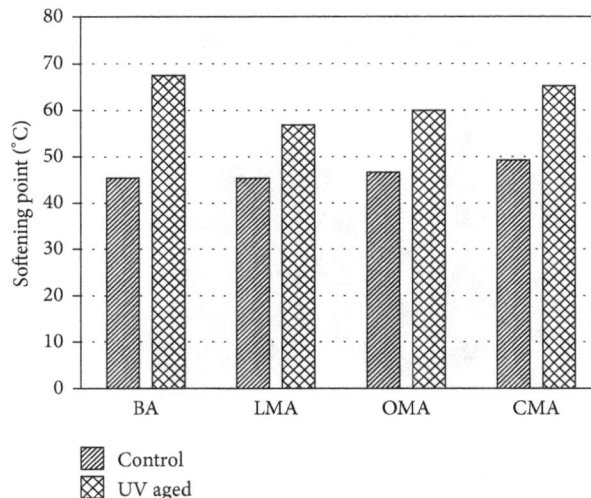

FIGURE 6: Softening point of UV aged and unaged asphalt.

point increments of LDHs modified asphalt and OMMT are almost the same which are much lower than that of CB modified asphalt. This indicates that LHDs and OMMT could significantly prevent asphalt ageing during UV radiation. And the antiageing effect of CB is limited compared to the other two UV blocking materials.

4.2. Penetration. Penetration reflects the hardness of asphalt and indicates the ageing index of asphalt after UV ageing. Smaller penetration means more serious ageing. From Figure 7, it can be seen that the penetration value declines for all four asphalt samples after UV ageing. The penetration values of the three modified asphalt samples are much higher compared to base asphalt. The penetration decrement of LDHs modified asphalt is the smallest, and the penetration decrement of CB modified asphalt is the highest in the three modified asphalt samples. The penetration results are in concordance with the softening point results. The anti-UV ageing effects of LDHs and OMMT are better than CB.

4.3. Ductility. Results of ductility are presented in Table 2. Incorporation of LDHs, OMMT, and CB decreased the ductility of asphalt by 16%, 26%, and 37%, respectively. But after UV ageing, ductility values of three anti-UV ageing materials of modified asphalt were higher than aged base asphalt. For LDHs modified asphalt was highest and for CB it was smallest. The rate of the loss of ductility was reduced greatly by the incorporation of UV barrier materials. Ultraviolet blocking materials dispersed in asphalt could reflect and absorb UV radiation and reduced the loss rate of light component in asphalt, which leads to a higher residual ductility during UV ageing process.

4.4. Rheological Properties

4.4.1. Strain Sweep Test. There are various test methods to measure the rheological properties: temperature sweep test,

TABLE 2: Ductility of UV aged and unaged asphalt.

Asphalt	Original ductility (cm)	Ductility after UV ageing (cm)
BA	170	41
LMA	142	62
OMA	126	52
CMA	107	45

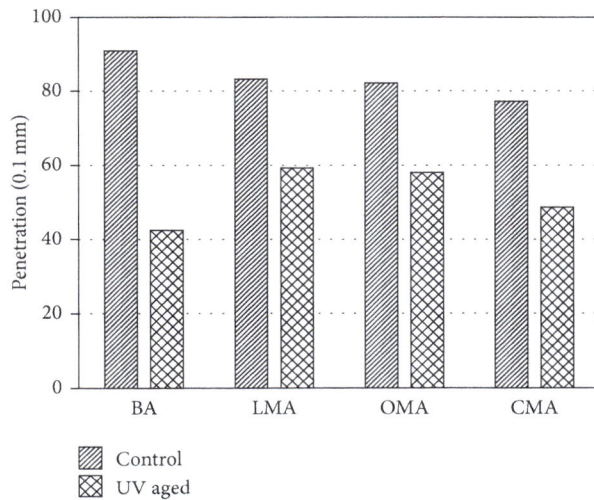

FIGURE 7: Penetration of UV aged and unaged asphalt.

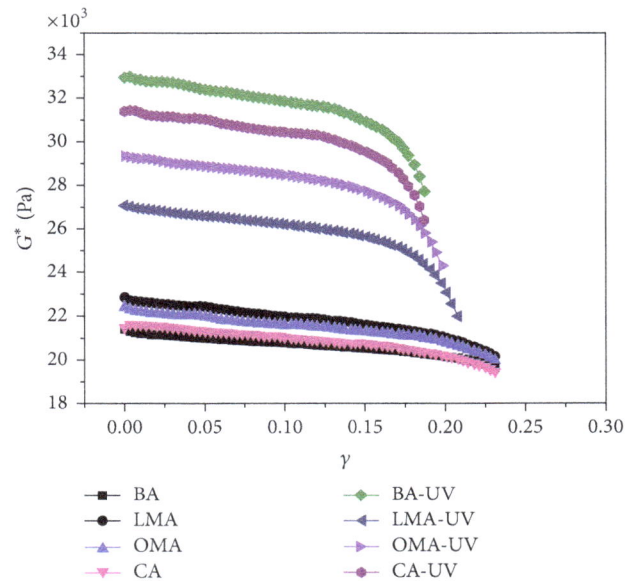

FIGURE 8: G^* of UV aged and unaged asphalt versus strain.

frequency sweep test, and sheer creep test. The shear stress applied in these tests must be low enough to keep the measurement inside the region of linear viscoelastic (LVE). In this LVE region, the sum of responses to an individual stress is equal to the sum of stresses [40]. The LVE strain limit represents the deformation where molecular structure of materials begins to experience irreversible changes [41]. Tests on asphalt must be done inside the LVE region to ensure test repeatability. The dynamic shear modulus is relatively independent of the strain applied at sufficiently small strains. Therefore, the strain level or stress level must be selected so that the resulting response (G^*) is within the LVE limit. The LVE strain limit was defined as the strain value at which the complex modulus has decreased to 95% of its initial value by Airey et al. [42].

The strain sweep tests were performed firstly with the purpose of determining the limits of the LVE region and stress range. G^* and stress (σ) of UV aged and unaged asphalt versus strain are shown in Figures 8 and 9. Tendencies of G^* and σ of unaged asphalt stay at a similar level. The LVE strain limits of several kinds of unaged asphalt are similar, and the LVE strain limits of several kinds of unaged asphalt are smaller than aged asphalt. But for aged asphalt and unaged asphalt, 0.05% and 0.5% strain amplitude are sufficiently small for the following temperature sweep tests.

4.4.2. High Temperature Sweep. Dynamic viscoelastic performances are an important part of asphalt rheological properties, which depends strongly on temperature and

loading frequency [43]. There are several parameters such as G^* and δ which can represent the principal viscoelastic performance. G^* represents the ratio between maximum stress and maximum strain, which shows the deformation resistance ability of asphalt [44]. In other words, the higher complexity the modulus has, the higher the deformation resistance ability becomes. G^* of both base and modified asphalt before and after ageing from 30°C to 80°C is shown in Figure 10. For asphalt ageing, the higher complexity the modulus has, the higher the ageing degree becomes. It can be seen that G^* of base asphalt is lower than G^* of modified asphalt before UV ageing. It means that asphalt ability for deformation resistance is increased with additives. G^* of LMA and OMA is higher than CMA, which means that LDHs and OMMT have a stronger effect for improving the deformation resistance than CB. G^* of modified asphalt is lower than G^* of base asphalt after UV ageing, which shows that UV ageing resistance of asphalt is increased by modifiers; besides it can be seen that G^* of LMA-UV and OMA-UV is lower than CA-UV. The results indicate that the anti-UV ageing effects of LDHs and OMMT are better than CB.

δ describes the time between the applied stress and the generated strain, which shows viscosity and elasticity characteristics of asphalt [45]. With the decrease of δ, the time is shortened. Lower phase angle means higher ageing degree of asphalt. δ of base and modified asphalt before and

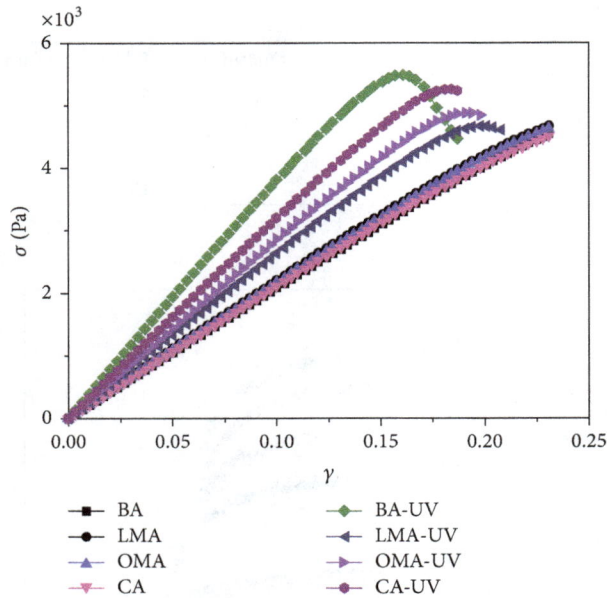

FIGURE 9: Stress of UV aged and unaged asphalt versus strain.

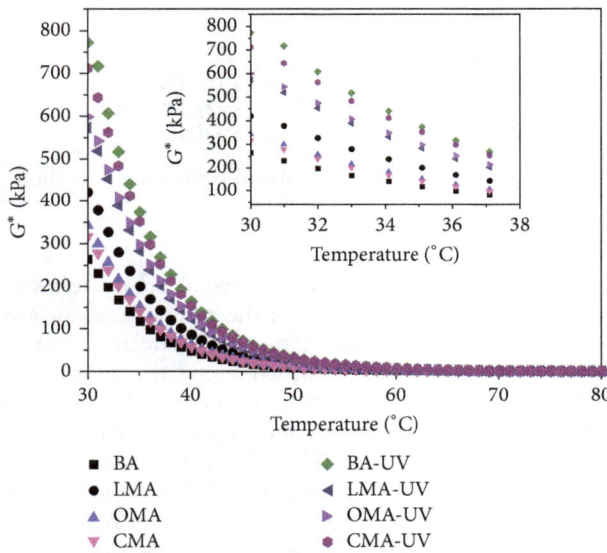

FIGURE 11: δ of UV aged and unaged asphalt from 30°C to 80°C.

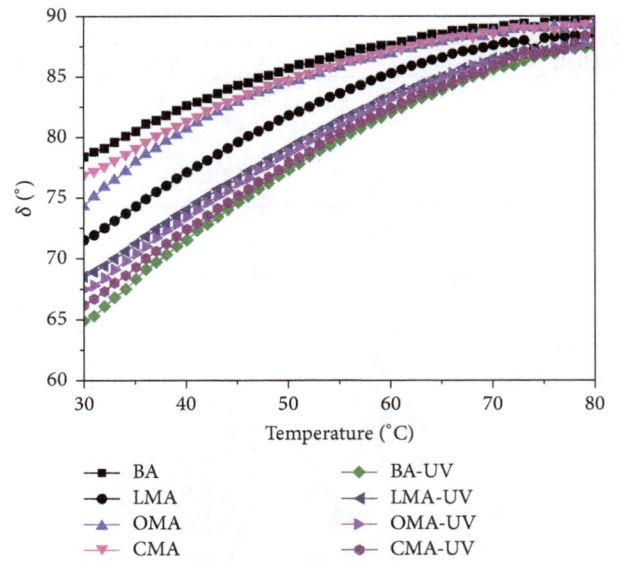

FIGURE 10: G^* of UV aged and unaged asphalt from 30°C to 80°C.

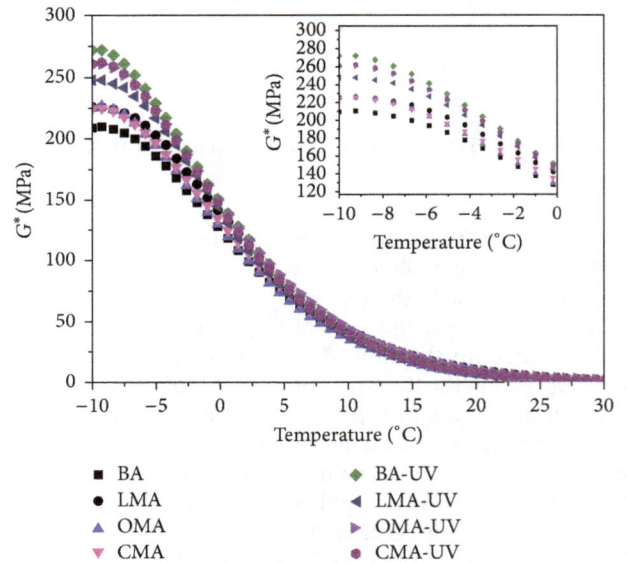

FIGURE 12: G^* of UV aged and unaged asphalt from −10°C to 30°C.

after ageing from 30°C to 80°C is shown in Figure 11. It can be seen that δ of modified asphalt is higher than δ of base asphalt before UV ageing, which indicates that the time lag is shortened by modifiers. δ of asphalt after UV ageing is BA-UV > CA-UV > OMA-UV > LMA-UV. The results indicate that the anti-UV ageing effect of asphalt is improving with the additives and the anti-UV ageing effects of LDHs and OMMT are better than CB, which is in concordance with results of complex modulus.

4.4.3. Low Temperature Sweep.

G^* of both base and modified asphalt before and after ageing from −10°C to 30°C is

illustrated in Figure 12. It can be seen that modifiers used in this research increase G^* of bitumen at low temperature. The phenomenon could be explained as asphalt ability of deformation resistance is increased with modifiers. G^* of unaged modified bitumen is slightly higher than unaged base bitumen. The results show that UV ageing resistance of asphalt is increased with additives modifiers and G^* of CMA-UV is higher than LMA, which means that the anti-UV ageing effects of LDHs are better than CB.

δ of base and modified asphalt before and after ageing from −10°C to 30°C is shown in Figure 13. δ of base and modified asphalt before ageing is similar, which means that the ratio between viscous component and elastic component of asphalt is not changed by modifiers. δ of modified asphalt

TABLE 3: $I_{C=O}$ and $I_{S=O}$ of UV aged and unaged asphalt.

Asphalt	$I_{C=O}$	$I_{S=O}$
BA	0	0.00127
BA-UV	0.0482	0.0954
LMA-UV	0.0257	0.0623
OMA-UV	0.0318	0.0817
CMA-UV	0.0367	0.0785

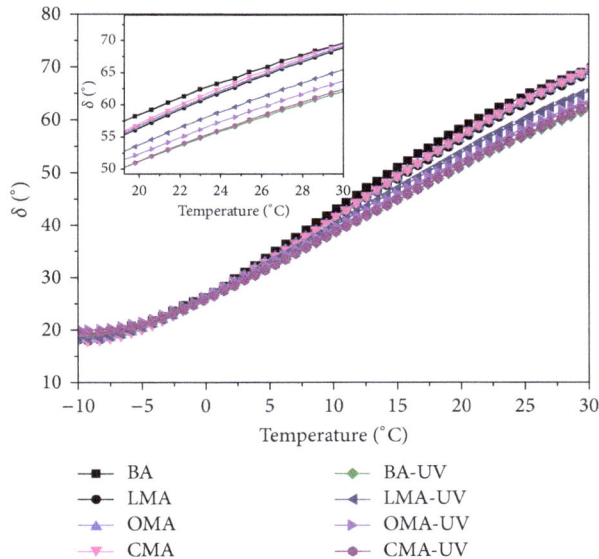

FIGURE 13: δ of UV aged and unaged asphalt from $-10°$C to $30°$C.

Spectroscopy (FTIR) test were conducted to evaluate anti-UV ageing effects of the three UV blocking materials. According to the discussion above, conclusions can be drawn as follows:

(1) The modifications of the asphalt with LDHs, OMMT, and CB could all improve the anti-UV ageing ability of asphalt. The physical properties of base asphalt were more influenced by UV ageing than those of modified asphalt. The increment of carbonyl groups and sulfoxide groups of modified asphalt is less than that of base asphalt.

(2) The complex modulus of asphalt before UV ageing is increased by adding the UV blocking additives, while the complex modulus of base asphalt after UV ageing is higher than modified asphalt. The phase angle of base asphalt after UV ageing is lower than that of modified asphalt. These results showed that the ageing extents of the three anti-UV ageing asphalt samples were much lower than that of base asphalt.

(3) LDHs could decrease the ageing rate of asphalt most effectively, followed by OMMT. CB showed the worst anti-UV ageing effect.

Conflicts of Interest

The authors declare that they have no conflicts of interest.

Acknowledgments

This work was financially supported by the National Basic Research Program of China (973 Program no. 2014CB932104), the Natural Science Foundation of China (no. 51508433), and the National Key Scientific Apparatus Development Program from the Ministry of Science and Technology of China (no. 2013YQ160501).

after ageing is slightly higher than base bitumen. It indicated that UV resistance of modified bitumen is increased. Variation of δ of CB before and after UV ageing is slightly bigger than LMA, which means that the anti-UV ageing effects are better than CB.

4.5. FTIR. With the increase of ageing degree, the content of carbonyl groups and sulfoxide groups is promoted [46]. From Table 3, it can be seen that the peak areas of carbonyl and butadiene double bonds of the BA are increased after UV ageing. Increasing $I_{C=O}$ and $I_{S=O}$ indicate absorbing more oxygen and more serious ageing index of asphalt after UV ageing. $I_{C=O}$ and $I_{S=O}$ of LMA, OMA, and CA are decreased compared to BA after UV ageing process. $I_{C=O}$ and $I_{S=O}$ of LDHs are minimums and $I_{C=O}$ and $I_{S=O}$ of carbon black are maximums. The effect of improving the UV ageing resistance of asphalt is highest with LDHs, medium with OMMT, and lowest with CB under the specific mixing amount. FTIR results show that the ageing extent of LMA is the lowest.

5. Conclusions

In this research, LDHs, OMMT, and CB were used as modifiers to prepare anti-UV ageing asphalt. Softening point, penetration, ductility, DSR test, and Fourier Transform Infrared

References

[1] Z. Chen, S. Wu, Y. Xiao, W. Zeng, M. Yi, and J. Wan, "Effect of hydration and silicone resin on Basic Oxygen Furnace slag and its asphalt mixture," *Journal of Cleaner Production*, vol. 112, pp. 392–400, 2016.

[2] P. Cui, S. Wu, Y. Xiao, M. Wan, and P. Cui, "Inhibiting effect of Layered Double Hydroxides on the emissions of volatile organic compounds from bituminous materials," *Journal of Cleaner Production*, vol. 108, pp. 987–991, 2015.

[3] P. Pan, S. Wu, Y. Xiao, and etal., "A review on hydronic asphalt pavement for energy harvesting and snow melting," *Renewable & Sustainable Energy Reviews*, vol. 48, pp. 624–634, 2015.

[4] X. Lu and U. Isacsson, "Effect of ageing on bitumen chemistry and rheology," *Construction and Building Materials*, vol. 16, no. 1, pp. 15–22, 2002.

[5] Y. Ruan, R. Davison, C. J. Glover, and etal., "Oxidation and viscosity hardening of polymer-modified asphalts," *Energy & Fuels*, vol. 17, pp. 991–998, 2003.

[6] L. Xiaohu and U. Isacsson, "Chemical and rheological evaluation of ageing properties of SBS polymer modified bitumens," *Fuel*, vol. 77, no. 9-10, pp. 961–972, 1998.

[7] M. Poirier and H. Sawatzky, "Changes ih chemical component type composition and effect on rheoloqical properties of asphalts," *Petroleum Science and Technology*, vol. 10, no. 4-6, pp. 681–696, 2007.

[8] M. N. Siddiqui and M. F. Ali, "Studies on the aging behavior of the Arabian asphalts," *Fuel*, vol. 78, no. 9, pp. 1005–1015, 1999.

[9] Y. Ruan, R. R. Davison, and C. J. Glover, "The effect of long-term oxidation on the rheological properties of polymer modified asphalts," *Fuel*, vol. 82, no. 14, pp. 1763–1773, 2003.

[10] P. Cong, X. Wang, P. Xu, J. Liu, R. He, and S. Chen, "Investigation on properties of polymer modified asphalt containing various antiaging agents," *Polymer Degradation and Stability*, vol. 98, no. 12, pp. 2627–2634, 2013.

[11] E. Kalkornsurapranee, N. Vennemann, C. Kummerlöwe, and C. Nakason, "Novel thermoplastic natural rubber based on thermoplastic polyurethane blends: Influence of modified natural rubbers on properties of the blends," *Iranian Polymer Journal*, vol. 21, no. 10, pp. 689–700, 2012.

[12] M. N. Siddiqui and M. F. Ali, "Investigation of chemical transformations by NMR and GPC during the laboratory aging of Arabian asphalt," *Fuel*, vol. 78, no. 12, pp. 1407–1416, 1999.

[13] F. Durrieu, F. Farcas, and V. Mouillet, "The influence of UV aging of a styrene/butadiene/styrene modified bitumen: comparison between laboratory and on site aging," *Fuel*, vol. 86, no. 10-11, pp. 1446–1451, 2007.

[14] K. Yamaguchi, I. Sasaki, I. Nishizaki, S. Meiarashi, and A. Moriyoshi, "Effects of film thickness, wavelength, and carbon black on photodegradation of asphalt," *Journal of the Japan Petroleum Institute*, vol. 48, no. 3, pp. 150–155, 2005.

[15] Z. Feng, J. Yu, H. Zhang, D. Kuang, and L. Xue, "Effect of ultraviolet aging on rheology, chemistry and morphology of ultraviolet absorber modified bitumen," *Materials and Structures*, vol. 46, no. 7, pp. 1123–1132, 2012.

[16] S. Xu, J. Yu, C. Zhang, and Y. Sun, "Effect of ultraviolet aging on rheological properties of organic intercalated layered double hydroxides modified asphalt," *Construction and Building Materials*, vol. 75, pp. 421–428, 2015.

[17] H. Wu, L. Li, J. Yu, S. Xu, and D. Xie, "Effect of layered double hydroxides on ultraviolet aging properties of different bitumens," *Construction and Building Materials*, vol. 111, pp. 565–570, 2016.

[18] S. Xu, J. Yu, W. Wu, L. Xue, and Y. Sun, "Synthesis and characterization of layered double hydroxides intercalated by UV absorbents and their application in improving UV aging resistance of bitumen," *Applied Clay Science*, vol. 114, pp. 112–119, 2015.

[19] H. L. Zhang, C. Z. Zhu, J. Y. Yu, C. Shi, and D. Zhang, "Influence of surface modification on physical and ultraviolet aging resistance of bitumen containing inorganic nanoparticles," *Construction and Building Materials*, vol. 98, pp. 735–740, 2015.

[20] V. B. Bojinov, I. P. Panova, and D. B. Simeonov, "Design and synthesis of polymerizable, yellow-green emitting 1,8-naphthalimides containing built-in s-triazine UV absorber and hindered amine light stabilizer fragments," *Dyes and Pigments*, vol. 78, no. 2, pp. 101–110, 2008.

[21] A. Rehab and N. Salahuddin, "Nanocomposite materials based on polyurethane intercalated into montmorillonite clay," *Materials Science and Engineering A*, vol. 399, no. 1-2, pp. 368–376, 2005.

[22] A. Gultek, T. Seckin, Y. Onal, and M. Galip Icduygu, "Preparation and phenol captivating properties of polyvinyl pyrrolidone-montmorillonite hybrid materials," *Journal of Applied Polymer Science*, vol. 81, no. 2, pp. 512–519, 2001.

[23] Y. Qu, Y. Su, J. Sun, and etal., "Preparation of poly(styrene-block-acrylamide)/organic montmorillonite nanocomposites via reversible additionfragmentation chain transfer," *Journal of Applied Polymer Science*, vol. 110, pp. 387–391, 2008.

[24] C. Ouyang, S. Wang, Y. Zhang, and Y. Zhang, "Improving the aging resistance of styrene-butadiene-styrene tri-block copolymer modified asphalt by addition of antioxidants," *Polymer Degradation and Stability*, vol. 91, no. 4, pp. 795–804, 2006.

[25] J. Yu, L. Wang, X. Zeng, S. Wu, and B. Li, "Effect of montmorillonite on properties of styrene-butadiene-styrene copolymer modified bitumen," *Polymer Engineering and Science*, vol. 47, no. 9, pp. 1289–1295, 2007.

[26] J. Y. Yu, P. C. Feng, H. L. Zhang, and S. P. Wu, "Effect of organomontmorillonite on aging properties of asphalt," *Construction and Building Materials*, vol. 23, no. 7, pp. 2636–2640, 2009.

[27] H. Zhang, J. Yu, H. Wang, and L. Xue, "Investigation of microstructures and ultraviolet aging properties of organomontmorillonite/SBS modified bitumen," *Materials Chemistry and Physics*, vol. 129, no. 3, pp. 769–776, 2011.

[28] H. Zhang, J. Yu, and S. Wu, "Effect of montmorillonite organic modification on ultraviolet aging properties of SBS modified bitumen," *Construction and Building Materials*, vol. 27, no. 1, pp. 553–559, 2012.

[29] Q. Wang and D. O. Hare, "Recent advances in the synthesis and application of layered double hydroxide (LDH) nanosheets," *Chemical Reviews*, vol. 112, no. 7, pp. 4124–4155, 2012.

[30] Y. Huang, Z. Feng, H. Zhang, and J. Yu, "Effect of layered double hydroxides (LDHs) on aging properties of bitumen," *Journal of Testing and Evaluation*, vol. 40, no. 5, pp. 734–739, 2012.

[31] X. Liu, S. Wu, L. Pang, Y. Xiao, and P. Pan, "Fatigue properties of layered double hydroxides modified asphalt and its mixture," *Advances in Materials Science and Engineering*, vol. 2014, Article ID 868404, 6 pages, 2014.

[32] S. P. Wu, J. Han, L. Pang, M. Yu, and T. Wang, "Rheological properties for aged bitumen containing ultraviolate light resistant materials," *Construction and Building Materials*, vol. 33, pp. 133–138, 2012.

[33] A. Gomes, D. Cocke, D. Tran, and etal., "Layered double hydroxides in energy research: advantages and challenges," in *Energy Technology 2015: Carbon Dioxide Management and Other Technologies*, pp. 309–316, USA, 2015.

[34] E. P. Giannelis, "Polymer layered silicate nanocomposites," *Advanced Materials*, vol. 8, no. 1, pp. 29–35, 1996.

[35] G. Liu, *Characterization and Identification of Bituminous Materials Modified with Montmorillonite Nanoclay*, TU, Delft, Delft University of Technology, 2011.

[36] W. Zhu, D. E. Miser, W. G. Chan, and M. R. Hajaligol, "HRTEM investigation of some commercially available furnace carbon blacks," *Carbon*, vol. 42, no. 8-9, pp. 1841–1845, 2004.

[37] F. Cataldo, "The impact of a fullerene-like concept in carbon black science," *Carbon*, vol. 40, no. 2, pp. 157–162, 2002.

[38] N. Pieri, "Étude du vieillissement simulé et in situ des bitumes routiers par IRTF et fluorescence UV en excitation-emission synchrones," *d'Aix-Marseille 3*, 1994.

[39] J. Lamontagne, P. Dumas, V. Mouillet, and J. Kister, "Comparison by Fourier transform infrared (FTIR) spectroscopy of different ageing techniques: application to road bitumens," *Fuel*, vol. 80, no. 4, pp. 483–488, 2001.

[40] F. Morea, J. O. Agnusdei, and R. Zerbino, "Comparison of methods for measuring zero shear viscosity in asphalts," *Materials and Structures/Materiaux et Constructions*, vol. 43, no. 4, pp. 499–507, 2010.

[41] O. V. Laukkanen, T. Pellinen, and M. Makowska, "Exploring the observed rheological behaviour of in-situ aged and fresh bitumen employing the colloidal model proposed for bitumen," *Multi-Scale Modeling and Characterization of Infrastructure Materials*, pp. 185–197, 2013.

[42] G. Airey, B. Rahimzadeh, and A. Collop, "Linear viscoelastic performance of asphaltic materials," *Road Materials and Pavement Design*, vol. 4, no. 3, pp. 269–292, 2003.

[43] X. Lu and U. Isacsson, "Rheological characterization of styrene-butadiene-styrene copolymer modified bitumens," *Construction and Building Materials*, vol. 11, no. 1, pp. 23–32, 1997.

[44] F. Xiao, S. N. Amirkhanian, M. Karakouzian, and M. Khalili, "Rheology evaluations of WMA binders using ultraviolet and PAV aging procedures," *Construction and Building Materials*, vol. 79, pp. 56–64, 2015.

[45] F. Xiao, S. Amirkhanian, H. Wang, and P. Hao, "Rheological property investigations for polymer and polyphosphoric acid modified asphalt binders at high temperatures," *Construction and Building Materials*, vol. 64, pp. 316–323, 2014.

[46] V. Mouillet, J. Lamontagne, F. Durrieu, J.-P. Planche, and L. Lapalu, "Infrared microscopy investigation of oxidation and phase evolution in bitumen modified with polymers," *Fuel*, vol. 87, no. 7, pp. 1270–1280, 2008.

Permissions

All chapters in this book were first published in AMSE, by Hindawi Publishing Corporation; hereby published with permission under the Creative Commons Attribution License or equivalent. Every chapter published in this book has been scrutinized by our experts. Their significance has been extensively debated. The topics covered herein carry significant findings which will fuel the growth of the discipline. They may even be implemented as practical applications or may be referred to as a beginning point for another development.

The contributors of this book come from diverse backgrounds, making this book a truly international effort. This book will bring forth new frontiers with its revolutionizing research information and detailed analysis of the nascent developments around the world.

We would like to thank all the contributing authors for lending their expertise to make the book truly unique. They have played a crucial role in the development of this book. Without their invaluable contributions this book wouldn't have been possible. They have made vital efforts to compile up to date information on the varied aspects of this subject to make this book a valuable addition to the collection of many professionals and students.

This book was conceptualized with the vision of imparting up-to-date information and advanced data in this field. To ensure the same, a matchless editorial board was set up. Every individual on the board went through rigorous rounds of assessment to prove their worth. After which they invested a large part of their time researching and compiling the most relevant data for our readers.

The editorial board has been involved in producing this book since its inception. They have spent rigorous hours researching and exploring the diverse topics which have resulted in the successful publishing of this book. They have passed on their knowledge of decades through this book. To expedite this challenging task, the publisher supported the team at every step. A small team of assistant editors was also appointed to further simplify the editing procedure and attain best results for the readers.

Apart from the editorial board, the designing team has also invested a significant amount of their time in understanding the subject and creating the most relevant covers. They scrutinized every image to scout for the most suitable representation of the subject and create an appropriate cover for the book.

The publishing team has been an ardent support to the editorial, designing and production team. Their endless efforts to recruit the best for this project, has resulted in the accomplishment of this book. They are a veteran in the field of academics and their pool of knowledge is as vast as their experience in printing. Their expertise and guidance has proved useful at every step. Their uncompromising quality standards have made this book an exceptional effort. Their encouragement from time to time has been an inspiration for everyone.

The publisher and the editorial board hope that this book will prove to be a valuable piece of knowledge for researchers, students, practitioners and scholars across the globe.

List of Contributors

Hui Wang
Key Laboratory of Highway Construction & Maintenance Technology in Loess Region, Shanxi Transportation Research Institute, Taiyuan, Shanxi 030006, China

Haoqi Tan, Tian Qu and Jiupeng Zhang
School of Highway, Chang'an University, Xi'an, Shaanxi 710064, China

Xuedong Guo, Mingzhi Sun, Wenting Dai and Shuang Chen
School of Transportation, Jilin University, Changchun 130022, China

Tao Ma, Yongli Zhao and Xiaoming Huang
School of Transportation, Southeast University, Nanjing, Jiangsu 210096, China

Kai Cui
State Engineering Laboratory of Highway Maintenance Technology, Changsha University of Science and Technology, Changsha 410114, China

Marco Pasetto
Department of Civil, Environmental and Architectural Engineering, University of Padua, Via Marzolo 9, 35131 Padua, Italy

Nicola Baldo
Polytechnic Department of Engineering and Architecture, University of Udine, Via del Cotonificio 114, 33100 Udine, Italy

Danhua Wang
School of Computer Engineering, Nanjing Institute of Technology, 1 Hongjin Road, Nanjing, Jiangsu 211167, China

Xunhao Ding, Linhao Gu and Tao Ma
School of Transportation, Southeast University, 2 Sipailou, Nanjing, Jiangsu 210096, China

Jie Ji
School of Civil Engineering and Transportation, Beijing University of Civil Engineering and Architecture and Beijing Urban Transportation Infrastructure Engineering Technology Research Center, Beijing 100044, China

Hui Yao
Department of Civil and Environmental Engineering, Michigan Technological University and School of Traffic and Transportation, Changsha University of Science and Technology, 1400 Townsend Drive, Houghton, MI 49931, USA

Di Wang and Zhi Suo
Beijing Urban Transportation Infrastructure Engineering Technology Research Center and Beijing Collaborative Innovation Center for Metropolitan Transportation, Beijing 100044, China

Luhou Liu
School of Civil Engineering and Transportation, Beijing University of Civil Engineering and Architecture and Beijing Cooperative Innovation Research Center on Energy Saving and Emission Reduction, Beijing 100044, China

Zhanping You
Department of Civil and Environmental Engineering, Michigan Technological University, 1400 Townsend Drive, Houghton, MI 49931, USA

Wenliang Wu, Zhi Li, Xiaoning Zhang and Minghui Li
School of Civil Engineering and Transportation, South China University of Technology, Guangzhou 510640, China

Moein Hasaninia
Department of Civil Engineering, Iran University of Science and Technology (IUST), Narmak, Tehran, Iran

Farshad Haddadi
Department of Civil Engineering, Florida International University (FIU), Miami, FL, USA

Mingfeng Chang and Binhui Zheng
School of Materials Science and Engineering, Chang'an University, Shaanxi, Xi'an 710061, China

Pingming Huang, Jianzhong Pei and Jiupeng Zhang
School of Highway, Chang'an University, Shaanxi, Xi'an 710064, China

Xiaohu Yan and Zaiqin Wang
College of Water Conservancy and Hydropower Engineering, Hohai University, Nanjing 210098, China Yangtze River Scientific Research Institute, Wuhan 430010, China

Meijuan Rao
State Key Laboratory of Silicate Materials for Architecture, Wuhan University of Technology, Wuhan 430070, China

Mingxia Li
Yangtze River Scientific Research Institute, Wuhan 430010, China

Yafei Li, Jing Chen and Jin Yan
Research and Consulting Department of Engineering Structure and Materials Research Center, China Academy of Transportation Sciences, 240 Huixinli, Chaoyang District, Beijing 100029, China

Meng Guo
College of Architecture and Civil Engineering, Beijing University of Technology, Beijing 100124, China

Jianbing Lv, Xu Zhancheng, Yin Yingmei, Zhang Jiantong and Sun Xiaolong
Guangdong University of Technology, Guangzhou 510006, Guangdong, China

Wu Chuanhai
Guangdong Hualu Traffic Technology Co. Ltd, Guangzhou 510006, Guangdong, China

Tian Xiaog, Ren Zhang, Zhen Yang, Yantian Chu, Shaohua Zhen and Yichao Xv
School of Traffic & Transportation Engineering, Changsha University of Science & Technology, 960 Wanjiali Road, Tianxin District, Changsha, Hunan 410114, China

Chaohui Wang, Xiaolong Sun and Xuancang Wang
School of Highway, Chang'an University, Middle-Section of Nan'er Huan Road, Xi'an, CN 710064, China

Qiang Li and Kevin C. P. Wang
School of Civil and Environmental Engineering, Oklahoma State University, Stillwater, OK 74078, USA

Xiaolong Zou
School of Architecture and Civil Engineering, Xi'an University of Science and Technology, Xi'an, Shaanxi, China
Key Laboratory for Special Area Highway Engineering of Ministry of Education, Chang'an University, Xi'an, Shaanxi, China
Guangxi Key Lab of Road Structure and Materials, Guangxi Transportation Research & Consulting Co., Ltd., Nanning, Guangxi, China

Aimin Sha, Yuqiao Tan and Xiaonan Huang
Key Laboratory for Special Area Highway Engineering of Ministry of Education, Chang'an University, Xi'an, Shaanxi, China

Biao Ding
CCCC First Highway Consultants Co., Ltd., Xi'an, Shaanxi, China

Hanbing Liu, Mengsu Zhang, Yubo Jiao and Liuxu Fu
College of Transportation, Jilin University, Changchun 130025, China

Biao Ding and Zixin Peng
CCCC First Highway Consultants Co., Ltd., Xi'an, Shaanxi 710065, China

Xiaolong Zou
School of Architecture and Civil Engineering, Xi'an University of Science and Technology, Xi'an, Shaanxi, China
Guangxi Key Lab of Road Structure and Materials, Guangxi Transportation Research & Consulting Co., Ltd., Nanning, Guangxi, China
Key Laboratory for Special Area Highway Engineering of Ministry of Education, Chang'an University, Xi'an, Shaanxi, China

Xiang Liu
School of Highway, Chang'an University, Xi'an, Shaanxi, China

Xing-jun Zhang
School of Petrochemical Engineering, Lanzhou University of Technology, Lanzhou 730050, China
Gansu Province Highway Maintenance Engineering Research Center, Lanzhou 730070, China

Hui-xia Feng
School of Petrochemical Engineering, Lanzhou University of Technology, Lanzhou 730050, China

Xiao-min Li and Bo Li
Key Laboratory of Road & Bridge and Underground Engineering of Gansu Province, Lanzhou Jiaotong University, Lanzhou 730070, China

Xiao-yu Ren
Gansu Province Highway Maintenance Engineering Research Center, Lanzhou 730070, China
Key Laboratory of Road & Bridge and Underground Engineering of Gansu Province, Lanzhou Jiaotong University, Lanzhou 730070, China

Zhen-feng Lv
School of Highway, Chang'an University, Xi'an 710064, China

Mojtaba Mohammadafzali and Hesham Ali
Department of Civil and Environmental Engineering, Florida International University, Miami, FL, USA

James A. Musselman
Oldcastle Materials Group, FL, USA

Gregory A. Sholar and Wayne A. Rilko
State Material Office, Florida Department of Transportation, Gainesville, FL, USA

Mohamadtaqi Baqersad, Amirmasoud Hamedi, Mojtaba Mohammadafzali and Hesham Ali
Department of Civil and Environmental Engineering, Florida International University, 10555 West Flagler Street, Miami, FL 33174, USA

Jinxuan Hu, Shaopeng Wu, Quantao Liu, María Inmaculada García Hernández and Wenbo Zeng
State Key Laboratory of Silicate Materials for Architectures, Wuhan University of Technology, Wuhan, China

Wenhua Xie
Wuhan Youfeng Moulding Co. Ltd., Wuhan, China

Index

www.ingramcontent.com/pod-product-compliance
Lightning Source LLC
Chambersburg PA
CBHW080651200326
41458CB00013B/4820